W9-BXY-609

JERUSALEM EINSTEIN CENTENNIAL SYMPOSIUM
14-23 MARCH 1979

ORGANIZED BY

The Israel Academy of Sciences and Humanities
The Hebrew University of Jerusalem
The Van Leer Jerusalem Foundation
The Jerusalem Foundation
Aspen Institute for Humanistic Studies

ALBERT EINSTEIN

Historical and Cultural Perspectives

THE CENTENNIAL SYMPOSIUM
IN JERUSALEM

EDITED BY

Gerald Holton and Yehuda Elkana

DOVER PUBLICATIONS, INC.
Mineola, New York

Copyright

Published in Canada by General Publishing Company, Ltd., 30 Lesmill Road, Don Mills, Toronto, Ontario.
Published in the United Kingdom by Constable and Company, Ltd., 3 The Lanchesters, 162–164 Fulham Palace Road, London W6 9ER.

Bibliographical Note

This Dover edition, first published in 1997, is an unabridged republication of the book first published by Princeton University Press in 1982.

Library of Congress Cataloging-in-Publication Data

Jerusalem Einstein Centennial Symposium (1979)
Albert Einstein : historical and cultural perspectives : the centennial symposium in Jerusalem / edited by Gerald Holton and Yehuda Elkana. – Dover ed.
p. cm.
Originally published: Princeton, N. J. : Princeton University Press, 1982.
Includes bibliographical references and index.
ISBN 0-486-29879-5 (pbk.)
1. Einstein, Albert, 1879–1955–Anniversaries, etc. 2. Relativity (Physics)–Congresses. 3. Science–Social aspects–Congresses. 4. Physicists–Biography–Congresses. I. Holton, Gerald James. II. Elkana, Yehuda, 1934– . III. Akademyah ha-le 'umit ha-Yiśre 'elit le-mada' im. IV. Title.
QC16.E5.J48 1997
530'.092–dc21
[B] 97-22479
CIP

Manufactured in the United States of America
Dover Publications, Inc., 31 East 2nd Street, Mineola, N.Y. 11501

Contents

GERALD HOLTON Introduction: Einstein and the Shaping of Our Imagination vii

HISTORICAL PERSPECTIVES ON EINSTEIN'S SCIENTIFIC CONTRIBUTIONS

ARTHUR I. MILLER The Special Relativity Theory: Einstein's Response to the Physics of 1905 3

PETER G. BERGMANN The Quest for Unity: General Relativity and Unitary Field Theories 27

MARTIN J. KLEIN Fluctuations and Statistical Physics in Einstein's Early Work 39

MAX JAMMER Einstein and Quantum Physics 59

RECEPTION OF EINSTEIN'S SCIENTIFIC IDEAS

P. A. M. DIRAC The Early Years of Relativity 79

BANESH HOFFMANN Some Einstein Anomalies 91

LOREN R. GRAHAM The Reception of Einstein's Ideas: Two Examples from Contrasting Political Cultures 107

EINSTEIN'S IMPACT ON SCHOLARSHIP AND TWENTIETH-CENTURY CULTURE

ROMAN JAKOBSON Einstein and the Science of Language 139

ERIK H. ERIKSON Psychoanalytic Reflections on Einstein's Centenary 151

NATHAN ROTENSTREICH Relativity and Relativism 175

YEHUDA ELKANA The Myth of Simplicity 205

YARON EZRAHI Einstein and the Light of Reason 253

EINSTEIN AND DEVELOPMENTS IN THE JEWISH WORLD

ISAIAH BERLIN Einstein and Israel 281

YITZHAK NAVON On Einstein and the Presidency of Israel 293

URIEL TAL Jewish and Universal Social Ethics in the Life and Thought of Albert Einstein 297

FRITZ STERN Einstein's Germany 319

EINSTEIN AND THE NUCLEAR AGE

PAUL DOTY Einstein and International Security 347

BERNARD T. FELD Einstein and the Politics of Nuclear Weapons 369

WORKING WITH EINSTEIN: REMINISCENCES BY ASSOCIATES AND FRIENDS

PETER G. BERGMANN Reminiscences 397

BANESH HOFFMANN Reminiscences 401

NATHAN ROSEN Reminiscences 405

BORIS SCHWARZ Musical and Personal Reminiscences of Albert Einstein 409

ERNST G. STRAUS Reminiscences 417

Index of Names 425

Subject Index 431

Gerald Holton

INTRODUCTION: EINSTEIN AND THE SHAPING OF OUR IMAGINATION

THE ONE-HUNDREDTH anniversary of Albert Einstein's birth occasioned an outpouring of celebrations and scholarly meetings all over the world, and in the most diverse forums. These were touching and sometimes overwhelming testimonies to the continuing role that Einstein's ideas and ideals still play, both in the aspirations of contemporary scientists and in the popular imagination.

Those who attended several of these occasions in 1979 and 1980 will, I believe, generally agree on the special significance of the international meeting held in Jerusalem from 14 to 23 March 1979—the Jerusalem Einstein Centennial Symposium. The organizing idea was the ambitious one of using the occasion of the centenary to elucidate some of the main influences of Einstein's work on our time and over the whole spectrum—from science to humanistic studies, from the problems of international security in the Nuclear Age to the special meaning Einstein had and has for the nation where the meeting took place. Einstein was of course first of all a builder of a new view of the physical universe; but in addition to contributing to many branches of physical science, he also published widely on social and philosophical issues. He challenged current philosophies, both of science and of the state. He waged a constant fight for individual liberty and dignity and against persecution and war. His views on science and society, or views attributed to him, reverberate to this day in fields as distant from his own direct scientific contributions as psychology, linguistics, the analysis of modern art, and the study of the impact of science and technology on ethics.

In addition, the planning committee designed the Jerusalem program so as to take advantage of the individual testimonies of some of the foremost persons whose thoughts and actions in various fields had been shaped by Einstein's presence or legacy. This opportunity was not only irresistible—it amounted to a historic duty. One can hope that many of the papers given at the Symposium will be read for a long time to come as firsthand analyses of Einstein's impact on the total culture of our century. (How useful it would be if today's historians had available to them the record of some analogous meeting of distinguished figures of the first half of the eighteenth century on the occasion of Newton's first centenary!)

It may help to set the stage for reading these essays if I sketch out the program as a whole. The Symposium opened on the evening of March 14, Einstein's birthday proper, with the city luminous under the nearly full moon. Sir Isaiah Berlin delivered his paper after the ceremonial opening. There followed a day of historical and sociological analyses of the place of Einstein's science in the science of his own time and of the reception of his ideas by his contemporaries. Presentations included those by the historians of science Loren Graham, Max Jammer, Martin J. Klein, and Arthur I. Miller and the physicists Peter Bergmann and Banesh Hoffmann. On the third day, the focus began to shift to Einstein's influence in fields outside his own direct scientific work, including the reflections of Erik H. Erikson on psychology, Roman Jakobson on the science of language, and Nathan Rotenstreich on philosophy.

Fritz Stern's and Uriel Tal's papers followed when the discussion turned to Einstein and developments in the Jewish world. Thereafter came the presentations on Einstein and the culture of our time, including those by Yehuda Elkana and Yaron Ezrahi; papers by Paul Doty and Bernard T. Feld on the effects of modern science and technology on policy in the Nuclear Age; and a special evening round table entitled "Working with Einstein" (reminiscences by associates and assistants of Einstein). After these largely historical and humanistic sessions—the fruits of which constitute the present volume—there followed three days of morning and afternoon meetings on current physics and cosmology, involving many of the foremost scientists. By prior agreement among the organizers, the latter essays are to be published in a separate book,[1] except for P. A. M. Dirac's paper, which is included here.

Throughout the two weeks of the Symposium, other occasions had also been arranged, from an exhibit of some of Einstein's papers at Hebrew University—an institution he helped to found and to which all his papers are to go, according to his last will—to receptions given by the President of the State and the Mayor of the City, to three large lectures for the public of Jerusalem given by Professors S. Sambursky, J. Shaham, and H. Harari. President Yitzhak Navon's remarks fitted well into the plan of the historical portion of the Symposium, and we were glad to receive his permission to publish them here.

The idea of holding such a meeting, addressed to an audience of scientists, scholars, cultural leaders, statesmen, and students, originated in discussions between the President of the Israel Academy of Sciences and Humanities, Professor Aryeh Dvoretzky, and the President of the Aspen Institute for Humanistic Studies, Mr. Joseph E. Slater.[2] Throughout the planning, the chief guide was Einstein's own example of reaching out not only to scholars in science and outside but also to all others who are touched by the fruits of the scientific imagination. He wrote, "All religions, arts, and sciences are branches of the same tree. All these aspirations are directed towards ennobling man's life, lifting it from the sphere of mere physical existence and leading the individual toward freedom." And, on the occasion of a day of celebration, he added, "A memory of the best of the past is proper to stimulate the well-disposed of today to a courageous effort."

∞

The essays that follow deal in various ways with the question of what Einstein's "influence" has been on the world in which he began to launch his ideas soon after the turn of our century. Rather than repeat or summarize the essays in this introduction, I prefer to supplement their coverage and argument. But one's first thought must of course go to science itself. And there it is significant to note how long it took, by present standards, for his seminal, early work to be understood even by his fellow physicists. Six years elapsed after the first publication of the special theory of relativity before it established itself sufficiently to merit a textbook (Max von Laue's *Das Relativitätsprinzip*), and for some years after that the theory continued to be confused by most scientists with the electrodynamics of H. A. Lorentz. Einstein's ideas on quantum physics, published from 1905 on, were also generally neglected or discounted for years. R. A. Millikan, on accepting his Nobel Prize for 1923, confessed that the validity of Einstein's "bold, not to say reckless" explanation of the photoelectric effect forced itself on him slowly, "after ten years of testing . . . [and] contrary to my own expectation." The transcripts of the questions asked in scientific meetings in the decade after 1905 contain many passages that demonstrated to the historian of science the large intellectual effort required at the time to enter fully into the meaning of the new physics.

Today, virtually every student who wishes can learn at least the elements of relativity or quantum physics before leaving high school, and the imprint of Einstein's work on the different areas of physical science is so large and varied that a scientist who tries to trace it would be hard put to it to know where to start. A modern dictionary of scientific terms contains thirty-five entries bearing his name, from "einstein: A unit of light energy used in photochemistry" and "Einstein-Bose statistics" to "Einstein tensor" and "Einstein viscosity equation."[3] It is ironic that now, a quarter of a century after his death, there is in many branches of the physical sciences more awareness of his generative role than would have been credited during the last decade or two of his life. His ideas became essential for laying out conceptual paths for contemporary work in astronomy or cosmology, for unifying gravitation with the quantum field theory of gauge fields, or even for understanding new observations that were not possible in his time but were predicted by him (as in his 1936 paper on the optical lens formed by gravitational fields).

Apart from changing science itself, Einstein has reached into the daily life of virtually every person on the globe in direct or indirect ways through the incorporation of his ideas on physics into a vast range of technical devices and processes. I need cite only some of the most obvious ones. Every photoelectric cell can be considered one of his intellectual grandchildren. Hence, we are in his debt whenever photo emission or absorption is used, in the home or on the job, to capture an image by means of a television camera, or to project the optical soundtrack of a motion picture, or to set the page of a book or newspaper by photocomposition, or to make a telephone call over a modern fiber cable, or (eventually) to replace the oil-fired heater by an array of photovoltaic cells. In each case, if a law required a label on the appliance giving its intellectual content or pedigree, such a display would list prominently: "Einstein, *Annalen der Physik* **17** (1905), pp. 132-148; **20** (1906), pp. 199-206," etc.

One would find an entry of this sort also on the laser, whose beam was probably used to lay out the highway on which one travels to the office or to site the office building itself ("Einstein, *Physikalische Zeitschrift* **18** (1917), pp. 121-128," etc.). Or again, the same kind of answer comes if one lists key ideas that helped to make possible modern electric machinery, such as power generators, or precision clocks that allow the course of planes and ships to be charted. Einstein appears also, if one looks for the ancestry of the ideas, in quantum and statistical physics, by which solid-state devices operate, from calculators and computers to the transistor radio and the ignition system; and once more, even when one takes one's vitamin pill or other pharmaceutical drug, for it is likely that its commercial production involved diffusion processes, first explained in Einstein's papers on Brownian movement and statistical mechanics ("Einstein, *Annalen der Physik* **17** (1905), pp. 549-560," etc.).

As Edward M. Purcell remarked in his lecture at the Centennial Symposium at Princeton a few days before the Jerusalem Symposium, since the magnetism set up by electric currents is a strictly relativistic effect, derivable from Coulomb's law of electrostatics and the kinematics of relativity, and nothing more, it requires no elaboration to discuss "special relativity in engineering": "This is the way the world *is*. And it does not really take gigavolts or nanoseconds to demonstrate it; stepping on the starter will do it!" It is not too much to say that even in our most common experiences, that unworldly theoretician's publications help to explain what happens to us all day—indeed, from the moment we open our eyes on the light of the morning, since the act of seeing is initiated by a photochemical reaction ("Einstein, *Annalen der Physik* **37** (1912), pp. 832-838; **38** (1912), pp. 881-884," etc.).

The proverbial man in the street is quite blissfully ignorant of all that, and has preferred to remain so, even while expecting fully that, mysteriously yet automatically, a stream of practical, benign "spin-offs" continues from the pursuit of pure science. But the philosopher, the writer, the artist, and many others outside the scientific laboratories could not help but be caught up to some extent by the wave that spread beyond science and technology, at first slowly, then with astonishing intensity. As the best scientists were coming to understand what Einstein had done, the trumpets began to sound. Even Max Planck, a person conservative in thought and expression, enthused by 1912: "This new way of thinking . . . well surpasses in daring everything that has been achieved in speculative scientific research, even in the theory of knowledge. . . . This revolution in the physical *Weltanschauung*, brought about by the relativity principle, is to be compared in scope and depth only with that caused by the introduction of the Copernican system of the world."[4] At the same time, on the other end of the philosophical spectrum, the followers and heirs of Ernst Mach rushed to embrace relativity as a model triumph of positivistic philosophy. In the inaugural session of the Gesellschaft für positivistische Philosophie in Berlin (11 November 1912), relativity was interpreted as an antimetaphysical and instrumentalist tract and was hailed as "a mighty impulse for the development of the philosophical point of view of our time." When in London on 6 November 1919 the result of the British eclipse expedition was revealed to bear out one of the predictions of general relativity theory, the discussion of implications rose to fever pitch among scholars and laymen, beginning with declarations such as that in *The Times* of London (8 November 1919): the theory had served "to overthrow the certainty of ages, and to require a new philosophy, a philosophy that will sweep away nearly all that has hitherto been accepted as the axiomatic basis of physical thought." It became evident that, as Newton had "demanded the muse" after the *Principia*, now it would be Einstein's turn.

In fact, Einstein did his best to defuse the euphoria and excess of attention that engulfed and puzzled him from that time on. When asked to explain the mass enthusiasm, his answer in 1921 was that "it seemed psychopathological." The essence of the theory was chiefly "the logical simplicity with which it explained apparently conflicting facts in the operation of natural law," freeing science of the burden of "many general assumptions of a complicated nature."[5] That was all. As for being labeled a great revolutionary, as his friends and opponents did equally, Einstein took every opportunity to disavow it. He saw himself essentially as a continuist, had specific ideas on the way scientific theory developed by evolution,[6] and attempted to keep the discussion limited to work done and yet to be done in science. He did not get much help, however. Thus, the physicist J. J. Thomson reported that the Archbishop of Canterbury, Randall Davidson, had been told by Lord Haldane "that relativity was going to have a great effect upon theology, and that it was his duty as Head of the English Church to make himself acquainted with it. . . . The Archbishop, who is the most conscientious of men, has procured several books on the subject and has been trying to read them, and they have driven him to what it is not too much to say is a state of *intellectual desperation.*" On Einstein's first visit to England in June 1921, the Archbishop of Canterbury therefore sought him out to ask what effect relativity would have on religion. Einstein replied briefly and to the point: "None. Relativity is a purely scientific matter and has nothing to do with religion."[7] But of course this did not dispose of the question. Later that year, even the scientific journal *Nature* felt it necessary to print opposing articles on whether "Einstein's space-time is the death-knell of materialism."[8]

Although the crest of the flood, and the worst excesses, have now passed, debates of this sort continue. More constructively, since modern philosophy is concerned in good part with the nature of space and time, causality, and other conceptions to which relativity and quantum physics have contributed, Einstein has had to be dealt with in the pages of philosophers, from Henri Bergson and A. N. Whitehead to the latest issues of the professional journals. As John Passmore observed correctly, it appeared in this century that "physics fell heir to the responsibility of metaphysics."[9]

Some philosophers and philosopher-scientists have confessed—and in the framework of this collection this is of special interest—that Einstein's work started them off on their speculations in the first place, thus giving some direction to their very careers. One example is P. W. Bridgman, who disclosed that the effort to clarify in his mind the issues in relativistic electrodynamics, when first asked to teach that course, drew him to the task of writing the influential book *The Logic of Modern Physics* (1927). Another case is Karl Popper, who in his recent autobiography reveals that his falsification criterion originated in his interpretation of a passage in Einstein's popular exposition of relativity, which Popper says he read with profound effect when he was still in his teens.

Philosophy was no doubt destined to be the most obvious and often the earliest and most appropriate field, outside science itself, that the radiation from Einstein's work would reach. But soon there were others, as several essays in this collection show, even though the connections made or asserted were not always valid. From Einstein's wide-ranging output, relativity was invoked most frequently. Cultural anthropology, in Claude Lévi-Strauss's phrase, had evolved the doctrine of cultural relativism "out of a deep feeling of respect toward other cultures than our own"; but this doctrine often invited confusion with physical relativity. Much that has been written on "ethical relativity" and on "relativism" is based on a seductive play with words. And painters and art critics have helped to keep alive the rumor of a supposed genetic connection of visual arts with Einstein's 1905 publication.

Here again, Einstein protested when he could and, as so often, without effect. One art historian submitted to him a draft of an essay entitled "Cubism and the Theory of Relativity," which argued for such a connection—for example, that in both fields "attention was paid to relationships, and allowance was made for the simultaneity of several views."[10] Politely but firmly, Einstein tried to put him straight, and he explained the difference between physical relativity and vulgar relativism so succinctly as to invite an extensive quotation:

> The essence of the theory of relativity has been incorrectly understood in it [your paper], granted that this error is suggested by the attempts at popularization of the theory. For the description of a given state of facts one uses almost always only one system of coordinates. The theory says only that the general laws are such that their form does not depend on the choice of the system of coordinates. This logical demand, however, has nothing to do with how the single, specific case is represented. A multiplicity of systems of coordinates is not needed for its representation. It is completely sufficient to describe the whole mathematically in relation to one system of coordinates.
>
> This is quite different in the case of Picasso's painting, as I do not have to elaborate any further. Whether, in this case, the representation is felt as artistic unity depends, of course, upon the artistic antecedents of the viewer. This new artistic "language" has nothing in common with the Theory of Relativity.[11]

Einstein might well have added here, as he did elsewhere, that the existence of a multiplicity of frames, each one as good as the next for solving some problems in mechanics, went back to the seventeenth century (Galilean relativity). As to the superposition of different aspects of an object on a canvas, that had been done for a long time; thus Canaletto drew various parts of a set of buildings from different places and merged them in a combined view on the painting (for example, *Campo S. S. Giovanni e Paolo*), the view becoming thereby an impossible *veduta*.

It was therefore doubly wrong to invoke Einstein as authority in support of the widespread misunderstanding that physical relativity meant that all frameworks, points of view, narrators, fragments of plot or thematic elements are created equal, that each of the polyphonic reports and contrasting perceptions is as valid or expedient as any other, and that all of these, when piled together or juxtaposed, *Rashomon*-like, somehow constitute the real truth. If anything, twentieth-century relativistic physics has taught the contrary: that under certain conditions we can extract from different reports, or even from the report originating in one frame properly identified, all the laws of physics, each applicable in any framework, each having therefore an invariant meaning, one that does not depend on the accident of which frame one inhabits. It is for this reason that, by comparison with classical physics, modern relativity is simple, universal, and, one may even say, "absolute." The cliché became, erroneously, that "everything is relative," whereas the whole point is that out of the vast flux one can distill the very opposite: "some things are invariant."

The cost of the terminological confusion has been so great that a brief elaboration on this point will be relevant. Partly because he saw himself as a continuist rather than as an iconoclast, Einstein was reluctant to present this new work as a new *theory*. The term "relativity theory," which made the confusions in the long run more likely, was provided by Max Planck in 1907. For a time Einstein referred to it in print as the "so-called relativity theory," and until 1911 he avoided using the term altogether in the titles of his papers on the subject. In informal correspondence Einstein seemed happier with the term *Invariantentheorie*, which is of course much more true to its method and aim. How much nonsense we might have been spared if Einstein had adopted that term, even with all its shortcomings! To a correspondent who suggested just such a change, Einstein replied (letter to E. Zschimmer, 30 September 1921): "Now to the name relativity theory. I admit that it is unfortunate, and has given occasion to philosophical misunderstandings. . . . The description you proposed would perhaps be better; but I believe it would cause confusion to change the generally accepted name after all this time."

To come back to Einstein's careful disavowal of a substantive genetic link between modern art and relativity: far from abandoning the quest for it, his correspondent forged onward enthusiastically and published three such essays instead of one. Newton did not always fare better at the hands of eighteenth-century literati and divines who thought they were following in his footsteps. Poets rush in where scientists no longer fear to tread. And why not, if the apparent promises are so great? In April 1921, at the height of what Einstein on his first journey to the United States all too easily diagnosed as a pathological mass reaction, William Carlos Williams published a poem entitled "St. Francis Einstein of the Daffodils,"[12] containing such lines as "April Einstein / . . . has come among the daffodils / shouting / that flowers and men / were created / relatively equal. . . ." Declaring simply that "relativity applies to everything"[13] and that "Relativity gives us the clue. So, again, mathematics comes to the rescue of the arts," Williams felt encouraged to adopt a new variable measure for his poems—calling it "a *relatively* stable foot, not a rigid one"[14]—that proved of considerable influence on other poets.

Williams was of course not alone. Robert Frost, Archibald MacLeish, E. E. Cummings, Ezra Pound, T. S. Eliot, and some of their disciples (and outside the English-speaking world, others such as Thomas Mann and Hermann Broch) referred directly to Einstein or to his work. Some were repelled by the vision thought to be opened by the new science, but there were at least as many who seemed to be in sympathy with Jean-Paul Sartre's remark that "the theory of relativity applies in full to the universe of fiction."[15] Perhaps the most cheerful of the attempts to harness science and literature to common purpose is Lawrence Durrell's entertaining set of novels, *The Alexandria Quartet*, of which its author says by way of preface: "Modern literature offers us no Unities, so I have turned to science and am trying to complete a four-decker novel whose form is based on the relativity proposition. Three sides of space and one of time constitute the soup-mix recipe of a continuum."[16] The intention is to use the properties of space and time as determining models for the structure of the book. Durrell says "the first three parts . . . are to be deployed spatially . . . and are not linked in a serial form. . . . The fourth part alone will represent time and be a true sequel."

For that alone one would not have had to wait for Einstein. But more seems to be hoped for; that, and the level of understanding, is indicated by the sayings of Pursewarden recorded in the novel. Pursewarden—meant to be one of the foremost writers in the English language, his death mask destined to be placed near those of Keats and Blake—is quoted as saying, "In the Space and Time marriage we have the greatest Boy meets Girl story of the age. To our great-grandchildren this will be as poetical a union as the ancient Greek marriage of Cupid and Psyche seems to us." Moreover, "the Relativity proposition was directly responsible for abstract painting, atonal music, and formless . . . literature."[17]

Throughout the novel it is evident that Durrell has taken the trouble to read up on relativity, but chiefly out of impressionistic popularizations such as *The Mysterious Universe* by James Jeans, even though Durrell readily confessed that "none of these attempts has been very successful."[18] There is something touching and, from the point of view of an intellectual historian, even a bit tragic about the attempt. In his study, *A Key to Modern British Poetry*, Durrell revealed his valid concern to show that as a result of "the far-reaching changes in man's ideas" about the outer and inner universe, "language has undergone a change in order to keep in line with cosmological inquiry (of which it forms a part)."[19] Yet on page after page the author demonstrates that he has been misled by the simplifications of H. V. Routh and Jeans; he believes that Rutherford and Soddy suggested that the "ultimate laws of nature were not simply causal at all," that "Einstein's theory joined up subject and object," that "sofar as phenomena are concerned . . . the uniformity of nature disappears," and so forth.[20] The terrible but clarifying remark of Wolfgang Pauli comes to mind, who said about a physical theory that seemed to him doomed: "It is not even wrong."

If I have spelled out some of the misunderstandings by which Einstein's work, for better or worse, has been thought to have found its way into contemporary culture, I had to do so in justice to any occasion associated with his name. But the examples of incorrect interpretation prepare us to appreciate that much more the correct ones. Most of the essays in this book deal with those. I should confess that my own favorite example of the successful transmutation of scientifically based conceptions in the writer's imagination is a novel, and a controversial one. William Faulkner's *The Sound and the Fury* is more like an earthquake than a book. Immediately on publication in 1929 it caused universal scandal; for example, not until Judge Curtis Bok's decision in 1949 was this, among Faulkner's other novels, allowed to be sold in Philadelphia. On the surface it seems unlikely that this book—even a friendly reviewer characterized it as "designedly a silo of compressed sin"—has any resonance with the ideas of modern physics, by intent or otherwise. At the time he poured himself into the book, Faulkner was still almost unknown, largely self-taught, eking out a meager living as a carpenter, hunter, and coal carrier on the night shift of a power station, his desk the upturned wheelbarrow on which he would write while kneeling on the floor. Yet, even there, he was not isolated if he read even a small part of the flood of articles in newspapers, periodicals, and popular books in the 1920s dealing with the heady concepts of relativity theory—such as the time dilation experienced by a clock traveling through space, the necessity to recognize the meaninglessness of absolute time and space—and the recent quantum physics, with its denial of the comforts of classical causality. Particularly in America, Einstein was quoted down to the level of local evening papers and *Popular Mechanics*, resulting in wide circulation of such haunting epigrams as his remark, made in exasperation to Max Born (1926), that "God does not throw dice." Could any of this have reached Faulkner?

In the second of the four chapters of *The Sound and the Fury* we follow Quentin Compton of Jefferson, Mississippi, as he lives through a day in June 1910. It is the end of his freshman year at college and the culmination of a short life wrenched by the degeneration and guilt, the fixations and tribal racism of his whole haunted family—from his father Jason, drinking himself to death, to his idiot brother Benjamin, whose forty-acre pasture has been sold to send Quentin to college. The only resource of human affection he has known came from black laborers and servants, although they have been kept in the centuries-old state of terror, ignorance, and obeisance. But the Comptons are doomed. As the day unfolds, Quentin moves toward the suicide he knows he will commit at midnight.

It is all too easy to discover theological and Freudian motives woven into the text, and one must not without provocation drag an author for cross-examination into the physics laboratory when he has already suffered through interrogations at the altar and on the couch. But Faulkner asked for it. Let me select here from a much more extensive body of evidence in the novel itself.

Quentin's last day on earth is a struggle against the flow of time. He attempts to stem the flow, first by deliberately breaking the cover glass of the pocket watch passed down to him from grandfather and father, then twisting off the hands of the watch, and then launching on seemingly random travel, by streetcar and on foot, across the whole city. His odyssey brings him to the shop of an ominous, cyclopean watch repairer. Quentin forbids him to tell him the time but asks if any of the watches in the shop window "are right." The answer he gets is "No." But wherever he then turns, all day and into the night, he encounters chimes, bells ringing the quarter hours, a factory whistle, a clock in the Unitarian steeple, the long, mournful sound of the train tracing its trajectory in space and time, "dying away, as though it were running through another month." Even his stomach is a kind of space-time metronome. "The business of eating inside of you space too space and time confused stomach says noon brain says eat o'clock All right I wonder what time it is what of it." Throughout, Quentin carries the blinded watch with him, the watch that never knew how to tell real time and cannot even tell relative time. But it is not dead: "I took out my watch and listened to it clicking away, not knowing it could not even lie."[21] And in the streetcar, the clicking away of time is audible to him only while the car has come to a stop.

Quentin has taken a physics course that freshman year and uses it to calculate how heavy the weights must be that he buys to help drown himself. It is, he says wryly to himself, "the only application of Harvard," and as he reflects on it: "The displacement of water is equal to the something of something. Reducto absurdum of all human experience, and two six-pound flat-irons weigh more than one tailor's goose. What a sinful waste, Dilsey would say. Benjy knew it when Damuddy died. He cried."

As midnight approaches, before he is ready to put his "hand on the light switch" for the last time,[22] Quentin is overcome by torment, caused by the shamed memory of his incestuous love for his sister Candace, by her loss, and by his own sense of loss even of the meaningfulness of that double betrayal. In anguish he remembers his father's terrible prediction after he had made his confession:

> You cannot bear to think that some day it will no longer hurt you like this now were getting at it. . . . you wont do it under these conditions it will be a gamble and the strange thing is that man who is conceived by accident and whose every breath is a fresh cast with dice already loaded against him will not face that final main which he knows beforehand he has assuredly to face without essaying expedients. . . . that would not deceive a child until some day in very disgust he risks everything on a single blind turn of a card no man ever does that under the first fury of despair or remorse or bereavement he does it only when he has realized that even the despair or remorse or bereavement is not particularly important to the dark diceman. . . . it is hard believing to think that a love or a sorrow is a bond purchased without design and which matures willynilly and is recalled without warning to be replaced by whatever issue the gods happen to be floating at the time.

This was not the God Newton had given to his time—Newton, of whom, just two centuries before Faulkner's soaring outcry, the poet James Thomson had sung in 1729 that "the heavens are all his own, from the wide role of whirling vortices, and circling spheres, to their great simplicity restored." Nor, of course, was it Einstein's God, a God whose laws of nature are both the testimony of His presence in the universe and the proof of its saving rationality. But this, it seems to me, defines the dilemma precisely. If the poet neither settles for the relief of half-understood analogies nor can advance to an honest understanding of the rational structure of that modern world picture, and if he is sufficiently sensitive to this impotency, he must rage against what is left him: time and space are then without meaning; so is the journey through them; so is grief itself, when the very gods are playing games of chance, and all the sound and the fury signify nothing. And this leads to recognizing the way out of the dilemma, at least for a few. At best, as in the case of Faulkner, this rage itself creates the energy needed for a grand fusion of the literary imagination with perhaps only dimly perceived scientific ideas. There are writers and artists of such inherent power that the ideas of science they may be using are dissolved, like all other externals, and rearranged in their own glowing alchemical cauldron.

It should not, after all, surprise us; it has always happened this way. Dante and Milton did not use the cosmological ideas of their time as tools to demarcate the allowed outline or content of their imaginative constructs. Those students of ours who, year after year, write us dutifully more or less the same esssay, explaining the structure of the *Divine Comedy* or *Paradise Lost* by means of astronomy, geography, and the theory of optical phenomena—they may get the small points right, but they miss the big one, which is that the good poet is a poet surely because he can transcend rather than triangulate. In Faulkner, in Eliot's *The Waste Land*, in Woolf's *The Waves*, in Mann's *Magic Mountain* it is futile to judge whether the traces of modern physics are good physics or bad, for these trace elements have been used in the making of a new alloy. It is one way of understanding Faulkner's remark on accepting his Nobel Prize in 1950: the task was "to make out of the material of the human spirit something which was not there before."[23] And insofar as an author *fails* to produce the feat of recrystallization, I suspect this lack would not be cured by more lessons on Minkowski's space-time, or Heisenberg's indeterminacy principle, or even thermodynamics, although these lessons could occasionally have a prophylactic effect that might not be without value.

Here we suddenly remember that, of course, the very same thing is true for scientists themselves. The most creative ones, almost by definition, do not build their constructs patiently by assembling blocks that have been precast by others and certified as sound. On the contrary, they too melt down the ready-made materials of science and recast them in a way that their contemporaries tend to think is outrageous. That is why Einstein's own work took so long to be appreciated even by his best fellow physicists, as I noted earlier. His physics looked to them like alchemy, not because they did not understand it at all, but because, in one sense, they understood it all too well. From their thematic perspective, Einstein's was anathema. Declaring, by simple postulation rather than by proof, Galilean relativity to be extended from mechanics to optics and all other branches of physics; dismissing the ether, the playground of most nineteenth-century physicists, in a peremptory half-sentence; depriving time intervals of inherent meaning; and other such outrages, all delivered in a casual, confident way in the first, short paper on relativity—those were violent and "illegitimate" distortions of science to almost every physicist. As for Einstein's new ideas on the quantum physics of light emission, Max Planck felt so embarrassed by it when he had to write Einstein a letter of recommendation seven years later that he asked that this work be overlooked in judging the otherwise promising young man.

Moreover, the process of transformation characterizes not only science itself and the flow of ideas from high science to high literature. It also works across the boundaries in other ways. The most obvious example is Einstein's importation into his early physics of an epistemology that he himself thought, with some enthusiasm, to be based on Ernst Mach's kind of positivism. Mach had begun to find him out on this point even while Einstein was still signing his letters to Mach as "Your devoted Student."

It seems clear to me that without this process of transformation, willing or unwilling, of ideas from science and from philosophy, physics would not have come into its twentieth-century form. (A similar statement may well be made for the cases of Copernicus, Kepler, Galileo, and Newton.) The case of Einstein suggests, therefore, that the accomplishments of the major innovator—and not only in science—depends on his ability to persevere in four ways: by giving his loyalty primarily to his own belief system rather than to the current faith; by perceiving and exploiting the man-made nature and plasticity of human conceptions; by demonstrating eventually that the new unity he has promised does become lucid and convincing to lesser mortals active in his field—that he has it all "wrong" in the right way; and, in those rare cases, by even issuing ideas that lend themselves, quite apart from misuse and oversimplification, to further adaptation and transformation in the imagination of similarly exalted spirits who live on the other side of disciplinary boundaries.

∞

It remains to deal with one more, somewhat different mechanism by which Einstein's imprint came to be felt far beyond his own field of primary attention: the power of his personal intervention on behalf of causes ranging from the establishment of a homeland for a persecuted people (discussed in Isaiah Berlin's paper) to his untiring efforts, over four decades, for peace and international security (analyzed in Paul Doty's essay). In retrospect we can see that he had the skill, at strategic periods of history, to lend his ideas and prestige to the necessary work of a Chaim Weizmann or a Bertrand Russell. Even the most famous of these personal interventions, the call on President Roosevelt in 1939 to initiate a study whether the laws of nature allow anyone to produce an atomic weapon, was of that sort, although it has perhaps been misunderstood more widely than anything else Einstein did. He was, after all, correct in his perception that the Germans, who were pushing the world into a war, had all the skill and intention needed to start production of such a weapon if it was feasible. In fact, they had a headstart, and but for some remarkable blunders, they might have fulfilled the justified fears, with incalculable consequences on the course of civilization.

To highlight these personal interactions, I select one as more or less paradigmatic of the considerable effect Einstein had even in some brief or seemingly casual discussion with the right person. The illustration is particularly apt because, in response to our invitation, the Swiss psychologist Jean Piaget had expressed the hope that he could come to speak about it at the Symposium. This proved not possible after all; but the record allows a certain amount of reconstruction. Piaget's work entered its most important phase with the publication in 1946 of *The Child's Conception of Time.* The book begins with a plain acknowledgment: "This work was promoted by a number of questions kindly suggested by Albert Einstein more than fifteen years ago [1928, at a meeting in Davos]. . . . Is our intuitive grasp of time primitive or derived? Is it identical with our intuitive grasp of velocity? What if any bearing do these questions have on the genesis and development of the child's conception of time? Every year since then we have made a point of looking into these questions. . . . The results [concerning time] are presented in this volume; those bearing on the child's conception of motion and speed are reserved for a later work."[24] Throughout his later writings, Piaget remarks on this debt: "It was the author of the theory of relativity who suggested to us our work,"[25] or "Einstein once suggested we study the question from the psychological viewpoint and try to discover if there existed an intuition of speed independent of time."[26] In addition, Piaget refers explicitly to notions of relativity and other aspects of Einstein's work.[27]

Einstein came to have an immense correspondence with leaders in virtually every type of endeavor. Much of that has been preserved, thanks largely to the devoted labors and care of his longtime secretary, Miss Helen Dukas. These documents have now been catalogued for scholarly research, and there is hope that the task of actual publication of the letters, draft manuscripts, and other materials, which Princeton University Press has accepted, will begin soon. Working in this archive has made it already clear that it will not be difficult to reconstruct instances of the same sort as Einstein's brief but seminal interaction with Piaget.[28]

∞

Looking back at the variety of ways in which Einstein came to impress the imagination of his time and ours, we can discern some rough categories, spread out, as it were, in a spectrum from left to right. At the center portion, corresponding to the largest intensity, one finds the widespread but unfocused and mostly uninformed fascination, manifested in a variety of ways, from enthusiastic mass gatherings to glimpse the man, to the outpouring of popularizations with good intentions, to responses that betray the vague discomfort aroused by the ideas. A good example of the last is an editorial entitled "A Mystic Universe" in the *New York Times* of 28 January 1928 (p. 14): "The new physics comes perilously close to proving what most of us cannot believe. . . . Not even the old and much simpler Newtonian physics was comprehensible to the man on the street. To understand the new physics is apparently given only to the highest flight of mathematicians. . . . We cannot grasp it by sequential thinking. We can only hope for dim enlightenment." The editorial writer then notes that the ever-changing scene in physics does offer some "comfort":

> Earnest people who have considered it their duty to keep abreast of science by readapting their lives to the new physics may now safely wait until the results of the new discoveries have been fully tested out by time, harmonized and sifted down to a formula that will hold for a fair term of years. It would be a pity to develop an electronic marriage morality and find that the universe is after all ether, or to develop a wave code for fathers and children only to have it turn out that the family is determined not by waves but by particles. Arduous enough is the task of trying to understand the new physics, but there is no harm in trying. Reshaping life in accordance with the new physics is no use at all. Much better to wait for the new physics to reshape our lives for us as the Newtonian science did.

Similarly, in Tom Stoppard's play *Jumpers* a philosopher is heard to ask: "If one can no longer believe that a twelve-inch ruler is always a foot long, how can one be sure of relatively less certain propositions?"

Near this position, as we said, are the enthusiastic misapplications, usually achieved by an illicit shortcut of meaning from, say, the true statement that the operational definition of length is "framework" dependent, to the invalid deduction that mental phenomena in a human observer have thereby been introduced into the very definitions of physical science. (To be sure, the layman has not always been served by the explanation on this point given by the scientists themselves; for example, in such pronouncements as "the object of research is no longer nature in itself but rather nature exposed to man's questioning, and to this extent man here also meets himself."[29]) The irony here is that the first lessons we learned from relativity physics were that short circuits in signification must be avoided, for they were just what burdened down classical physics, and that attention must be paid as never before to the meaning of the terms we use.

When we now glance further toward the left, or blue, end of the spectrum, the expressions of resignation or futility become more explicit. Indeed, among some of the most serious intellectuals there seems, on this point, a sense of despair. By the very nature of their deep motivation they must feel most alienated from a universe whose scientific description they can hardly hope to understand except in a superficial way. The much-admired humanistic scholar Lionel Trilling spoke for many when he stated the dilemma frankly and eloquently:

> The operative conceptions [of science] are alien to the mass of educated persons. They generate no cosmic speculation, they do not engage emotion or challenge imagination. Our poets are indifferent to them. . . .
>
> This exclusion of most of us from the mode of thought which is habitually said to be the characteristic achievement of the modern age is bound to be experienced as a wound given to our intellectual self-esteem. About this humiliation we all agree to be silent, but can we doubt that it has its consequences, that it introduced into the life of mind a significant element of dubiety and alienation which must be taken into account in any estimate that is made of the present fortunes of mind?[30]

Einstein, who had intended originally to become a science teacher, came to understand this syndrome, and the obligation it put on him. He devoted a good deal of time to popularization of his own. His avowed aim was to simplify short of distortion. In addition to a large number of essays and lectures, he wrote, and repeatedly updated, a short book on relativity that he promised in the very title to be *gemeinverständlich.*[31] It is, however, overly condensed for most nonscientific beginners. Later, Einstein collaborated with Leopold Infeld in a second attempt to reach out to the population at large by means of a book-length treatment of modern physics. As the preface acknowledged, the authors no longer attempted "a systematic course in elementary facts and theories." Rather, they aimed at a historical account of how the ideas of relativity and quanta entered science, "to give some idea of the eternal struggle of the inventive human mind for a fuller understanding of the laws governing physical phenomena."[32] In fact, there is to this day no generally agreed source, the *reading* of which by itself will bring a large fraction of its nonscientific audiences to a sound enough understanding of these ideas, even for those who truly want to attain it and are willing to pay close attention. I believe it is a fact of great consequence that it takes a much larger effort, and one starting earlier than most people undertake. To make matters worse, so little has been found out about how scientific literacy is achieved or resisted that little blame can be spun off on the would-be students, young or old.

Going now further along the spectrum in the same direction, we encounter outright hostility and opposition to Einstein's work, either on scientific or on ideological grounds. Almost all scientists, even those initially quite reluctant, became eventually at least reconciled, save (to this day) for Einstein's famous refusal to regard the statistical interpretation as fundamental. Even on that point, as P. A. M. Dirac's paper in this volume notes, the last word may not have been said. On the other hand, the opposition to Einstein's work on grounds other than scientific has had a longer history. Thus, a number of studies now exist that show the lengths to which various totalitarian groups, for various reasons, felt compelled to go in their attacks.

Turning now to the other, more "positive" half of the spectrum, we see there the gradual acceptance and elaboration of Einstein's work within the corpus of physical science; its penetration into technology (largely unmarked) and into the more thoughtful philosophies of science; Einstein's effect through his personal intervention, causing some historic redirections of research; and its passage into the scientific world picture of our time, as it tries to achieve a unification that eluded Einstein. And beyond that, at the end of the spectrum, where the number of cases is small but the color deep and vibrant, we perceive the examples of creative transformation beyond science. Those are the works of the few who found that scientific ideas, or rather *metaphors* embodying such ideas, released in them a fruitful response with an authenticity of its own, far removed from textbook physics.

This last is the oldest and surely still the most puzzling interplay between the science and the rest of culture. Evidently, the mediation occurs through a sharing of an analogy or metaphor—irresistible, despite the dangers inherent in the obvious differences or discontinuities. We know that such a process exists, because any major work of science itself, in its nascent phase, is connected analogically rather than fully logically, both with the historic past in that science and with its supporting data. The scientist's proposal may fit the facts of nature as a glove fits a hand, but the glove does not uniquely imply the hand, nor the hand the glove.

Einstein spoke insistently over the decades about the need to recognize the existence of such a discontinuity, one that in his early scientific papers asserted itself first in his audacious method of postulation. In essay after essay, he tried to make the same point, even though it had little effect on the then reigning positivism. Typical are the phrases in his Herbert Spencer Lecture of 1933.[33] The rational and empirical components of human knowledge stand in "eternal antithesis," for "propositions arrived at by purely logical means are completely empty as regards reality." In this sense, the "fundamentals of scientific theory," being initially free inventions of the human mind, are of "purely fictional character." The phenomenic-analytic dichotomy makes it inherently impossible to claim that the principles of a theory are "deduced from experience" by "abstraction," that is, by logically complete claims of argument. As he put it soon afterwards, the relation between sense experience and concept "is analogous not to that of soup to beef, but rather that of check number to overcoat."[34]

If this holds for the creative act in science itself, we should hardly be surprised to find the claim to be extended to more humanistic enterprises. The test, in both instances, is of course whether the freely invented check token yields the intended use of a suitable overcoat. The existence of both splendid scientific theories and splendid products of the humanistic imagination shows that despite all their other differences, they share the ability to build on fundamentals of a "purely fictional character." And even the respective fundamentals, despite all their differences, can share a common origin. That is to say, at a given time the cultural pool contains a variety of themata and metaphors, and some of these have a plasticity and applicability far beyond their initial provenance. The innovator, whether a scientist or not, necessarily dips into this pool for his fundamental notions and in turn may occasionally deposit into it new or modified themata and metaphors of general power.

Examples of such science-shaped metaphors, each of these by no means a "fact" of the external world, yet revealing immense explanatory energy, are easy to find: Newton's "innate force of matter (*vis insita*)" and the Newtonian clockwork universe; Faraday's space-filling electric and magnetic lines of force; Niels Bohr's examples of complementarity in physics and in daily life; Mendeleev's neat tableau setting for the families of elements and Rutherford's long parent-daughter-granddaughter chains of decaying atoms; Minkowski's space-time "World," of which our perceptible space and time are like shadows playing on the wall of Plato's cave; and of course the imaginative scenes Einstein referred to—the traveler along the light beam, the calm experimenter in the freely falling elevator, the dark, dice-playing God, the closed but unbounded cosmos, the Holy Grail of complete unification of all forces of nature. So it continues in science.

The allegorical use of such conceptions may, as we have noted, help to shape works of authenticity outside the sciences. And the process works both ways. Thus Niels Bohr acknowledged that his reading in Kierkegaard and William James helped him to the imaginative leap embodied in his physics, Einstein stressed the influence on his early scientific thinking of the philosophical tracts of that period, and Heisenberg noted the stimulus of Plato's *Timeaus*, read in his school years. No matter if such "extraneous" elements are eventually suppressed or forgotten, or even have to be overcome; at an early point they can encourage the mind's struggle.

∞

We conclude, then—and, I trust, in the spirit of Einstein—that in pursuing the evident and documentable cases of "impact" of one person or field on another, we have been led to a more mysterious fact, namely, the mutual adaptation and resonance of the innovative mind with portions of the total set of metaphors current at a given time. The philosopher José Ortega y Gasset was one of those who struggled with this idea. In 1921-1922, evidently caught up by the rise of the new physics, he began an essay on "The Historical Significance of the Theory of Einstein."[35] There he remarked quite correctly that the most relevant issue was not that

> the triumph of the theory will influence the spirit of mankind by imposing on it the adoption of a definite route. . . . What is really interesting is the inverse proposition: the spirit of man has set out, of its own accord, upon a definite route, and it has therefore been possible for the theory of relativity to be born and to triumph. The more subtle and technical ideas are, the more remote they seem from the ordinary preoccupations of men, the more authentically they denote the profound variations produced in the historical mind of humanity.[36]

But to this day, attempts to go much beyond that point, it seems to me, have not been very successful. Perhaps the tantalizing task will have borne fruit by the time scholars meet for discussion at Einstein's Second Centennial Symposium.

ACKNOWLEDGMENTS

It is a pleasure to acknowledge the help received from Professor Dvoretzky and Mr. Slater and their associates throughout the many months of planning before the conference and in the editorial work afterward; from Professor Elkana and his staff at the Van Leer Jerusalem Foundation; from Dr. Katherine Sopka and Ms. Joan Laws of my own staff; from the organizations that provided the essential financial support for assembling the Symposium itself; from Abraham Pais, Heinz Maier-Leibnitz, Christian Møller, Gershom Scholem, Robert K. Merton, Margaret Gowing, Alfred Kastler, and John A. Wheeler, who were among the persons acting as chairmen of the sessions at which these papers were given; from Miss Helen Dukas and the Estate of Albert Einstein, whose cooperation over many years has been essential to my work; and from Herbert S. Bailey and his associates at the Princeton University Press, whose care and interest helped to bring this volume into print.

NOTES

1. Yuval Ne'eman, ed., *To Fulfill a Vision: Jerusalem Einstein Centennial Symposium on Gauge Theories and Unification of Physical Sources*, Reading, Mass., Addison-Wesley, in press.

2. Ultimately, the Symposium was organized jointly by the Israel Academy of Sciences and Humanities, the Hebrew University of Jerusalem, the Van Leer Jerusalem Foundation, the Jerusalem Foundation, and the Aspen Institute for Humanistic Studies. Financial support for the Symposium came from the Joseph H. Hazen Foundation, from The Rockefeller Foundation, and from the five organizing groups listed above.

The operational organizing committee, appointed by the Academy, consisted of Aryeh Dvoretzky, Yehuda Elkana, Gerald Holton, Max Jammer, Yuval Ne'eman, Nathan Rosen, Nathan Rotenstreich, and Igal Talmi.

The list of Honorary Sponsors of the Symposium included the International Council of Scientific Unions, the International Union of Pure and Applied Physics, International Union of the History and Philosophy of Science, Committee on General Relativity and Gravitation, Institute for Advanced Study (Princeton), Académie Royale des Sciences, des Lettres et des Beaux-Arts de Belgique, Royal Danish Academy of Sciences and Letters, Institut de France-Académie des Sciences, Deutsche Forschungsgemeinschaft, Accademia Nazionale dei Lincei, The Japan Academy, Royal Netherlands Academy of Arts and Sciences, The Norwegian Academy of Science and Letters, Royal Swedish Academy of Sciences, Swiss Academy of Sciences, The Royal Society of London, National Academy of Sciences (U.S.A.), Council for Higher Education in Israel, Association for the Advancement of Science in Israel, Bar-Ilan University, Ben-Gurion University of the Negev, Haifa University, Technion-Israel Institute of Technology, Tel-Aviv University, and Weizmann Institute of Science.

3. Daniel N. Lapedes, ed., *McGraw-Hill Dictionary of Scientific and Technical Terms*, 2d ed., New York, McGraw-Hill, 1978, pp. 512-513. As another measure in the continuing, albeit sometimes only ritualistic, reference made in the ongoing research literature to Einstein's publica-

tions, Eugene Garfield has found that during the period 1961-1975 the serious scientific journals *in toto* carried no less than 40 million citations to previously published articles. Of these, 58 cited articles stand out by virtue of having been published before 1930 *and* cited over 100 times each; and among these 58 enduring classics, ranging from astronomy and physics to biomedicine and psychology, 4 are Einstein's. See E. Garfield, *Current Contents* **21** (1976), pp. 5-9.

4. Quoted in Ernst Lechner, *Physikalische Weltbilder*, Leipzig, Theodore Thomas Verlag, [1912], p. 84.

5. Quoted in the *New York Times*, 3, 4, and 5 April 1921.

6. For discussion and documentation, see G. Holton, "Einstein's Search for the '*Weltbild*,'" *Proceedings of the American Philosophical Society* **125**, no. 1 (1981), pp. 1-15.

7. Quoted from J. J. Thomson, *Reflections and Recollections*, London, G. Bell and Sons, Ltd., 1936, p. 431 (italics in original). See also Philipp Frank, *Einstein: His Life and Times*, New York, Alfred A. Knopf, 1947, p. 190. Frank's book is one of the good sources for documentation on the reception and rejection of Einstein's theories by various religions and philosophic and political systems, ranging from Cardinal O'Connell's assessment that "those theories [Einstein's as well as Darwin's] became outmoded because they were mainly materialistic and therefore unable to stand the test of time" (p. 262), to the attempt of a Nazi scientist to overcome his aversion sufficiently to "recommend Einstein's theory of relativity to National Socialists" as a weapon in the fight against "materialistic philosophy" (p. 351).

8. These articles, and excerpts from some other publications dealing with the influence of Einstein's work, have been gathered in L. Pearce Williams, ed., *Relativity Theory: Its Origins and Impact on Modern Thought*, New York, John Wiley and Sons, 1968. It is a useful compendium, and I am indebted to it for a number of illustrations to be referred to below.

9. John Passmore, *A Hundred Years of Philosophy*, rev. ed., New York, Basic Books, 1966, p. 332.

10. As reported by Paul M. LaPorte, "Cubism and Relativity, with a Letter of Albert Einstein," *Art Journal* **25**, no. 3 (1966), p. 246.

11. Ibid. See also C. H. Waddington, *Behind Appearances: A Study of the Relations between Painting and the Natural Sciences in this Century*, Edinburgh, Edinburgh University Press, 1969, Cambridge, Mass., MIT Press, 1970, pp. 9-39. At the Jerusalem Symposium, Professor Meyer Schapiro presented an extensive and devastating critique of the frequently proposed relation between modern physics and modern art.

12. *Contact* **4** (1923), p. 3. I am indebted to Carol Donley's draft paper, "Einstein, Too, Demands the Muse" for this lead and others in the following paragraphs.

13. *Selected Essays of William Carlos Williams*, New York, Random House, 1954, p. 283.

14. Ibid., p. 340.

15. J.-P. Sartre, "François Mauriac and Freedom," in *Literary and Philosophical Essays*, New York, Criterion Books, 1955, p. 23.

16. Lawrence Durrell, *Balthazar*, New York, E. P. Dutton, 1958, Author's Note, p. 9.

17. Ibid., p. 142.

18. For a good review of details, to which I am indebted, see Alfred M. Bork, "Durrell and Relativity," *Centennial Review* 7 (1963), pp. 191-203.

19. L. Durrell, *A Key to Modern British Poetry*, Norman, University of Oklahoma Press, 1952, p. 48.

20. Ibid., pp. 25, 26, 29.

21. All quotations are from William Faulkner, *The Sound and the Fury*, London, Chatto and Windus, 1961, pp. 81-177. I thank Dr. J. M. Johnson for a draft copy of her interesting essay, "Albert Einstein and William Faulkner," and have profited from some passages even while differing with others.

22. The chapter is shot through with references to light, light rays, even to travel "down the long and lonely light rays."

23. In *Les Prix Nobel en 1950*, Stockholm, Imprimierie Royale, 1951, p. 71.

24. Jean Piaget, *The Child's Conception of Time*, New York, Ballantine Books, 1971, p. vii.

25. Jean Piaget, *Genetic Epistemology*, New York, Columbia University Press, 1970, p. 69; see also p. 7.

26. Jean Piaget, *Psychology and Epistemology*, New York, Grossman Publishers, 1971, p. 82; see also pp. 10, 110. A similar statement is to be found in Piaget's *Six Psychological Studies*, New York, Vintage Books, 1968, p. 85.

27. For example, Jean Piaget, with Bärbel Inhelder, *The Child's Conception of Space*, New York, W. W. Norton, 1967, pp. 232-233; *The Child's Conception of Time*, London, Routledge and Kegan Paul, 1969, pp. 305-306; *Biology and Knowledge*, Chicago, University of Chicago Press, 1971, pp. 308, 337, 341-342. I wish to express my thanks to Dr. Katherine Sopka for help in tracing these references.

28. All this is quite apart from the role of model or personal culture hero that Einstein played in the lives of a great many individuals whom he never met. Indeed, several participants at the Symposium reminisced that they had tried to correspond with Einstein at a point in their lives when they were deciding to become scientists. To this day, his picture can be found in wide circulation and in the most unlikely places, from the tee shirt of a student in high school to the workbench of a cobbler in Shanghai.

29. Werner Heisenberg, "The Representation of Nature in Contemporary Physics," *Daedalus*, Summer 1958, pp. 103-105.

30. Lionel Trilling, *Mind in the Modern World: The 1972 Jefferson Lecture in the Humanities*, New York, Viking Press, 1972, pp. 13-14.

31. Albert Einstein, *Über die spezielle und die allgemeine Relativitätstheorie, gemeinverständlich*, Braunschweig, Vieweg, 1917. It was often translated and to this day is perhaps his most widely known work.

32. Albert Einstein and Leopold Infeld, *The Evolution of Physics*, New York, Simon and Schuster, 1938.

33. Albert Einstein, *On the Method of Theoretical Physics*, Oxford, Clarendon Press, 1933.

34. Albert Einstein, "Physics and Reality," *Journal of the Franklin Institute* **221** (1936), pp. 349-382.

35. First published in English in 1933, in his *The Modern Theme*, New York, W. W. Norton.

36. Ibid., pp. 135-136.

Historical Perspectives on Einstein's Scientific Contributions

Arthur I. Miller

THE SPECIAL RELATIVITY THEORY: EINSTEIN'S RESPONSE TO THE PHYSICS OF 1905

IMAGINE THAT YOU are on the editorial board of a prestigious physics journal and that you receive a paper that is unorthodox in style and format. Its title has little to do with most of its content; it has no citations to current literature; a significant portion of its first half seems to be philosophical banter on the nature of certain basic physical concepts taken for granted by everyone; the only experiment explicitly discussed could be explained adequately using current physical theory and is not considered to be of fundamental importance. Yet, with a minimum of mathematics, the little-known author deduces exactly a result that has heretofore required several drastic approximations. Furthermore, you are struck by certain of the author's general principles, and you feel that they promise additional simplifications. So you decide to publish the paper. This could well have been the frame of mind of the most eminent theoretical physicist on the Curatorium of the *Annalen der Physik*, Max Planck, when he received from the editor's office Albert Einstein's 1905 paper "On the Electrodynamics of Moving Bodies"—the relativity paper.[1]

The kind of title Einstein had given his paper customarily signaled a discussion of the properties of moving bulk matter, either magnetic or dielectric. Einstein analyzed neither of these topics. In fact, the paper's first quarter contains a philosophical analysis of the notions of time and length. The paper's second half dispatches quickly certain problems of such fundamental importance that they generally rate separate papers—for example, the characteristics of radiation reflected from a moving mirror—and he concludes with certain results from the dynamics of electrons that generally appear at the beginning of papers in which electrons are discussed. The only experiment developed in detail is at the paper's beginning and concerns the generation of current in a closed circuit as a result of the circuit's motion relative to a magnet, that is, electromagnetic induction.

The phenomenon of electromagnetic induction had ushered the Western world into the age of technology because it is fundamental to electrical dynamos. Everyone knew dynamos worked, but there remained fundamental problems concerning their operation. This essay discusses the connection that Einstein realized in 1905 between problems concerning huge electrical dynamos, radiation, moving electrical bulk media, the dynamics of electrons, and the nature of space and time. In order to set the stage for Einstein's bold approach to the physics of 1905, let us review the treatment of these topics by scientists and philosophers of whom Einstein has acknowledged he was aware before 1905.[2]

THE NATURE OF SPACE AND TIME

In his *Science of Mechanics* (1883) the philosopher-scientist Ernst Mach leveled a devastating critique at the Newtonian notions of absolute space and time. According to Newton, absolute space was the ultimate receptacle in which all phenomena occurred, and absolute time flowed independent of the motion of clocks. Mach considered these notions to be "metaphysical obscurities" because they were unavailable to our sense perceptions. Consequently, Mach disagreed with Immanuel Kant, who by 1781 had elevated Newton's notions of absolute space and time to knowledge that we possessed before all else, that is, a priori intuitions. According to Kant, these intuitions serve as basic organizing principles that enable our minds to construct knowledge from the potpourri of sense perceptions. Thus, for example, we are driven irresistibly toward a three-dimensional Euclidean geometry and the law of causality, and then to such higher-order organizing principles as Newton's physics. Although the discovery of non-Euclidean geometries in 1827 had dealt the Kantian view a serious blow, Kant's emphasis on the role of a priori organizing principles was nevertheless considered important to an understanding of how exact laws of nature are possible. A priori organizing principles played an important role in the neo-Kantian frameworks of such influential philosopher-scientists as Hermann von Helmholtz, Heinrich Hertz, and Henri Poincaré, whose writings impressed Einstein no less than did Mach's "incorruptible skepticism."[3]

Although Mach and Poincaré probed the relation between time and sense perceptions, in their work time remained absolute because there was no reason for it to depend on motion. But Mach and Poincaré insisted on replacing motion relative to Newton's absolute space with motion relative to the distant stars or, even better, motion relative to the substance that electrodynamicists assumed to fill Newton's cosmic receptacle—the ether.[4] This brings us to electromagnetism.

ELECTROMAGNETIC THEORY

Newton's mechanics of 1687 had unified terrestrial and extraterrestrial phenomena. The next great synthesis occurred not quite two hundred years later, when James Clerk Maxwell unified electromagnetism and optics. Whereas in the Newtonian mechanics disturbances propagated instantaneously through empty space, in Maxwell's theory disturbances propagated at a large but finite velocity through an ether, like ripples in a pond. In 1892 there appeared the result of over two decades of elaborations and purifications of Maxwell's theory—the electromagnetic theory of that master of theoretical physics, Hendrik Antoon Lorentz.

Lorentz assumed that the sources of the electromagnetic fields were as yet undiscovered electrons, which moved about in an all-pervasive, absolutely resting ether.[5] The five fundamental equations of Lorentz's theory are:[6]

$$\vec{\nabla} \times \vec{E} = -\frac{1}{c}\frac{\partial \vec{B}}{\partial t} \tag{1}$$

$$\vec{\nabla} \times \vec{B} = \frac{1}{c}\frac{\partial \vec{E}}{\partial t} + \frac{4\pi}{c}\rho\vec{V} \tag{2}$$

Maxwell-Lorentz Equations

$$\vec{\nabla} \cdot \vec{E} = 4\pi\rho \tag{3}$$

$$\vec{\nabla} \cdot \vec{B} = 0 \tag{4}$$

$$\vec{F} = \rho\vec{E} + \rho\frac{\vec{V}}{c} \times \vec{B} \quad \text{Lorentz Force Equation} \tag{5}$$

where $\vec{E}$ and $\vec{B}$ are the electric and magnetic fields, respectively, and ϱ is the electron's volume density of charge. Since Lorentz's fundamental equations are written relative to a reference system at rest in the ether, which we shall call S, then c is the velocity of light measured in S, and $\vec{v}$ is the electron's velocity relative to S. The Maxwell-Lorentz equations possess the property expected of a wave theory of light, namely, that relative to S the velocity of light is independent of the source's motion and is always c. But this may not necessarily be the result of measuring the velocity of light in a reference system moving with a uniform linear velocity relative to the ether, that is, in an inertial reference system. Therefore, the reference systems in the ether are preferred reference systems. To be sure, despite much effort, experiments had not revealed that the earth's motion through the ether had any effect on optical or electromagnetic phenomena.

Concerning the velocity of light, Newtonian mechanics predicted that the velocity of light emitted from a moving source should differ from the velocity of the light emitted from a source at rest by the amount of the source's velocity; consequently, the velocity of light c' from a source moving with velocity v is given by Newton's law for the addition of velocities,

$$\vec{c}' = \vec{c} + \vec{v}. \qquad (6)$$

On the other hand, according to the wave theory of light, the quantity c' measured by an observer at rest in the ether is

$$\vec{c}' = \vec{c}, \qquad (7)$$

and Lorentz's equations agreed with this requirement. But the effect of the ether on the measuring apparatus was expected to yield a result in agreement with (6), where c' is the velocity of the light relative to the earth and v is the ether's velocity relative to the earth. However, experiments accurate to second-order in the ratio v/c, where v is the velocity of a body that is moving relative to the ether and c is the velocity of light, led to (7). To this order of accuracy, optical and electromagnetic phenomena occurred on the moving earth as if the earth were at rest in the ether. Therefore, to second-order accuracy in v/c, Newtonian mechanics and electromagnetism are inconsistent with optical phenomena occurring in inertial reference systems.

In an 1895 monograph titled *Treatise on a Theory of Electrical and Optical Phenomena in Moving Bodies*,[7] Lorentz responded fully to the failure of the first-order experiments to detect any effects of the earth's motion on optical and electromagnetic phenomena; these experiments were called ether-drift experiments. For regions of the ether that are free of matter, or within neutral matter that is neither magnetic nor dielectric, the Lorentz equations in the ether-fixed reference system are the set of equations (S) (see Fig. 1). Applying the modified space and time transformations to (S), Lorentz obtained their analogues in the inertial reference system S_r.

We can appreciate Lorentz's achievement at a glance because to first-order accuracy in the quantity v/c, the Maxwell-Lorentz equations have the same form in the inertial system S_r as in the ether-fixed system S, and thus the same physical laws pertain to both these reference systems; in other words, to this order of accuracy neither optical nor electromagnetic experiments could reveal the motion of the system S_r. Lorentz called this stunning and desirable result the "theorem of corresponding states." It rested on the hypothesis of the mathematical "local time coordinate" t_L; the real or physical time was still Newton's absolute time. Hence, to order v/c, the velocity of light in S_r was the same as in S, that is, $c' = c$. To this order of accuracy, then, Lorentz's theorem of corresponding states removed the inconsistency between Newton's prediction and that of electromagnetic theory in favor of electromagnetic theory.

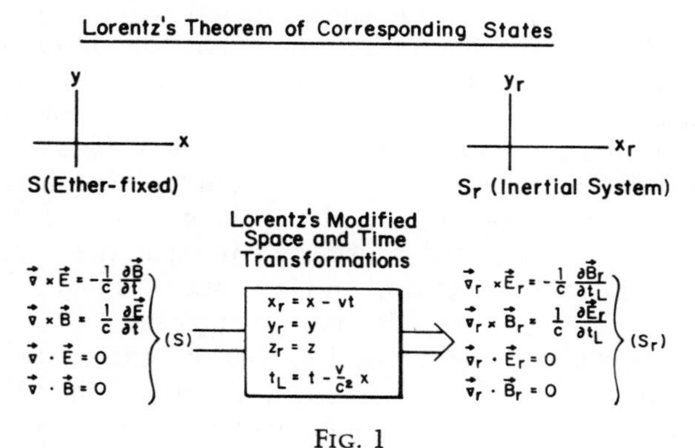

FIG. 1

Lorentz's modified space and time transformations contain the "local time coordinate" t_L and Lorentz referred to the electromagnetic field quantities $\vec{E}_r = \vec{E} + \vec{v}/c \times \vec{B}$ and $\vec{B}_r = \vec{B} - \vec{v}/c \times \vec{E}$ as "new" vectors.

But Lorentz had not yet explained the only reliable experiment accurate to second-order accuracy in v/c, namely, the 1887 experiment of Albert A. Michelson and Edward Williams Morley, in which light had been found to take the same time to race back and forth along each of two orthogonal rods of equal length that were at rest on the moving earth. He discussed this experiment in the final chapter of the 1895 treatise, which contained two other experiments that, as the chapter's title indicated, could "not be explained without further ado."[8] In order to explain the Michelson-Morley experiment, Lorentz indulged in a physics of desperation. From Newton's law for the addition of velocities, which the theorem of corresponding states was supposed to have obviated, he proposed the hypothesis that the dimensions of the rod in the direction of the earth's motion contracted by an amount $\sqrt{(1 - v^2/c^2)}$. In short, Lorentz's contraction hypothesis was admittedly ad hoc. This blemish on Lorentz's theory was emphasized in the philosophic-scientific criticism of that titan of international science, Henri Poincaré. Nevertheless, Poincaré was impressed with Lorentz's theorem of corresponding states because he eschewed absolute motion.[9]

So successful had been Newton's physics that many scientists had attempted to reduce all of physical theory to it; that is, they pursued a mechanical world-picture. For example, they attempted to simulate the contiguous actions of the ether with increasingly complex mechanical models. But these attempts paled before the successes of Lorentz's theory. Thus, in 1900 Wilhelm Wien suggested the "possibility of an electromagnetic foundation for mechanics," that is, pursuance of an electromagnetic world-picture based on Lorentz's electromagnetic theory.[10] A far-reaching implication of this program was that the electron's mass originated in its own electromagnetic field and should therefore be a velocity-dependent quantity. From studying the behavior of fast electrons that had been injected transversely into parallel electric and magnetic fields, Walter Kaufmann gave data for a dependence of the electron's mass on its velocity that increased without limit as the electron's velocity approached that of light.[11] Kaufmann's colleague at Göttingen, Max Abraham, developed the first field-theoretical description of an elementary particle.[12] Depending on whether his rigid-sphere electron experienced a force transverse or parallel to its motion, and with certain severe restrictions placed on the electron's acceleration, Abraham predicted that it had transverse (m_T) and longitudinal (m_L) masses:

$$m_T = \frac{m_o^e}{2\beta^3}\left[(1+\beta^2)\log\left(\frac{1+\beta}{1-\beta}\right) - 2\beta\right] \tag{8}$$

$$m_L = \frac{m_o^e}{\beta^3}\left[\frac{2\beta}{1-\beta^2} - \log\left(\frac{1+\beta}{1-\beta}\right)\right] \tag{9}$$

where $m_o{}^e = e^2/2Rc^2$ is the electron's electrostatic (that is, rest) mass and $\beta = v/c$. The transverse mass of Abraham's theory agreed with Kaufmann's data, and the goal of an electromagnetic world-picture appeared to be within reach. However, Abraham's theory offered no explanation for the Michelson-Morley experiment, and by 1904 it was in violent disagreement with the new optical experiments of Lord Rayleigh and D. B. Brace, which were accurate to second-order in v/c.[13]

Prompted by the new second-order data and by Kaufmann's measurements, as well as by Poincaré's criticisms, Lorentz proposed his own theory of the electron in which the contraction hypothesis was deemed no longer to be ad hoc, because it became one of several hypotheses that could explain more than one experiment.[14] Lorentz's electron can be likened to a balloon smeared with a uniform distribution of charge. While at rest, Lorentz's electron is assumed to be a sphere; but moving, it undergoes a Lorentz contraction, and its mass becomes a two-component quantity:

$$m_T = \frac{4}{3}\frac{m_o^e}{\sqrt{1-\beta^2}} \tag{10}$$

$$m_L = \frac{4}{3}\frac{m_o^e}{(1-\beta^2)^{3/2}} \tag{11}$$

where $m_o{}^e = e^2/2Rc^2$ is the electron's electrostatic (that is, rest) mass and $\beta = v/c$.

Lorentz's m_T agreed with Kaufmann's data as well as did Abraham's. But Abraham immediately leveled a severe fundamental criticism at Lorentz's theory: Lorentz's deformable electron was unstable because it could explode under the enormous repulsive forces among its constituent parts.[15] From newly discovered Lorentz-Poincaré correspondence we know that Poincaré had recognized this problem independently and then cracked it with his unmatched arsenal of mathematics. Poincaré's resulting papers were the penultimate effort toward an electromagnetic world-picture based on Lorentz's electromagnetic theory.[16] They included such advanced notions of mathematics as group-theoretical methods and four-dimensional spaces. Using a term familiar from fundamental studies in geometry, Poincaré renamed Lorentz's 1904 theorem of corresponding states that embraced all extant data—and, it was *hoped*, future ones as well—the "principle of relativity."[17] Einstein, we know, had not encountered Poincaré's 1905 version of Lorentz's theory of the electron when he wrote the relativity paper.

To summarize, by 1905 physicists believed that electromagnetic theory was proceeding in the correct direction. Many of them felt sure that with a little more tinkering, Lorentz's theory of the electron could serve as the cornerstone for a unified field-theoretical view of nature. Lorentz's was a dynamical theory that *explained* such effects as the presumed contraction of length, the observed variation of mass with velocity, and the fact that the measured velocity of light always turned out to be the same—all explained as resulting from the interaction of electrons with the ether. The stage was set for a great new era in science to emerge from what everyone considered to be the cutting edge of scientific research. But, as we shall see, this turned out not to be the case. We move next to an area of science and engineering whose basic problems were deemed unimportant for progress in basic physical theory: the area of electromagnetic induction. In German-speaking countries problems in this area combined technology and basic research. They received a particularly interesting treatment because, as the intellectual historian J. T. Merz has written, the "German man of science was a philosopher."[18]

ELECTRICAL DYNAMOS

In 1831 Michael Faraday discovered that relative motion between a wire loop and a magnet produces a current in the wire. Faraday interpreted this result as follows: the magnet affects the wire loop through its lines of force, which emanate from the magnet's north pole and enter through its south pole; consequently, relative motion between the loop and the magnet results in the loop's cutting the lines of force. Faraday's law states that the rate at which the lines of force are cut determines the strength of the current induced; furthermore, the direction and magnitude of the induced current depend on only the relative velocity between the loop and magnet.

But Faraday's interpretation of electromagnetic induction differed when circuit and magnet were rotating relative to each other. An apparatus of the sort in Fig. 2 was important to Faraday because by 1851 he had convinced himself that lines of force participated in the magnet's linear motion but not in its rotation. In Fig. 2 the wire loop makes sliding contact with the rotating magnet's periphery and touches one of the magnet's poles; hence, this sort of electromagnetic dynamo became known as a unipolar dynamo. Faraday's interpretation of unipolar induction is: if the loop rotates counterclockwise, a current appears in it owing to its cutting the lines of force. If the loop remains at rest, and the magnet rotates clockwise, a current of the same magnitude and direction appears in the loop; the current was thought to originate in the magnet owing to the magnet's rotating through its lines of force. When magnet and loop turn together in the same direction, there is no net current in the loop because the loop's current is canceled by the magnet's internal current. Thus, two different explanations were required for the current induced in the loop, depending on whether the loop or the magnet rotated.[19] Clearly, the experimental data were more easily understood in terms of relative motion between the wire loop and the magnet with its co-moving lines of force.

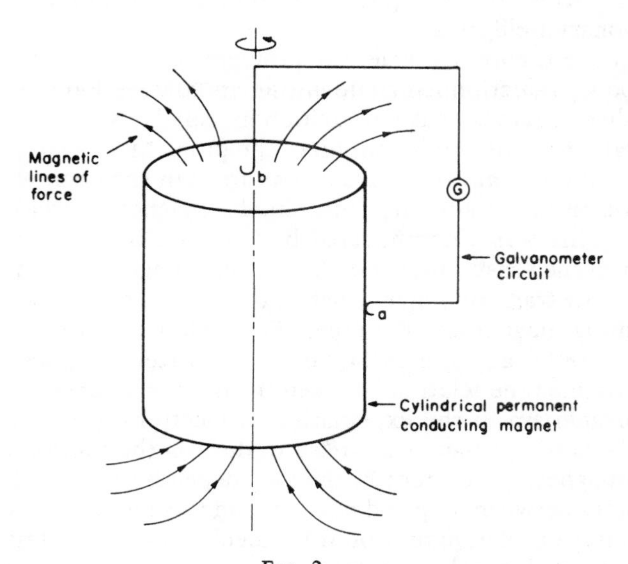

FIG. 2
An example of a unipolar dynamo.

Representing the magnetic field by lines of force found fertile ground in the German-speaking countries. Although by the end of the nineteenth century most British scientists considered lines of force to be useful chiefly for pedagogy, the Kantian philosophical position of scientists and engineers in German-speaking countries led them to consider lines of force as a fundamental *Anschauung*. In this context *Anschauung* refers to the intuition through pictures formed in the mind's eye from previous visualizations of physical processes in the world of perceptions; *Anschauung* is superior to viewing merely with senses.[20] In short, lines of force were seen everywhere. There ensued a controversy in the German-speaking scientific-engineering community on the merits of the *Anschauungen* of Faraday versus the rotating-line view, and experiments were offered to distinguish between them. One engineer emphasized that this controversy was "not an academic moot point" for two chief reasons. First, there was intense research and development toward unipolar direct-current and alternating-current generators owing to the relative simplicity of these machines vis-à-vis multipolar machines. Second, calculating the effect of a rotating magnet on stationary armature coils required knowing whether the magnetic field lines remained stationary; this problem appeared also in multipolar machines with stationary field magnets.[21]

On the theoretical side, the problem of how current arises in unipolar induction led some physicists to question the universal validity of Faraday's law of induction. Faraday's law was taken to be valid for closed wire circuits. The problem was how Faraday's law could explain the current that arose in the open wire in Fig. 2 as a result of slipping between the magnet's surface and the wire. This was a complicated boundary-value problem because it concerned assumptions on how electromagnetic quantities varied in the transition layer (that is, the surface of slip) between the spinning magnet and the ether. The widely read texts of August Föppl (1894) and the 1904 edition rewritten by Max Abraham emphasized the relation between the laws of mechanics and electromagnetism in the explanation of electromagnetic induction: the laws of mechanics are used to discuss the motion of circuit and magnet; the current in the loop depends on only the relative velocity between loop and magnet, and this result agrees with the principle of relative motion from mechanics.[22] This principle asserts that physical phenomena occur as a result of the relative motion of material bodies and that the laws of mechanics are the same for all inertial systems. Abraham stressed that experiments supported the principle of relative motion for Lorentz's theory only to first-order accuracy in v/c. Whereas Föppl's use of complicated notions of surfaces of slip allowed him to express only faith in the validity of Faraday's law, Abraham avoided any problems of unipolar induction by adhering strictly to Hertz's electromagnetic theory, in which the principle of relative motion held exactly. Then he calculated the current for the one case of the loop turning where Faraday's law could be applied with its line-cutting interpretation; the inverse case followed by replacing v with $-v$ in Faraday's law.

In Abraham's view, only the induced current, and not the lines of force, was measurable; that is, the equations of electromagnetic theory dictated what was to be observed, and lines of force were merely auxiliary quantities. In 1904 Lorentz enlarged his electromagnetic theory to describe moving bulk magnetic and dielectric matter, and his results agreed with empirical data better than did Hertz's. Lorentz's extension required his postulating three different sorts of electrons: conduction electrons responsible for electric current; polarization electrons that produce dielectric properties; and magnetization electrons to explain magnetic properties of matter. Using the local time coordinate, Lorentz showed that the effect of a moving magnet on a stationary current loop—that is, the electric field due to the moving magnet—had its origin in the moving magnet developing dielectric properties. But this explanation for electromagnetic induction involved a complicated intermingling of two different kinds of electrons in a substance that was assumed to possess only conduction and magnetization electrons. Lorentz himself found this explanation puzzling.[23]

Besides appending the hypotheses of additional electrons to an already overburdened superstructure, Lorentz by the end of 1904 was still unable to explain the stability of his electron. Then there was Max Planck's 1900 explanation for the radiation from hot substances, which brings us to the final ingredient necessary for describing Einstein's unique view of the physics of 1905—problems concerning radiation.

RADIATION

Since the late nineteenth century, the light emitted from a cavity within a hot substance had fascinated scientists because its characteristics are independent of the substance's constitution.

To his horror, Planck found that the successful empirical formula he offered to describe cavity radiation required that energy be exchanged in only discrete amounts between the cavity radiation and the hot substance's constituent electrons.[24] Yet, according to electromagnetic theory, processes involving radiation should be continuous. Furthermore, if only certain discontinuous exchanges of energy were allowed, then the constituent electrons' states of motion were similarly restricted. Consequently, Planck's formula violated both mechanics and electromagnetism. Einstein recalled that in 1904 this situation was a "second fundamental crisis" (in addition to the instability of Lorentz's electron).[25] He was alone in this assessment.

EINSTEIN'S VIEW OF PHYSICAL THEORY

Einstein's opinion of Planck's theory of cavity radiation received confirmation from the application of his 1904 results on the behavior of atoms in gases—that is, statistical mechanics—to Planck's radiation formula. He found that light exhibited particulate properties in regions of space whose size was that of the electron's volume.[26] Thus, current electromagnetic theory was insufficient for discussing the nature of the electron. In addition, his research on particles in solution (Brownian motion) convinced him of the insufficiency of thermodynamics and mechanics in microscopic volumes. These two results led Einstein to conclude that scientists of 1905 were "out of [their] depth":[27] neither the mechanical nor the electromagnetic world-picture could succeed. Einstein's first two papers in the now famous volume 17 of *Annalen der Physik* explored the consequences of particles of light, that is, light quanta[28] and Brownian motion.[29] In the third paper of that 1905 trilogy—the relativity paper—Einstein again dealt with the characteristics of light.

Einstein's studies of the great philosopher-scientists von Helmholtz, Hertz, and Poincaré had shown him the power of a fundamental analysis within a neo-Kantian framework. But when it came to what to analyze and where to begin anew, he was in a quandary. Theories based on assumptions concerning the constitution of matter were inadequate for a fundamental analysis. At this point, Einstein was aided by his predilection for visual thinking, which was reinforced when he encountered the writings of Ludwig Boltzmann and von Helmholtz, in particular, at the

Eidgenössische Technische Hochschule in Zurich. Before Einstein, however, no one had combined visual thinking so effectively with the thought-experiment, that is, an experiment capable of being performed in the mind.

In 1895, after ten years of obstinate pondering, Einstein had conceived of a thought-experiment that finally revealed the "germ of the special relativity theory." The essence of this experiment is: (1) Current physics asserts that an observer who is moving alongside a light wave whose source is in the ether should be able to discern the effects of his motion by, for example, measuring the velocity of light. (2) But to Einstein it was "intuitively clear" that the laws of optics could not depend on the state of the observer's motion. Statements (1) and (2) are mutually contradictory, and to Einstein this thought-experiment contained a paradox.[30]

In their own ways, Lorentz and Poincaré were also attempting to resolve this paradox—by one degree of accuracy at a time, that is—by proposing, as we saw, hypotheses such as the local time and the ad hoc contraction of moving bodies. But Einstein's 1904 results on the nature of light had convinced him that theories of matter could lead neither to exact explanations for ether-drift experiments nor to a consistent world-picture.

I turn next to the basic scientific-philosophic considerations that led Einstein to discover that the key to the paradox lay, as he said later, in the "axiom of the absolute character of time, viz., of simultaneity, [which] unrecognizedly was anchored in the unconscious."[31] My research, based on Einstein's writings, has resulted in the following reconstruction of Einstein's thinking toward this momentous discovery.[32] Of all the ether-drift experiments

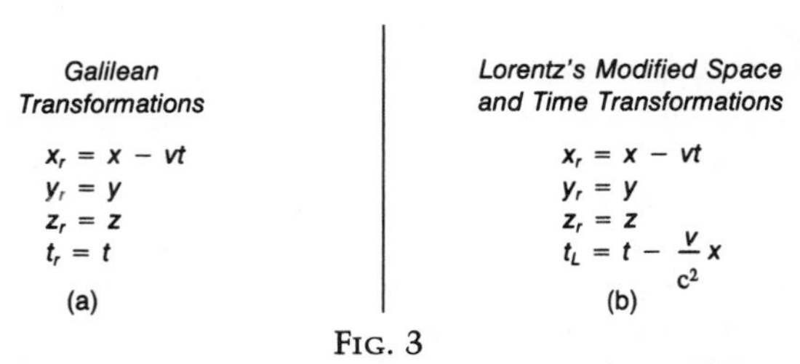

FIG. 3

(a) The coordinates (x_r, y_r, z_r, t_r) and (x, y, z, t) refer to the two *inertial* reference systems S_r and S.

(b) The coordinates $(x_r, y_r, z_r, t_r = t)$ and (x, y, z, t) refer to the inertial reference system S_r and to the ether-fixed reference system S; and t_L is the mathematical "local time coordinate."

that had been performed, those accurate to first order in v/c were explainable by Lorentz's theorem of corresponding states, which was based on the set of modified space and time transformations that included the mathematical local time coordinate (see Fig. 3b). Although the Maxwell-Lorentz equations remained unchanged under the modified space and time transformations of Fig. 3b, the same was not true of the equations of mechanics that were transformed between inertial reference systems according to the Galilean transformations (Fig. 3a). Thus, according to Lorentz's modified transformations, the laws of mechanics were not the same in every inertial reference system. But this result violated Newton's exact principle of relative motion, which states that no mechanical experiment could reveal an inertial system's motion. The mathematical statement of Newton's principle of relative motion lies in the Galilean transformations of Fig. 3a, where for mechanics the reference systems S_r and S are *both* inertial reference systems. Consequently, the transformation rules for the laws of mechanics and of electromagnetism depended on two different notions of time—one physical and the other mathematical—contrary to Lorentz's goal of unification. Thus, whereas most scientists in 1905 considered the tension between mechanics and electromagnetism to be rooted in the inability of mechanics to explain the measured velocity of light, Einstein delved deeper and found that current physics rendered mechanics and electromagnetism *incompatible.*

From the texts of Föppl and Abraham, Einstein could have learned well the intimate connection between mechanics and electromagnetism for interpreting electromagnetic induction. Faraday's law was basic also to Lorentz's electromagnetic theory, where the local time made it possible to calculate the moving magnet's effect on either a resting conductor or the open circuit in unipolar induction. But, as we recall, this calculation rested on certain special assumptions on the constitution of matter. Moreover, electromagnetic theory explained electromagnetic induction in two different ways, depending on whether the conductor or the magnet was moving, even though the physically measurable effect was a function of their relative velocity alone: when the conductor moved relative to the magnet, a current flowed owing to a force on the conductor's electrons; in the inverse case, a current flowed owing to an electric field at the conductor's site. For Einstein, two explanations for an effect that depended on only the relative velocity between magnet and conductor was more than a shortcoming of Lorentz's theory; it was, as he wrote in the relativity paper, an asymmetry that was "not inherent in the phenomena."[33] Einstein found the theoretical situation in electromagnetic induction so "unbearable," as he recalled in 1919,[34] that he focused on the necessity for an equivalence of viewpoints between observers on the wire loop and on the magnet, rather than on the source of the moving magnet's effect on the wire. He turned the fashionable programmatic research efforts of physics sideways by enlarging Newton's principle of relative motion to treat mechanics and electromagnetism on an equal footing, instead of attempting to reduce one to the other. In the relativity paper, he referred to this widened version of Newton's principle of relative motion as the "principle of relativity." Since, as we saw, the Galilean transformations could not explain the velocity of light, the modified transformations of Lorentz, with their apparently ubiquitous local time coordinate, would have to play a role in relating phenomena between reference systems. From Lorentz's modified transformations Einstein could have deduced a new result for the addition of velocities:

$$w_r = \frac{w - v}{1 - vw/c^2} \tag{12}$$

where w_r is the velocity of a moving point relative to S_r and w is its velocity relative to S. For the case of $w = c$, then $w_r = c' = c$, instead of Newton's addition law of velocities, $c' = c - v$. Thus Einstein could have realized that, to first order in v/c, the addition law for velocities from Lorentz's modified transformations produced a result that agreed with the intuition of his thought-experimenter. Since Lorentz's modified transformation differed from the Galilean transformation only in the local time coordinate, Einstein asked himself whether the local time might be the physical "time"? But this step required asserting that the times in inertial reference systems differed because the local time coordinate depends on their relative velocity. Yet the absoluteness of time had always been accepted. Furthermore, the thought-experimenter's intuition demanded an examination of the mathematical relation between Newton's principle of relative motion and Lorentz's theorem of corresponding states. After all, the spatial coordinates of Lorentz's modified transformation equations of 1895 were mathematically the same as the ones in the Galilean transformations, and the local time coordinate had been invented for use in electromagnetic theory. Einstein's imposing a Newtonian unity upon Lorentz's modified transformation equations meant also his asserting the equivalence of the reference systems S and S_r. This was a big step, for it meant rejecting Lorentz's ether, and with it the dynamical interpretations of an enormously successful and, for the most part, satisfying theory. To his surprise, Einstein had found that the notion of time was both the central point and the Achilles' heel of the electrodynamics of moving bodies.

For aid in analyzing the nature of time, Einstein benefited from the critical reasoning "especially . . . of David Hume's and Ernst Mach's philosophical writings." Their analyses of sense perceptions offered strong evidence that exact laws of nature could not be induced from empirical data. And their analyses of the limits imposed by sense perceptions on notions of causality and of time enabled Einstein to realize that the high value of the velocity of light, compared with the other velocities we encounter daily, had prevented our appreciating that "the absolute character of time, viz., of simultaneity, unrecognizedly was anchored in the unconscious."[35] Einstein's fundamental analysis of physical theory went far beyond science as it is normally conceived: from an analysis of electromagnetic induction into an analysis of sensations, and then into an analysis of thinking itself. He concluded that the customary sensation-based notions of time and simultaneity resulted in a physics burdened with asymmetries, unobservable quantities, and ad hoc hypotheses. Thus prevented from lapsing into a dogmatic slumber, Einstein took recourse in the neo-Kantian view that was predicated on the usefulness of organizing principles such as the second law of thermodynamics. He enlarged Newton's principle of relativity to include Lorentz's theory, and then he raised to axioms this principle and the basic principle of every wave theory of light: in the ether-fixed system S the velocity of light is independent of the source's motion and is always c.

Einstein's two principles do not attempt to explain anything—for example, why the measured velocity of light always turns out to be c. Whereas Lorentz's and Poincaré's principle of relativity was based on a theory of matter that was heavily dependent on experimental data and was contingent on negative results of future ether-drift experiments, Einstein's was independent of assumptions concerning electrons and was an axiom. As Einstein wrote in 1907,[36] the principle of relativity was the basis of a theory that specified the form that laws of physics should assume in order to be used again to investigate the constitution of matter;

it was a theory of principle. In short, Einstein moved boldly counter to the prevailing currents of theoretical physics by resolving problems in a Gordian manner, that is, by formulating a view of physics in which certain problems do not occur, a view in which the 1895 paradox becomes a mere fiction. Consequently, Einstein's relativity paper looked insignificant alongside the papers of Abraham, Lorentz, and Poincaré, which used the most *au courant* methods of mathematical physics in order to derive what Einstein took to be axiomatic.

Straightway in the relativity paper, Einstein emphasized that the basic problems confronting physical theory concerned not the constitution of matter but understanding the equivalence of viewpoints between moving observers. For this purpose, he began with the simplest thought-experiment illustrating the problems of electromagnetic induction: a magnet and conducting loop in relative inertial motion. No mathematics was necessary to demonstrate that Maxwell's electrodynamics led observers on the wire loop and the magnet to different interpretations for the physically measurable effect, the current induced in the conductor. Thus, Faraday's law was universally valid, though it had been misinterpreted.

In the second paragraph of the relativity paper Einstein mentioned a facet of the results of ether-drift experiments accurate to the first order in v/c (which he left unnamed) that may have been glimpsed only by Abraham,[37] but within the context of the electromagnetic world-picture: namely, the laws of electrodynamics and mechanics were valid in inertial reference systems to first-order accuracy in v/c. Einstein's masterstroke was to link the experiment of magnet and conductor, which he had just explained, to the ether-drift experiments. He reasoned that electromagnetic induction depends on the laws of mechanics and of electromagnetism, which cover optics also; he then "conjectured" that Newton's principle of relative motion covers these three disciplines to order v/c. He boldly continued: "We will raise this conjecture (whose intent will from now on be referred to as the 'Principle of Relativity') to the status of a postulate."[38] Employing the principle of relativity and the principle governing the velocity of light, Einstein went on to develop a view of physics so powerful that the only mention he made of problems involving unipolar induction was to dismiss them as "meaningless."[39] In the relativity paper the Lorentz contraction and most of the other hypotheses of Lorentz's theory surfaced only as "secondary consequences,"[40] and it was unnecessary for Einstein to review every extant ether-drift experiment, because in his view their results were *ab initio* a foregone conclusion.

With further simple thought-experiments the first part of the relativity paper demonstrated how imprecise were current notions of time and length, and the relative nature of these notions emerged. Most readers of 1905, however, emphasized the end of the paper, where Einstein deduced exactly a formula for the electron's transverse mass that was almost Lorentz's, and so they interpreted Einstein's work as a valuable generalization of Lorentz's theory of the electron. Soon Kaufmann himself pointed out that these two equations should be the same, and he was one of the first to use the name "Lorentz-Einstein theory."[41] But what did Einstein have to say in the relativity paper about the data in Kaufmann's 1902 and 1903 papers? Nothing, because it is reasonable to conjecture that Einstein knew well that they disagreed with his prediction for the electron's transverse mass.[42] Einstein considered this prediction to be of secondary importance anyway. In a 1907 paper reviewing the status of the principle of relativity,[43] Einstein discussed Kaufmann's data from late 1905, which, as Kaufmann had written in no uncertain terms, were *"not consistent with the Lorentz-Einstein fundamental assumption."*[44] Einstein acutely emphasized that the "systematic deviation" between the Lorentz-Einstein predictions and the data could indicate a hitherto "unnoticed source of error"; consequently, Einstein called for further experiments. Then he dismissed Kaufmann's data, because in his "opinion" they supported theories that did not "embrace a greater complex of phenomena," that is, the electron theories of Abraham and A. H. Bucherer, which could not explain optical experiments accurate to second order in v/c.[45]

A letter among Poincaré's correspondence, which I discovered in Paris, permits us to compare Einstein's bold defense of the principle of relativity in the face of disconfirming data with Lorentz's own response to the same data (see Fig. 4). On 8 March 1906, Lorentz wrote to Poincaré: "Unfortunately my hypothesis of the flattening of electrons is in contradiction with Kaufmann's new results, and I must abandon it. I am, therefore, at the end of my Latin. It seems to me impossible to establish a theory that demands the complete absence of an influence of translation on the phenomena of electricity and optics." What a remarkable confession, and what a clear-cut case of falsification. After all

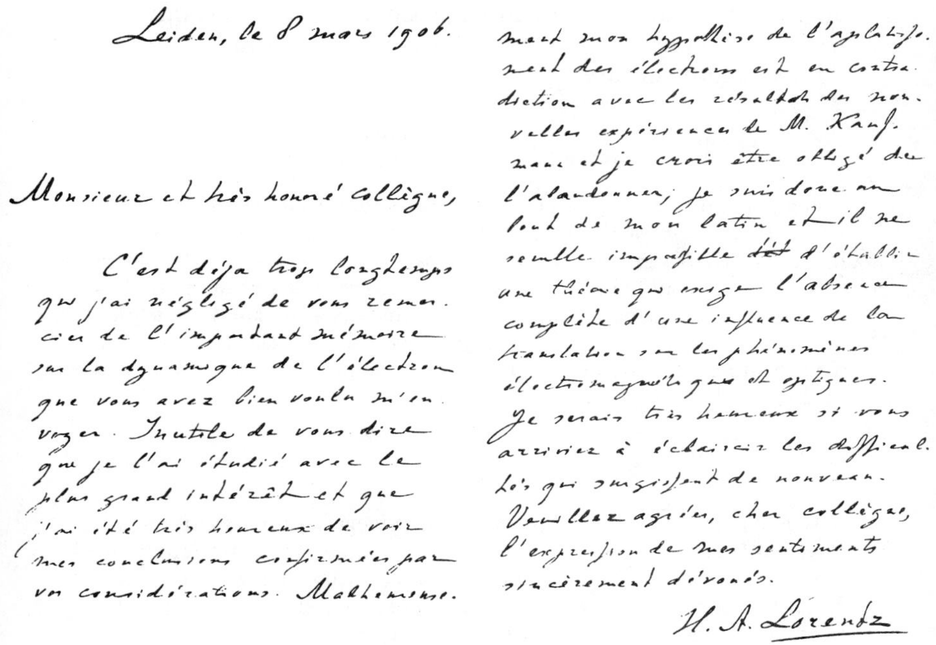

Leiden, le 8 mars 1906.

Monsieur et très honoré Collègue,

C'est déjà trop longtemps que j'ai négligé de vous remercier de l'important mémoire sur la dynamique de l'électron que vous avez bien voulu m'envoyer. Inutile de vous dire que je l'ai étudié avec le plus grand intérêt et que j'ai été très heureux de voir mes conclusions confirmées par vos considérations. Malheureusement mon hypothèse de l'aplatissement des électrons est en contradiction avec les résultats des nouvelles expériences de M. Kaufmann et je crois être obligé de l'abandonner; je suis donc au bout de mon latin et il me semble impossible d'établir une théorie qui exige l'absence complète d'une influence de la translation sur les phénomènes électromagnétiques et optiques. Je serais très heureux si vous arriviez à éclaircir les difficultés qui surgissent de nouveau.

Veuillez agréer, cher collègue, l'expression de mes sentiments sincèrement dévoués.

H. A. Lorentz

FIG. 4
Letter from H. A. Lorentz to H. Poincaré, 8 March 1906.

those years of work Lorentz was willing to abandon his theory on the basis of a report of a single experiment.

Einstein's intuition served him well here, for in fact, Kaufmann's data were incorrect. In 1908 the Lorentz-Einstein prediction for the electron's transverse mass was considered to have been vindicated by the new data of Bucherer.[46] Bucherer's results were not immediately accepted by the entire scientific community, but Poincaré and Lorentz did not call for additional tests. Incidentally, three decades later it was proved that Bucherer's data were also incorrect.[47] Undoubtedly having the Kaufmann episode of 1907 in mind, Einstein in 1946 described one of two criteria for assessing a scientific theory: "The first point of view is obvious: the theory must not contradict empirical facts. However evident this demand may in the first place appear, its application turns out to be quite delicate."[48]

In conclusion, although by mid-1905 high-velocity data dominated basic physical theory, Einstein needed only familiar low-velocity data, such as electromagnetic induction, in conjunction with the data of the thought-experimenter, to bring physics into the twentieth century.

ACKNOWLEDGMENTS

Figures 1 and 2 are reproduced from Harry Woolf, ed., *Some Strangeness in the Proportion: A Centennial Symposium to Celebrate the Achievements of Albert Einstein*, Reading, Mass., Addison-Wesley, 1980, by permission of the publisher. This essay is based on research supported by a grant from the National Science Foundation's History and Philosophy of Science Program. I am grateful to the Poincaré Estate, Paris, for permission to quote from their correspondence. I thank Professor Gerald Holton for his comments on this essay.

NOTES

1. A. Einstein, "Zur Elektrodynamik bewegter Körper," *Annalen der Physik* **17** (1905), pp. 891-921. As far as we know, the editorial policy of the *Annalen* was that an author's initial contributions were scrutinized by either the editor (in 1905, Paul Drude) or a member of the Curatorium; subsequent papers were published with no refereeing. Einstein had appeared in print in the *Annalen* five times by 1905; so his relativity paper was probably accepted on receipt. I thank Mr. Allan Needell for this information, at which he arrived as a result of studying the correspondence between Planck and Wilhelm Wien.

For the purpose of serious historical analysis I had to retranslate Einstein's relativity paper from the *Annalen* version for my book, *Albert Einstein's Special Theory of Relativity: Emergence (1905) and Early Interpretation (1905-1911)* (Reading, Mass., Addison-Wesley, 1981), where the retranslation appears in the Appendix (pp. 391-415). The hitherto most frequently quoted English translation of the relativity paper is in the Dover reprint volume, *The Principle of Relativity*, trans. W. Perrett and G. B. Jeffery (New York: Dover, n.d.; reprinted from London: Methuen, 1923). But that contains some substantive mistranslations, infelicities, and outdated Britishisms. For example, Einstein's second principle of the relativity theory is mistranslated as: "Any ray of light moves in the 'stationary' system of co-ordinates with the determined velocity *c*, whether the ray be emitted by a stationary or by a moving body." The correct translation is: "Any ray of light moves in the 'resting' coordinate system with the definite velocity *c, which is independent* of whether the ray was emitted by a resting or by a moving body" (italics added to indicate a key phrase that was omitted in the Dover translation).

The Dover translation was made from a retypeset version of Einstein's relativity paper that had appeared in a Teubner reprint volume, thereby adding to the misprints in the original *Annalen* version. In addition, the Dover translation does not distinguish between Einstein's footnotes and those added to the Teubner edition by Arnold Sommerfeld. This state of affairs is an example of the pitfalls inherent in using a translation that was not made from the original paper and of the importance of going back to the original papers.

Hereinafter all citations to Einstein's relativity paper are to the translation in my book.

2. I discuss these materials in *Einstein's Special Theory of Relativity.* See also G. Holton, *Thematic Origins of Scientific Thought: Kepler to Einstein*, Cambridge, Mass., Harvard University Press, 1973.

3. A. Einstein, "Autobiographical Notes," trans. P. A. Schilpp, in Schilpp, ed., *Albert Einstein: Philosopher-Scientist*, Evanston, Ill., Library of Living Philosophers, 1949, p. 21.

4. See, for example, H. Poincaré, *Science and Hypothesis* (1902), trans. unknown, New York, Dover, 1952.

5. H. A. Lorentz, "La théorie électromagnétique de Maxwell et son application aux corps mouvants," *Archives Néerlandaises des Sciences Exactes et Naturelles* **25** (1892), p. 363; reprinted in Lorentz, *Collected Papers*, 9 vols., The Hague, Nijhoff, 1935-1939, vol. 2, pp. 164-343.

6. Instead of using Lorentz's original units, I take the liberty to write (1)–(5) in absolute Gaussian (cgs) units, whose usefulness was emphasized first by Max Abraham in 1902 and 1903; see the papers cited in note 12. Abraham was also the first to use the nomenclature "Maxwell-Lorentz equations," which appears in Einstein's relativity paper of 1905.

7. H. A. Lorentz, *Versuch einer Theorie der elektrischen und optischen Erscheinungen in bewegten Körpern*, Leiden, Brill, 1895; reprinted in Lorentz, *Collected Papers*, vol. 5, pp. 1-137.

8. Lorentz originally proposed the contraction hypothesis in his "The Relative Motion of the Earth and the Ether," *Koninklijke Akademie van Wetenschappen te Amsterdam* **1** (1892), p. 74; reprinted in Lorentz, *Collected Papers*, vol. 4, pp. 219-223. For further discussion of Lorentz's contraction hypothesis, see my article, "On Lorentz's Methodology," *British Journal for the Philosophy of Science* **25** (1974), pp. 29-45, and Miller, *Einstein's Special Theory of Relativity.*

9. See my "A Study of Henri Poincaré's 'Sur la Dynamique de l'Electron,' " *Archive for History of Exact Sciences* **10**, nos. 3-5 (1973), pp. 207-328, and "Poincaré and Einstein: A Comparative Study," *Boston Studies in the Philosophy of Science*, vol. 31 (forthcoming).

10. W. Wien, "Über die Möglichkeit einer elektromagnetischen Begründung der Mechanik," in *Recueil de travaux offerts par les auteurs à H. A. Lorentz*, The Hague, Nijhoff, 1900, pp. 96-107; reprinted in *Annalen der Physik* **5** (1901), pp. 501-513.

11. W. Kaufmann, "Die magnetische und elektrische Ablenkbarkeit

der Becquerelstrahlen und die scheinbare Masse der Elektronen," *Nachrichten von der Gesellschaft der Wissenschaften zu Göttingen* (1901), pp. 143-155, and "Die elektromagnetische Masse des Elektrons," *Physikalische Zeitschrift* **4** (1902), pp. 54-57. For discussion, see my *Einstein's Special Theory of Relativity*, esp. chaps. 1, 7, and 12.

12. M. Abraham, "Dynamik des Elektrons," *Nachrichten von der Gesellschaft der Wissenschaften zu Göttingen* (1902), pp. 20-41, "Prinzipien der Dynamik des Elektrons," *Physikalische Zeitschrift* **4** (1902), pp. 57-63, and "Prinzipien der Dynamik des Elektrons," *Annalen der Physik* **10** (1903), pp. 105-179. For discussion, see my *Einstein's Special Theory of Relativity*, esp. chap. 1.

13. Lord Rayleigh, "Does Motion through the Aether Cause Double Refraction?" *Philosophical Magazine* **4** (1902), pp. 678-683; D. B. Brace, "On Double Refraction in Matter Moving through the Aether," *Philosophical Magazine* **7** (1904), pp. 317-329.

14. H. A. Lorentz, "Electromagnetic Phenomena in a System Moving with Any Velocity Less than That of Light," *Proceedings of the Royal Academy, Amsterdam* **6** (1904), p. 809; reprinted in Lorentz, *Collected Papers*, vol. 5, pp. 172-197, and in part in *The Principle of Relativity: A Collection of Original Memoirs on the Special and General Theory of Relativity by H. A. Lorentz, A. Einstein, H. Minkowski and H. Weyl*, trans. W. Perrett and G. B. Jeffery, New York, Dover, n.d., pp. 11-34.

15. M. Abraham, "Die Grundhypothesen der Elektronentheorie," *Physikalische Zeitschrift* **5** (1904), pp. 576-579.

16. H. Poincaré, "Sur la dynamique de l'électron," *Comptes rendus de l'Académie des Sciences* **140** (1905), pp. 1504-1508, and "Sur la dynamique de l'électron," *Rendiconti del circolo matematico di Palermo* **21** (1906), pp. 129-175. Both are reprinted in *Oeuvres de Henri Poincaré*, 11 vols., Paris, Gauthier-Villars, 1934-1954, vol. 9, pp. 489-493 and 494-550, respectively. I have discussed these papers in "Analysis of Henri Poincaré's 'Sur la Dynamique de l'Electron' "; "Poincaré and Einstein"; "On Some Other Approaches to Electrodynamics in 1905," in H. Woolf, ed., *Some Strangeness in the Proportion: A Centennial Symposium to Celebrate the Achievements of Albert Einstein*, Reading, Mass., Addison-Wesley, 1980, pp. 66-91; and *Einstein's Special Theory of Relativity*, esp. chap. 1.

17. H. Poincaré, "L'état actuel et l'avenir de la physique mathématique," paper delivered at the International Congress of Arts and Science, St. Louis, Mo., 24 September 1904, and published in *Bull. Sci. Mat.* **28** (1904), pp. 302-324; reprinted in H. Poincaré, *The Value of Science* (1905), trans. George Bruce Halsted, New York, Dover, 1958.

18. J. T. Merz, *A History of European Scientific Thought in the Nineteenth Century* (1904-1912), 2 vols., rpt. New York, Dover, 1954, vol. 1, p. 215.

19. For a discussion of Faraday's experiments with rotating magnets

and of the problem of unipolar induction, see my "Unipolar Induction: A Case Study of the Interaction Between Science and Technology," *Annals of Science* **38** (1981), pp. 155-189.

20. For discussions of the importance of *Anschauung* in the German-speaking physics community, see ibid., and my "Visualization Lost and Regained: The Genesis of the Quantum Theory in the Period 1913-27," in J. Wechsler, ed., *On Aesthetics in Science*, Cambridge, Mass., MIT Press, 1978, pp. 72-102.

21. C. L. Weber, "Über unipolare Induktion," *Elektrotechnische Zeitschrift* **16** (1895), pp. 513-514.

22. A. Föppl, *Einführung in die Maxwellsche Theorie der Elektrizität*, Leipzig, Teubner, 1894; M. Abraham and A. Föppl, *Theorie der Elektrizität: Einführung in die Maxwellsche Theorie der Elektrizität*, Leipzig, Teubner, 1904. For discussion of fundamental problems concerning electromagnetic induction, see my *Einstein's Special Theory of Relativity*, esp. chap. 3.

23. H. A. Lorentz, "Alte und neue Fragen der Physik," *Physikalische Zeitschrift* **11** (1910), pp. 1234-1257.

24. For discussions of Planck's work, see M. J. Klein, "Max Planck and the Beginnings of the Quantum Theory," *Archive for History of Exact Sciences* **1** (1962), pp. 459-479, and T. S. Kuhn, *Black-Body Theory and the Quantum Discontinuity, 1894-1912*, Oxford, Oxford University Press, 1978.

25. Einstein, "Autobiographical Notes," p. 37.

26. See my *Einstein's Special Theory of Relativity*, esp. chaps. 2 and 11.

27. A. Einstein, "Fundamental Ideas and Problems of the Theory of Relativity," address to acknowledge the Nobel Prize, delivered to the Nordic Assembly of Naturalists at Gothenburg, 11 July 1923, in *Nobel Lectures: 1901-1921*, New York, Elsevier, 1967, pp. 482-490, p. 484.

28. A. Einstein, "Über einen die Erzeugung und Verwandlung des Lichtes betreffenden heuristischen Gesichtspunkt," *Annalen der Physik* **17** (1905), pp. 132-148; translated by A. B. Aarons and M. B. Shepard in *American Journal of Physics* **33** (1965), pp. 367-374.

29. A. Einstein, "Die von der Molekularkinetischen Theorie der Wärme geforderte Bewegung von in ruhenden Flüssigkeiten suspendierten Teilchen," *Annalen der Physik* **17** (1905), pp. 549-560; reprinted in A. Einstein, *Investigations on the Theory of Brownian Movement*, trans. A. D. Cowper, with notes by R. Furth, New York, Dover, 1956.

30. Einstein, "Autobiographical Notes," p. 53.

31. Ibid.

32. For detailed development, see my *Einstein's Special Theory of Relativity*.

33. Einstein, "On the Electrodynamics of Moving Bodies," p. 392.

34. Holton, *Thematic Origins*, p. 364.

35. Einstein, "Autobiographical Notes," p. 53.

36. Einstein, "Über das Relativitätsprinzip und die aus demselben gezogenen Folgerungen," *Jahrbuch der Radioaktivität* **4** (1907), pp. 411-462.

37. Abraham and Föppl, *Theorie der Elektrizität.*

38. Einstein, "On the Electrodynamics of Moving Bodies," p. 392.

39. Ibid., p. 406.

40. Einstein, "Über das Relativitätsprinzip und die aus demselben gezogenen Folgerungen," p. 413.

41. W. Kaufmann, "Über die Konstitution des Elektrons," *Annalen der Physik* **19** (1906), pp. 487-553.

42. Miller, *Einstein's Special Theory of Relativity*, pp. 333-334.

43. Einstein, "Über das Relativitätsprinzip und die aus demselben gezogenen Folgerungen."

44. Kaufmann, "Über die Konstitution des Elektrons," p. 495.

45. Einstein, "Über das Relativitätsprinzip und die aus demselben gezogenen Folgerungen," p. 439.

46. A. H. Bucherer, "Messungen an Becquerelstrahlen. Die experimentelle Bestätigung der Lorentz-Einsteinschen Theorie," *Physikalische Zeitschrift* **9** (1908), pp. 755-762.

47. C. T. Zahn and A. A. Spees, in "A Critical Analysis of the Classical Experiments on the Variation of Electron Mass," *Physical Review* **53** (1938), pp. 511-521, demonstrated that the resolution of Bucherer's velocity filters was inadequate to distinguish between the various competing electron theories. In fact, this shortcoming pervaded every other experimental determination of the electron's mass in the period 1908-1915.

48. Einstein, "Autobiographical Notes," p. 21.

Peter G. Bergmann

THE QUEST FOR UNITY: GENERAL RELATIVITY AND UNITARY FIELD THEORIES

A CENTENARY JUBILEE is a good time to evaluate an outstanding individual's intellectual contributions. Enough time has elapsed to free us of the fashions of the moment; yet that time is sufficiently short so that some of us who came directly under that person's influence are still alive. Albert Einstein, who earned the greatest fame of all physicists in his own lifetime, has come to serve as a focus for a vast range of endeavors, both humane and scientific, on the occasion of his one-hundredth birthday. Having had the privilege of being associated with Einstein in my youth, I am very happy to have this occasion to express my deep gratitude for the instruction and stimulation that I received from him. Albert Einstein's memory will last as long as there are human beings who strive for a more perfect society and for a deeper comprehension of the physical universe.

Einstein's contributions to physics are many and are being discussed by the participants of this celebration, just as at the many gatherings throughout the world. I shall address myself to but one, his quest for unity in science, which found expression in his formulation of the theory of relativity and in his research for a unitary field theory that would lead beyond it.

Let me begin with the special theory of relativity. Among the many puzzles confronting physicists at the turn of the century, one touched the very foundations of all natural science; it related to the nature of space and time. Most of physics was then dominated by mechanics, which dealt with the interaction of physical bodies. The crowning achievement of mechanics had been the complete and quantitative explanation of the workings of the solar system, so that astronomers were able to predict such future events as eclipses with great accuracy and complete reliability. The laws of mechanics had been formulated by Isaac Newton. They concerned the *accelerations* of the interacting bodies, determined by the forces of interaction, which in turn depended only on the (instantaneous) configuration. If Newton's laws were valid, then it followed that in our universe there is no possibility of identifying a state of rest or, for that matter, one of absolute motion. As far as absolute properties of space and time were concerned, the laws of mechanics called for a set of states of nonrotational, uniform, rectilinear motion, all of equal stature, which are usually referred to as inertial frames of reference.

The then new physics of the electromagnetic field, brilliantly formulated by Michael Faraday, James Clerk Maxwell, and Hendrik A. Lorentz, differed from the laws of mechanics in that it introduced the notion of a pervasive field that filled the space between the particles. The laws of the field, however, involved a *velocity*, the speed with which any electromagnetic disturbance would spread in empty space; today we call this velocity the speed of light. The electromagnetic laws would seem to single out one state, the state of absolute rest, in which in the absence of matter the speed of propagation of electromagnetic waves is isotropic. As everybody knows, the search for that state, or frame, of absolute rest was unsuccessful: it appeared that the electromagnetic field is totally insensitive to the absolute motion of the earth through space. This experimental fact, confirmed in the meantime in all manner of ways, apparently presented an internal inconsistency, unless of course you assumed that the earth represented the state of absolute rest. And that would have been a regression to Ptolemaic ideas, unacceptable to nineteenth-century scientists.

Increasingly tortured proposals were considered by the outstanding theorists of the time. Einstein's contribution was revolutionary because it was formally simple, yet conceptually deep. In addition to permitting observers in different states of motion to have different scales of distance and of time, Einstein demonstrated that the simultaneity of distant events would be observer- or frame-dependent if one accepted the proposition that the speed of light cannot be exceeded by any signaling device. By an intricate argument—intricate not because of abstruse mathematics but because of a very delicate analysis of experimental procedures—he showed that once the notion of absolute time marks is dropped, two moving observers *both* can perceive the other's clocks to be slow, and both can perceive the other's yardsticks to be contracted. The paradox was resolved by a profound modification of classical space and time concepts. A very few years later Hermann Minkowski discovered the natural mathematical formulation of Einstein's new physics, the four-dimensional space-time model. The relationship between space and time measurements of two observers moving differently was analogous to a rotation in four dimensions, except for a few signs that differed from an ordinary rotation.

Through his revision of the space-time concepts Einstein had succeeded in removing from physics the apparent contradiction between the (classical) principle of relativity of mechanics and the laws of electrodynamics. To this extent, unity was restored, but a new contradiction had been created. Newtonian mechanics involved at its foundations the notion of absolute simultaneity; the forces between distant bodies—for instance, between the sun and the earth—depended on their instantaneous distance from each other, which in relativity would differ for different observers. If the new theory of space and time was to prevail, mechanics needed to be modified.

Relativistic mechanics was designed to bridge the gap partially by making the mass velocity-dependent—hence the proportionality between mass and energy—and by modifying the force. These changes had no effect, however, on the dependence of the action at finite distances on absolute simultaneity. A relativistic theory of gravitation would require replacing the Newtonian action by the intermediary of fields. Hence the need for a relativistic gravitational field.

It is possible to introduce relativistic field equations for a gravitational field with relatively little effort. Einstein was troubled by two considerations, however. First, there were several ways of doing so, and there seemed to be very little grounds for choosing one over the others. The second consideration involved a peculiar property of gravitation, namely, the universality of gravitational acceleration. In a gravitational field all bodies undergo the same acceleration—on the surface of the earth, for instance, 9.8 m/sec/sec.

In an electric field the force acting on a body depends on that body's electric charge, and its acceleration, on the ratio of the body's charge to its mass, *e/m*. No analogous parameter enters into the expression for acceleration caused by a gravitational field. This fact was already ascertained by Galileo and certainly recognized by Newton; but it remained a curiosity. It was Einstein who recognized the implications. If gravitational acceleration is the same for all bodies, then it vanishes, locally, for an observer who himself undergoes the same acceleration. One is led naturally to the notion of a free-falling frame of reference, rather than the inertial frame of reference. The difference between the two concepts is this: whereas an inertial frame of reference presumably extends over the whole universe, a free-falling frame is defined only locally, in a sufficiently small region. An astronaut or cosmonaut will perceive no gravitational field in his free-falling vehicle, but distant objects appear to be accelerated relative to himself. Thus the local uniformity of gravitational acceleration precludes the determination of inertial frames of reference by local means, replacing these frames by constructs that cannot be extended globally.

This line of reasoning leads to the general theory of relativity, Einstein's theory of the gravitational field. When the new theory was completed some sixty years ago, it replaced the space-time of the special theory of relativity by a yet more general geometric concept, that of a Riemannian space-time, which locally has properties resembling those of the special theory but on a larger scale is much more involved, being a curved manifold.

I do not wish to give you the impression that the progress from the special to the general theory was straightforward or logically inescapable. Far from it. If inertial frames cannot be determined by local observations, it might be possible to preserve the concept by relying on observations of distant objects. This is in fact what astronomers do. But reliance *in principle* on distant objects runs counter to the spirit of a field theory, which relies on physical interaction of fields in the neighborhood of the particle, not at a distance, thus circumventing the embarrassment of instantaneous action at a distance. From a logical point of view, the progress toward general relativity depended on a number of choices to be made; its eventual adoption, first by Einstein himself and later by the community of physicists, depended on the esthetic appeal of the finished theory and on its confirmation by experiment and observation.

With regard to experimental confirmations, the universality of gravitational acceleration has been confirmed to an accuracy beyond 10^{-11}. As for relativistic effects—that is, gravitational effects that deviate from the predictions of classical mechanics and of the special theory of relativity—all quantitative observations that can be performed with today's technology have confirmed Einstein's theory well within the bounds of instrumental error, including such cases where competing modern theories predicted different results. These experiments are an ongoing enterprise.

The issue of the esthetic appeal of general relativity is closer to my principal theme. General relativity had restored a measure of unity to physics by modifying our ideas of space and time, which lie at the foundations of any dynamical conceptual construction. The new framework accommodated gravitation. Its essence was to be sought not in the properties of the single, local, free-falling frame of reference but in the relationship of that local frame to free-falling frames in adjacent regions. These relations were subject to field laws that were chosen according to principles of formal simplicity and the requirement that for weak fields the classical results should agree in lowest approximation with those of the new theory.

One major conceptual difficulty associated with the new theory—the interaction of the local field with a particle—was removed some twenty years after the theory's inception. Every mass serves as a source of the gravitational field, just as each charge is a source of the electromagnetic field. At the site of a particle the field becomes very large. If the particle is conceived of as a mass point, the field becomes infinite. But the force that affects the kinematic behavior of the particle is determined by the surrounding field. What if that field is finite? The first response, historically, was a holding operation. If the particle itself was small, if its mass was slight, then one could imagine the field as it would be if the particle under consideration did not exist. Einstein then postulated that such a small particle would travel on a so-called geodesic, a curve in space-time that corresponds to unaccelerated motion in special relativity, or in the local free-falling frame. This assumption was the point of departure for the geometric interpretation of the fact of uniform gravitational acceleration.

But what if the particle was not so small? How would one deal with the problem of a double star, for instance, in which the field caused by the "other" star could not reasonably be assumed to be larger than the field caused by the star under scrutiny? Eventually Einstein, Leopold Infeld, and Banesh Hoffmann developed an approach that was applicable in such cases. They found that the field laws outside a particle could not be satisfied unless the particle itself behaved "properly." Viewed from that angle, the behavior of sources of the gravitational field was determined by the field laws themselves. This was a property not shared by other field laws, and certainly not by those of electrodynamics. Thus general relativity turned out, after all, to be conceptually more nearly of one piece than any physical theory then known.[1]

With the laws of motion of particles having been absorbed into the logical structure of the field laws, mechanics, once the dominant structure of theoretical physics, was all but eliminated from it. The quest for unity had apparently reached its objective. But there were several hairs in that ointment.

Atomic physics, we know, is governed not by classical laws but by quantum rules. General relativity, however, is nonquantum. It satisfies strictly deterministic laws, whereas quantum laws are essentially statistical. Einstein could never bring himself to accept statistics as the definitive form of the laws of nature, even though as a young man he had made major contributions to quantum theory and to statistical mechanics. He always considered statistical approaches preliminary to a better understanding, which would be strictly causal.

The second drawback of general relativity was that it treated particles as singularities of the field, as infinities, and failed to explain their structural properties, such as their masses, charges, and other characteristics. Finally, as nature is not purely gravitational but allows for other forces as well, the gravitational and the nongravitational fields appear to be essentially different. From the point of view of general relativity, gravitation is needed in order to give space and time their geometric structure; all the other forces are gratuitous.

Unitary field theory was intended to remedy all these blemishes. From the early 1920s to the end of his life, Einstein developed ever new approaches to unitary field theory. At the time of his death, he was working, together with Bruria Kaufman, on the so-called asymmetric theory.

Riemannian geometry in four dimensions is a well-defined and fairly rigid structure that admits very little variation in the proposed dynamical laws. Somehow this mold must be broken if more physical fields than gravitation are to be accommodated within the geometric framework. Before I discuss a few of these attempts, permit me to address myself to a preliminary question: What is geometry?

I suspect that there is no answer to this question that will satisfy everybody. Basically, one might consider geometry to be any kind of mathematical structure that begins with the construction of a set of points that satisfies the minimal properties of continuity that justify one in speaking of a space. A space may, but need not, involve such concepts as volume and distance; it may, but need not, involve the existence of vector fields and the possibility of defining when two vectors at distinct locations are to be considered parallel to each other. These are but examples of properties that geometric spaces might possess. Many more have been investigated; in fact, many more have been used by physicists in their pursuits and endeavors to understand nature. Depending on the properties ascribed to a new model for space-time, its structures might lend themselves to interpretations that are reminiscent of fields known to physicists. How does such a "geometrization" contribute to unification? Einstein stated repeatedly that he did not consider geometrization of physics a foremost or even a meaningful objective, and I believe that his comments remain valid today. What really counts is not a geometric formulation or picturization but a real fusing of the mathematical structures intended to represent physical fields.

How can we visualize such a fusing? One possibility, suggested by the history of relativity itself, is that the decomposition of fields into gravitational, electromagnetic, "strong," and "weak" nuclear forces might depend on the frame used for their description—that, for instance, a field that appears in one frame to be purely gravitational is mixed gravitational and electromagnetic in another frame. This is possible if the variety of equivalent frames, or modes of description, is sufficiently large. There are other possibilities. Some fields might require additional fields complementing them before any meaningful differential operations can be defined. This situation obtains, for instance, in Hermann Weyl's geometry, on which I shall comment later.

In order to formulate and to survey such possibilities, a geometric formulation often is a real help. Essentially, mathematicians and physicists, too, proceed intuitively when they endeavor to create new concepts and relations. Geometry often helps them to "think in images." Thus, geometry may serve as a heuristic device. That role might not exhaust its possibilities, but it is a major one.

I cannot give you a complete listing of all attempts, by Einstein and by many others, to create generalizations of the four-dimensional Riemannian model of space-time. Though I have worked on unitary field theories myself, I cannot claim any comprehensive knowledge. One whole class of attempts may be characterized as maintaining the four-dimensionality of space-time but modifying or enriching the Riemannian structure. In this class belongs, for instance, Weyl's geometry. Weyl weakened Riemann's idea of an invariant distance at the infinitesimal level; he replaced it by the notion of relative distance. Only the ratio of two distances would have any invariant (frame-independent) meaning. With this weakening of the metric concept, one cannot form differential structures without introducing a pseudo-vector field that looks like the potentials of the electromagnetic field. Another enrichment, suggested originally by Elie Cartan, generalizes the notion of parallel transport of vectors. In Riemannian geometry, if you introduce a free-falling frame of reference, then a vector is parallel to a vector if in that frame the components are the same. In Cartan geometry, they may be rotated. Finally, in Einstein's asymmetric theory the dot product of two vectors (at the same point) is not symmetric in the two vectors, $a \cdot b \neq b \cdot a$. In all three of these examples the minimal geometric structures are richer than in Riemannian geometry and so are capable of accommodating a greater variety of physical fields.

In Weyl's geometry the gravitational and the electromagnetic structures are distinct in that they are not being converted into each other under changes of frame, but both are required to produce a harmonic whole. In Einstein's asymmetric theory there exists one type of change of frame that mixes the Riemannian with the other parts of the geometry. I do not see that kind of fusing in the Cartan geometry, but from a somewhat different point of view Cartan's geometry also hangs all together.

How can any enlargement of the geometry lead to an understanding of the properties of particles? That is a very difficult question to answer. The occurrence of singularities in a field theory represents a sort of breakdown of that theory: the field equations admit of solutions that go beyond the control of those equations, ruin the causal character of the field laws. And there seems to be little doubt that general relativity as we know it today leads to singularities under a variety of circumstances. There are no solutions that might be interpreted as particles that are everywhere finite. Once you are dealing with different field equations, you can hope that such solutions might exist. I might add that the theorems concerning the unavoidability of singularities in the standard theory were all discovered long after Einstein's death, mostly by Roger Penrose and Steven Hawking. I have not seen their methods of proof extended to any of the unitary theories, but this might well be possible.

If nonsingular solutions should exist, then one might investigate whether these can in some way be related to the properties or particles that occur in nature. There is a way to relate the ratio of the charge of an elementary particle to its mass to a pure number of the order of 10^{20}, depending on the kind of particle. A theory of elementary particles at the very least should yield numbers like this one.

Einstein hoped to obtain quantum rules in a similar fashion. If particles interact with one another, it is not likely that singularities can be avoided in the course of time unless the initial conditions are just right. I suspect that few practitioners of unitary field theory today would share these hopes; many of them would feel, I believe, that to achieve successes in other respects is worth their efforts, even if quantum theory will continue to flourish in its present form.

There are other kinds of unitary field theories, including some that today claim a great deal of interest. These use, in some way or another, an increase in the number of dimensions of space-time. One famous example is Theodor Kaluza's proposal. He increased the number of dimensions to five without changing the Riemannian character of the model. He was thus able to increase the number of components of the metric so as to accommodate the electromagnetic field as well. He set one extra component equal to a constant, because he had no use for it. To account for the observed four-dimensionality of space-time, he assumed that no field depended on the fifth coordinate.

Strangely enough, Kaluza's field, though conceived of as a single structure, the metric, separated quite naturally into the gravitational and the electromagnetic fields in a manner that did not at all depend on the frame used. To this extent, Kaluza's fusion of fields failed. But his idea continued to intrigue others, and several variants were tried in the course of the years. One, by Einstein, Valentin Bargmann, and myself, replaced Kaluza's assumption of strict independence from the fifth dependence by a weaker assumption: that the universe is closed in the fifth dimension, that it looks a bit like a tube, and that the dependence on the fifth coordinate, limited as it must be if the circumference of the tube is sufficiently small, has something to do with quantum phenomena. Alas, the idea did not work out.

Another idea, discovered and rediscovered several times over, was not to kill the supernumerary field component but to retain it and to assign to it such tasks as to serve as a cosmological parameter. Carl Brans's and Robert Dicke's so-called tensor-scalar theory is one of these attempts, though I believe that these authors were initially unaware of the preceding history of that idea.

There are other methods for increasing the dimensionality of space-time. One is to permit the coordinates of space-time to assume complex values. Penrose's twistor formalism is a case in point. Complexification is used by some authors as a mere technical device for discovering new solutions of Einstein's equations in the real domain. This is a productive approach, but it has little to do with unitary field theory. Others, and I believe Penrose is among them, take complex space-time seriously and hope to break new ground. Formally, a complex number is a pair of real numbers. A complex four-dimensional space or space-time is in that sense equivalent to a real eight-dimensional manifold. But if the pairing into sets of complex coordinates (or dimensions) is taken seriously, then the rules of algebra and of analysis applied to complex numbers are equivalent to the introduction of an additional invariant structure, the so-called complex structure, which must be reproduced under all changes of frame. Thus the structure of a complex space differs significantly from that of a real space having twice as many dimensions. Penrose hopes that by pursuing this line of inquiry he may succeed in understanding elementary particles, and perhaps also the quantum character of nature.

If complex numbers are good, hypercomplex numbers may be better. Hypercomplex numbers are one way of looking at algebras that have at least some of the properties of the algebra of ordinary numbers. Whereas complex numbers are equivalent to pairs of real numbers, hypercomplex numbers involve larger multiplets. Their rules of arithmetic cannot be as simple as those involving real and complex numbers. They will involve noncommutative products ($ij \neq ji$). Most systems of hypercomplex numbers also contain null divisors, nonzero elements whose product with some other nonzero number equals zero.

One particular type of hypercomplex algebras is known as Grassmann algebras. The product of any two of the basic elements of a Grassmann algebra is anticommutative, $ij + ji = 0$. Interest in Grassmann algebras and in fields formed with their help originated with mathematicians and with physicists who were impressed with the possibility of using them in elementary particle physics. It had been observed that there are collections of elementary particles that resemble each other even though some members of the set have integral spin, others half-odd spin. In quantum theory the state vectors, or wave functions, belonging to particles with integral spin are symmetric with respect to the permutation of particles; those belonging to particles with half-odd spin are antisymmetric. One type obeys Einstein-Bose statistics, the other, Fermi-Dirac statistics. Some elementary particle physicists believe that there must be some changes in frame that transform one kind of particle into the other field. Formally, such a scheme can be set up, and most conveniently with the help of Grassmann "numbers." These endeavors go under the name of supersymmetry. If they involve an attempt at unitary field theory, they are called supergravity.

There is some formal resemblance between complex field theories and supergravity, and I am impressed with the seriousness of these novel attempts to draw inspiration from elementary particle physics, an area in which many people are obtaining new and exciting insights. Supergravity meets one objection that has been raised against the search for unitary field theory: that it has been purely speculative, without nurture from the findings of experimental physics. Supersymmetry and supergravity are speculative, to be sure; but they are influenced by high-energy physics, and that to me is a very attractive feature. Many unsolved problems remain, of course. I certainly do not wish to give you the impression that I am all sold on supergravity. Rather, I should say, these many years after Albert Einstein's death a new generation of unitary field theorists is taking up the torch and proceeding along novel lines. They have good contact with other frontier areas of physics; one can only wish them well.

In twentieth-century theoretical physics, a number of major areas have emerged, each dominated by a closely reasoned and closely linked set of laws. These areas have emerged in response to the human quest for understanding, for comprehending the individual event as an instance of an overriding general principle. Albert Einstein created one such area, the theory of gravitation, and he did so by deepening our grasp of the nature of space and time, the scaffolding on which all of physical science takes place. He had hoped to expand and to strengthen this scaffolding so as to take in the physics of the atom and of the subatomic world as well, but this attempt did not succeed in his lifetime.

It behooves us to proceed, each of us in the manner we judge best, whether or not it resembles closely Einstein's own way. All our endeavors are supported by what he achieved, and our resolve is strengthened as we perceive not only Einstein's tenacity but also his creativity and flexibility. To most of us it is given to contribute but one small step or two toward man's understanding of nature. Let us be content with that. The quest for unity will never be sated. Each achievement will reveal new vistas and mysteries to be conquered.

ACKNOWLEDGMENTS

Research for this paper was supported in part by National Science Foundation Grant No. PHY78-06721.

NOTES

1. *Note added in proof.* Peter Havas has called my attention to the fact that the relationship between field equations and ponderomotive equations was known to Hermann Weyl and several other scientists quite early in the history of general relativity. Apparently, their understanding was not generally acknowledged, and the facts were rediscovered in the 1930s.

Martin J. Klein

FLUCTUATIONS AND STATISTICAL PHYSICS IN EINSTEIN'S EARLY WORK

"HE WAS ONE of that class of geniuses who early learn to trust themselves in an essential way, whatever moments of doubt they may have," who are soon "propounding the great questions and attempting the great answers."[1] Lionel Trilling's words about Keats provide an unexpectedly apt description of the young Einstein and his way of doing physics. Einstein recognized his unusual youthful confidence in his own thoughts about the natural world as he looked back on his early years while preparing his scientific autobiography. He had had no faith in his ability to select the centrally important problems in mathematics out of all "the more or less dispensable erudition." But in physics, he wrote, "I soon learned to scent out the paths that led to the depths and to disregard everything else, all the many things that fill up the mind and divert it from the essential."[2] The young man who created the special theory of relativity in 1905 evidently trusted his "nose" for the right paths to follow in exploring the profoundest issues in the science of his time.[3]

That same confidence in his own scientific judgment, that same readiness to pursue the consequences of his ideas no matter how unlikely they might appear to be, are also to be found in Einstein's papers on what he called the kinetic-molecular theory of heat, or what we would now call statistical physics. Many of these papers have remained largely unknown, despite their intrinsic interest and their historical importance. They are an indispensable aid in trying to understand and characterize Einstein's particular approach to physics. It is from this standpoint that I want to discuss a few of the principal features of Einstein's very early work in statistical physics.

∞

Even in his student years at Zurich, Einstein was struck by the powerful way in which mechanics could account for so much of the natural world. "What made the greatest impression upon the student," he wrote in his scientific autobiography, "was not so much the technical development of mechanics or the solution of complicated problems as the achievements of mechanics in areas that apparently had nothing to do with mechanics. . . ."[4] It was "above all the kinetic theory of gases" that caught his interest then, and he returned to it a year after his graduation from the Polytechnic. Einstein made a thorough study of Ludwig Boltzmann's book on the theory of gases and thought he saw a way of "supplying the keystone of a chain of argument begun by him."[5] This chain of argument was nothing less than the derivation of the second law of thermodynamics from the general laws of mechanics together with the theory of probability, and it was in no way limited to gases.

Einstein submitted this work to the *Annalen der Physik* in June 1902, a few days after he was appointed to the Patent Office at Bern.[6] This paper is impressive enough for its positive results, for its discussions of the fundamental distribution law, the equipartition theorem, and the statistical-mechanical interpretations of temperature and entropy. It is even more interesting for other reasons. The Einstein who appears in these pages is evidently something of an outsider in the community of physicists. He knew Boltzmann's book, but not the many papers in the *Wiener Berichte* in which Boltzmann had developed, elaborated, and qualified his ideas. Einstein's language, when he refers to the mechanical representation of a physical system or to the mechanical world picture, is the language of the previous generation, of Hermann von Helmholtz and Heinrich Hertz, of J. J. Thomson and J. Willard Gibbs.[7] But what is most remarkable in the paper is Einstein's awareness of the limitations of an approach too closely tied to mechanics and his eagerness to free his arguments from such restrictions. He calls his readers' attention to how little of the conceptual structure of mechanics has actually been used in the derivations—only the energy principle and Liouville's theorem—and how likely it is that the laws of thermodynamics would hold for systems more general than those governed by mechanics.[8] He emphasizes that the statistical expressions for temperature and entropy are independent of the "special form of the energy as the sum of potential and kinetic energies" and that they are probably more general than the "mechanical representation," with which he began, would suggest.[9]

Einstein developed these hints for freeing his statistical mechanics from its mechanical origins in a paper finished in January 1903.[10] In a letter to his friend Michele Besso he described this paper as "completely clear and simple" and declared that he was "entirely satisfied" with it after having repeatedly revised and improved his work.[11] The starting point of Einstein's new article retained only the bare essentials of the underlying mechanics—first-order linear differential equations for the variation with time of the state variables, a unique integral of the motion (the energy), and an analogue of Liouville's theorem. The equations did not have to be Hamilton's, and even the division of the energy into kinetic and potential terms was unnecessary. Einstein saw this generalization as another step in "the elimination of the force concept" from physics, as he pointed out in another letter to Besso.[12] He had evidently been much impressed by Hertz's program for mechanics.[13] Einstein's generalized statistical mechanics, based on a clearly formulated concept of ensembles, would be the foundation for a variety of his subsequent studies.[14]

This 1903 paper also deals with the problem of irreversibility, a problem totally omitted from Einstein's earlier work, which dealt only with systems in equilibrium. He proves that entropy will never decrease as a result of processes occurring within a closed system, but in order to obtain this strong result he had to make the drastic assumption that "more probable distributions always follow less probable ones, i.e. that *W* [the probability of the distribution] always increases until the distribution has become stationary and *W* has reached a maximum."[15] Einstein's recognition that one must assume that more probable states *always* follow less probable ones if one is to prove that entropy *always* increases was far from adequate as a basis for understanding the puzzle of irreversibility: namely, how the irreversibility of macroscopic phenomena can be explained when the underlying equations of motion are unaffected by time reversal. Boltzmann had gone a long way toward answering that question in the 1890s, but Einstein did not refer to this work at all. He would take up the problem in his own way a few years later.

∞

Einstein's works up to this time—the two papers just discussed and two earlier ones on intermolecular forces[16]—do not show the unique qualities that characterize his mature science. Those qualities appear for the first time in his next publication, "On the General Molecular Theory of Heat," written in March 1904.[17] This short article, which Einstein described as a report on some results supplementing his last paper, breaks new ground but does it so simply and directly that the boldness of the arguments and the unexpected nature of the results might easily go unnoticed. Einstein focused his attention on the single universal constant that appears in all the basic equations of statistical mechanics, the constant we now refer to as Boltzmann's constant and denote by the symbol k.[18] He offered a new interpretation of its significance, using the generality of his formulation of statistical theory, and suggested a fresh approach to its experimental determination. Let us examine his reasoning.

The single quantity that characterizes a physical system in Einstein's statistical mechanics is its structure function, which he denotes by $\omega(E)$. This function is defined so that $\omega(E)dE$ is the volume of that region in the system's phase space corresponding to the states in which the energy of the system lies in the interval between E and $E + dE$. If such a system is in contact with a heat bath at temperature T, the probability dW of finding the system with its energy in this interval is given by the equation

$$dW = C \exp(-E/kT)\, \omega(E)dE, \tag{1}$$

where C is determined by the condition that the probability be normalized. The average energy $\bar{E}$ of the system is then obtained from the equation

$$\bar{E} = \frac{\int_0^\infty E \exp\,(-E/kT)\, \omega(E)\, dE}{\int_0^\infty \exp\,(-E/kT)\, \omega(E)\, dE}. \tag{2}$$

The entropy S of the system is given by

$$S = k\, ln[\omega(\bar{E})], \tag{3}$$

and the energy and temperature are related by the equation

$$\frac{1}{kT} = \frac{d\, ln\, [\omega(E)]}{dE}, \tag{4}$$

where the derivative is evaluated at $E = \bar{E}$.

One interpretation of the constant k comes from a calculation of the average kinetic energy of a point molecule in an ideal gas of such molecules. This is readily found to be $\frac{3}{2}kT$, from which one can then show that k must be simply related to the gas constant R,

$$k = R/N_0, \tag{5}$$

where N_0 is Avogadro's number, the number of atoms in one gram atomic weight of any substance.[19] This interpretation of k, in which it appears as one of the ways of establishing the scale of atomic magnitudes—that is to say, as equivalent in importance to Avogadro's number—might seem fundamental enough, but Einstein evidently thought that there was more to be said about it.

He proceeded to calculate the fluctuations in the energy of a general system whose temperature is held at T by keeping it in contact with a heat bath. Those fluctuations about the average energy $\bar{E}$ are determined by the distribution law of (1). If one differentiates the expression for $\bar{E}$ given in (2) with respect to temperature, one finds directly that

$$\frac{d\bar{E}}{dT} = \frac{1}{kT^2}[\bar{E^2} - (\bar{E})^2], \tag{6}$$

or, equivalently, that

$$\overline{(\Delta E)^2} = kT^2\frac{d\bar{E}}{dT}, \tag{7}$$

where ΔE is the deviation of the energy from its average value, $E - \bar{E}$. The mean square fluctuation of the energy, $\overline{(\Delta E)^2}$, is "a measure of the thermal stability of the system," as Einstein put it; the greater the value of $\overline{(\Delta E)^2}$, the less stable the system. The most general significance of k, then, was that it "determines thermal stability," or, equivalently, that it sets the scale of fluctuation phenomena.

Such fluctuations or deviations of the variables from their average values are intrinsic to any statistical theory. They are, in fact, the peculiarly statistical part of such a theory. Thus it was not enough to show that the average values of the appropriate quantities in a macroscopic system obey equations having the same form as the laws of thermodynamics. Unless the fluctuations about these average values are small, the averages are a poor indication of what to expect. It is therefore not surprising that Boltzmann and Gibbs made calculations to show that fluctuations are indeed reassuringly small for systems of macroscopic size. This was an essential step in providing what Gibbs called a "rational foundation of thermodynamics."[20]

Both these masters of statistical mechanics went further than that, however, and argued that one could never expect to observe departures from average behavior. Thus Boltzmann wrote in his *Theory of Gases*: "Even in the smallest neighborhood of the tiniest particles suspended in a gas, the number of molecules is already so large that it seems futile to hope for any observable deviation, even in a very small time, from the limits that the phenomena would approach in the case of an infinite number of molecules."[21] And Gibbs, who devoted a chapter to a discussion of "anomalies" in the canonical ensemble and derived (7) among many other such results, came to an equally negative conclusion:

> It follows that to human experience and observation with respect to such an ensemble as we are considering, or with respect to systems which may be regarded as taken at random from such an ensemble, when the number of degrees of freedom is of such an order of magnitude as the number of molecules in the bodies subject to our observation and experiment, [ΔE] would be in general [a] vanishing quantity, since such experience would not be wide enough to embrace the more considerable divergencies from the mean values, and such observation not nice enough to distinguish the ordinary divergencies.[22]

Einstein took a very different approach to this question. Immediately after deriving the equation for the mean square fluctuation in energy, he remarked that this result would make possible an exact determination of the universal constant k, if only one could measure the energy fluctuations of some system. Although any system would do, since the result was so general, Einstein had to admit that "in the present state of our knowledge" there was no case at all that could be used for such a measurement. He did not, however, stop after that statement.

One could still explore the validity of his ideas on energy fluctuations for one particular system, unpromising as that system might seem to be for testing a result in statistical mechanics. The energy fluctuations in blackbody radiation—thermal radiation at equilibrium in an enclosure—would be too small to be measured if the enclosure were of macroscopic size. If, however, one considered an enclosure whose linear dimensions were of the order of the wave length at which the radiation spectrum has its maximum, then the energy fluctuations ought to be comparable to the energy itself. As Einstein pointed out, it was far from obvious that an enclosure containing radiation was a system to which his equations should apply. It was certainly not a mechanical system, not even in the generalized sense of his own theory. He went ahead nevertheless, arguing that his method should at least be adequate for order-of-magnitude estimates.

The problem was to estimate the linear dimension L of an enclosure for which the average fluctuations in energy are as big as the energy itself, thus requiring that

$$\overline{(\Delta E)^2} = (\bar{E})^2. \tag{8}$$

The average energy of blackbody radiation at temperature T in a cubic enclosure of side L is given by the Stefan-Boltzmann law,

$$\bar{E} = \alpha L^3 T^4, \tag{9}$$

where α is a constant. If one now evaluates the fluctuations from this with the help of (7) and then imposes the condition of (8), one is led directly to the result

$$L = \left(\frac{4k}{\alpha}\right)^{1/3} \frac{1}{T}. \tag{10}$$

From the qualitative discussion in the previous paragraph, this length L should be comparable to the wave length at which the blackbody radiation spectrum has its maximum. That wave length, however, was known on both experimental and theoretical grounds (Wien's displacement law) to be inversely proportional to the temperature. Furthermore, the coefficient of proportionality found experimentally between that wave length and the inverse temperature was of the same order of magnitude as the coefficient in (10), which Einstein estimated using the Stefan-Boltzmann constant α and the rough value of k available to him.

This agreement between the length L, calculated from the theory of fluctuations, and the wave length at the peak of the spectrum was impressive. Einstein thought it "ought not be ascribed to chance, because of the great generality of our assumptions." He was not yet able to obtain k from fluctuation measurements, but he had found an "extremely interesting relationship" between that constant, which fixed the scale of atomic quantities, and the wave lengths of blackbody radiation. And this result had been accomplished "without having recourse to special hypotheses."[23] Einstein was now convinced that fluctuations must be taken completely seriously.

∞

One might well ask what had led Einstein to explore the significance of that universal constant *k*, which is the focus of his 1904 paper. Although we know it as Boltzmann's constant, it never appears in the two volumes of Boltzmann's book on gases,[24] which came out only a few years before Einstein's work. It is also absent from Walther Nernst's contemporary work, *Theoretische Chemie*,[25] a thoroughgoing treatment of physical chemistry at the turn of the century, and one in which the subject is discussed "From the Standpoint of Avogadro's Rule and Thermodynamics," in the words of its subtitle. It must be remembered that the numerical values of atomic and molecular quantities were not at all well known at this time.[26] The estimates of atomic diameters and masses and of Avogadro's number made by Josef Loschmidt, J. D. van der Waals, and others were based on indirect arguments, approximate calculations, and crude models. One could still say in 1900, as James Clerk Maxwell had said in 1873, that the available knowledge of the basic atomic properties was "only as yet of the nature of a probable conjecture."[27] If none of the books that Einstein studied emphasized the atomic constants in general or the constant *k* in particular, should we conclude that he was uninfluenced in deciding to give it a prominent position in his thinking? Although this is certainly a possibility, I do not think it is a likely one.

In the opening paragraphs of his 1904 paper Einstein mentions an expression for the entropy of a system "which Boltzmann found for ideal gases and which is assumed by Planck in his theory of radiation." Although no reference to Planck's work is provided, Einstein surely meant the papers in which Planck derived the distribution law for blackbody radiation with the help of discrete elements of energy.[28] Planck's theory involved two constants, which he denoted by *k* and *h*: the *k* is just what we have been discussing, and *h* has been known ever since as Planck's constant.[29] The fundamental natural constants always had an almost mystical significance for Planck,[30] and he set particular importance on this aspect of his work. In the papers communicating his new theory Planck drew his readers' attention to these two universal constants and evaluated them from measurements on blackbody radiation with the help of the new distribution law. He also showed how, using his value of k, one could now calculate Avogadro's number (combining k with the gas constant R), and so the mass of a hydrogen atom. He also showed how one could then calculate the unit electrical charge—the charge of a singly charged ion or of the recently discovered electron (with the use of the Faraday constant).[31] These were, he asserted, direct consequences of his theory and could claim "not approximate but absolute validity" if the theory were correct.[32]

It seems that Einstein had read enough of Planck's work to be impressed by those results, no matter how perplexed he might have been by the derivation Planck had given for his distribution law. Einstein had already seen in his own earlier work how essential k was to the statistical theory. Planck's work underlined its importance for him. When Einstein referred to k as "determined by the magnitudes of the elementary units of matter and electricity," as he did early in his 1904 paper, he must have been thinking of Planck's results. And when he obtained a connection between k and the order of magnitude of the wave length at which the blackbody radiation spectrum has its maximum, he emphasized that this connection was found without any need to introduce special hypotheses. This assertion was in implied contrast with the very special hypothesis underlying Planck's theory.[33]

∞

In his first paper on the subject, Einstein had not been able to specify a system having observable fluctuations, but he did not give up the search for one. In May 1905 he reported the successful result of that search in "On the Motion of Particles Suspended in a Liquid at Rest That Is Required by the Kinetic-Molecular Theory of Heat."[34] Einstein showed that colloidal particles, even those large enough to be observed under a microscope, must carry out a completely chaotic perpetual motion as a result of the thermal motion of the molecules of the liquid in which the colloid is suspended. He calculated the average displacement to be expected of such particles in a given time interval and found that the phenomenon should be observable under reasonable experimental conditions.

Einstein had *invented* the Brownian motion. To say anything less, to describe this paper in the usual way, that is, as his *explanation* of the Brownian motion, is to undervalue it. The phenomenon he discussed had been discovered in 1827 and studied by a number of scientists during the three-quarters of a century that followed.[35] Many of Einstein's contemporaries were well aware of it, and some had even conjectured its molecular origins, although no theory yet existed. As for Einstein, his only reference to it was at the beginning of his paper: "It is possible that the movements to be discussed here are identical with the so-called 'Brownian molecular motion'; however, the information available to me regarding the latter is so lacking in precision that I can form no judgment in the matter." Einstein evidently knew that there was such a thing as Brownian motion, but he had no knowledge of the experiments reported in the literature. We see this fact confirmed in a letter he wrote at about this time to his friend Conrad Habicht, listing the works he had in progress in the spring of 1905. "The third [of these works] proves that on the assumption of the molecular theory of heat particles of the order of magnitude of 1/1000 mm, suspended in a liquid, must carry out an observable, irregular movement which is produced by the thermal motion. Movements of small, inanimate, suspended particles have in fact been observed by physiologists, and these movements are called by them the 'Brownian molecular motion.' "[36] Einstein clearly knew nothing about the experiments and theoretical proposals of the chemists and physicists who were working on that motion "observed by physiologists."[37] He was not providing a theoretical explanation of a familiar but puzzling phenomenon.

It is instructive to compare Einstein's paper with one published a year or so later by Marian von Smoluchowski.[38] Smoluchowski was not a largely self-taught, isolated individual like Einstein. He had taken his degree at Vienna and then worked in Paris, Glasgow, and Berlin before accepting a position at Lemberg. Smoluchowski knew about the Brownian motion and had been working on a theory of it for some five years when Einstein's paper appeared.[39] Although he published after Einstein and never contested his priority, Smoluchowski had independently arrived at essentially the same results earlier by a very different route. His paper shows a thorough knowledge of the literature and a close familiarity with the experimental methods used to study Brownian motion. Smoluchowski's work was a deliberate attempt to explain this motion theoretically by a direct analysis of the effects of molecular collisions on the suspended particle. Both Smoluchowski and Einstein saw that the Brownian motion was much more than just an interesting consequence of the kinetic theory.

"If the motion discussed here can actually be observed," Einstein wrote in the introduction to his paper, "together with the laws relating to it that one would expect to find, then classical thermodynamics can no longer be looked upon as applicable with precision to bodies even of dimensions distinguishable in a microscope: an exact determination of actual atomic dimensions is then possible. If on the other hand the prediction of this motion should prove to be incorrect, this would present a weighty argument against the kinetic-molecular conception of heat." Einstein saw his prediction of an observable fluctuation phenomenon as a crucial test of the whole program of statistical mechanics. If the fluctuations did not occur as predicted—if there were no "Brownian motion"—then there must be something fundamentally wrong with the idea that macroscopic bodies are composed of large numbers of particles obeying the laws of mechanics. But confirmation of Einstein's predictions would make possible a direct determination of k and thus of atomic magnitudes. Such confirmation would also demonstrate that there were processes going on in the world that could not be accounted for by thermodynamics, processes that contradicted the laws of thermodynamics. The "Brownian motion" would establish the limits of applicability of those laws.[40]

Einstein was certainly aware of the doubts about atomistic theories that were often expressed at this time by a number of influential physicists and chemists in Europe. He had studied the second volume of Boltzmann's lectures on the kinetic theory, which opens with Boltzmann's impassioned response to criticisms of the theory. Boltzmann expressed his fear that the theory of gases might be "temporarily thrown into oblivion because of a momentary hostile attitude toward it" and vowed, as "an individual struggling weakly against the stream of time," to do what he could so that "when the theory of gases is again revived, not too much will have to be rediscovered."[41] Einstein had also studied Wilhelm Ostwald's textbook of general chemistry. He admired Ostwald,[42] but he knew what the chemist thought about the use of "atomic hypotheses" in chemistry and how he had tried to eliminate these "hypothetical representations," these "superfluous accessories," from his science.[43]

More important to Einstein's development than Ostwald was Ernst Mach, whose "incorruptible skepticism and independence" he never stopped admiring.[44] During his years at the Polytechnic, Einstein read Mach, his *Mechanik* and also his *Wärmelehre*, and both books made a great impression on the young student.[45] Just as Einstein was absorbing the mechanical physics of the late nineteenth century and appreciating its successes in the kinetic theory of gases, he came upon Mach's sharply worded and forcefully stated criticism of this work. For Mach, atomism was an attempt to make "the idea of substance in its most naive and crudest form" into the very basis of physics. He thought that these "childish and superfluous" pictures made an "odd contrast" to the real spirit of current physics.[46] However useful the atomic hypothesis may have been as a guide to research, however valuable as a pedagogical device, one must be careful not to confuse the analogies to reality that it offered with reality itself.

Mach aimed some of his critical remarks directly at Boltzmann's efforts to explain the second law on the basis of statistical mechanics. "The mechanical interpretation of the second law, by making a distinction between ordered and disordered motions, by drawing a parallel between the increase of entropy and the increase of the disordered motion at the expense of the ordered, seems to be really *artificial.*" Mach completely agreed with a remark he quoted from his former student František Wald: " 'In my opinion the roots of this [entropy] law lie much deeper, and if the attempt to bring the molecular hypothesis and the entropy law into agreement succeeds, then this would be fortunate for the hypothesis but not for the entropy law.' "[47]

There was certainly something to be said for this attitude, as Einstein must have recognized. That the laws of thermodynamics could be derived from statistical mechanics was no proof that they must be derived on that basis. Unless one could find phenomena predicted by statistical mechanics but not by thermodynamics—phenomena that depended on departures from the averages that were related in accord with thermodynamics—there was no real need to accept statistical mechanics. Mach's cogent criticism of Boltzmann may well have spurred Einstein's search for an observable effect of fluctuations. Mach argued that no adequate proof of the existence of atoms had yet been given, and he concluded that atoms were only a fiction. Einstein accepted the criticism, but went on searching until he found an adequate proof.[48]

∞

An exposition of Einstein's theory of the motion of particles suspended in a fluid would be out of place here. I want to comment only on a few points that seem particularly characteristic of his style of thinking about physical problems. What Einstein actually calculated was the mean square displacement $\overline{(\Delta x)^2}$ that a suspended particle will undergo in any given direction during a time interval τ. His result is that $\overline{(\Delta x)^2}$ is proportional to τ. The coefficient of proportionality depends on the temperature of the fluid and the force with which the fluid opposes the motion of a solid (the particle) moving through it. That resistive force is assumed to be proportional to the velocity of the moving body with a coefficient f. The resulting equation has the form

$$\overline{(\Delta x)^2} = \frac{2kT}{f}\tau. \qquad (11)$$

(The coefficient f is taken to be that appropriate to a sphere of radius a moving through a fluid of viscosity η,

$$f = 6\pi\eta a; \qquad (12)$$

this is Stokes's law.)

Although Einstein made no point of it in his first papers on this subject, his result showed that it was hopeless to try to measure an average velocity of the suspended particles. The root mean square displacement in time τ, divided by that time, does not approach a limit as the time interval is decreased. This finding corresponds to the absolutely irregular, chaotic character of the motion, which is treated by considering the displacements in successive intervals τ as completely independent. The experimenters who had studied Brownian motion, however, had always tried to report their measurements in terms of average velocities, a procedure that Jean Perrin later described as "absolutely wrong."[49] It was not until 1907 that Einstein emphasized the importance of this aspect of his theory as a guide to what could and should be measured.[50]

Considering that Einstein's reason for undertaking this study was to test the "kinetic-molecular theory of heat," the appearance of k in his principal result, equation (11), is the crucial feature. How did it get there? The key step comes at the beginning of Einstein's paper, where he argues that each particle suspended in the fluid must contribute to an osmotic pressure in exactly the same way that a single molecule of a dissolved substance contributes to the osmotic pressure of the solution. "According to this theory [the kinetic-molecular theory of heat, or statistical mechanics] a dissolved molecule is differentiated from a suspended body *solely* by its dimensions, and it is not apparent why a number of suspended particles should not produce the same osmotic pressure as the same number of molecules." That pressure p will then be given by the usual law for ideal systems,

$$p = nkT, \tag{13}$$

where n is the number of suspended particles per unit volume, a result that Einstein justifies by a detailed derivation from his general statistical mechanics.

However involved the rest of the theory might be, that basic physical idea, at least, seems simple. It was an extension of the domain of applicability of statistical mechanics to include a system containing particles large enough to be visible. Each Brownian particle would then have an average kinetic energy of translational motion equal to $\frac{3}{2}kT$, just as if it were a molecule in a gas. Such an extended use of statistical mechanics seems simple only in retrospect. In 1905 it was not at all clear when one could or could not safely use the results of statistical mechanics, such as the energy formula just mentioned, a special case of the famous equipartition theorem. Einstein knew this better than anyone else, for he had shown just a few months earlier what disastrous conclusions follow when that theorem is applied to charged oscillators that interact with blackbody radiation.[51]

It is remarkable how closely the theory of Brownian motion—one of the most impressive successes of the equipartition theorem—followed after its catastrophic failure to account for the radiation spectrum. It is extraordinary that that success and that failure should have been worked out by the same man.

∞

Once Einstein had seen the power of the fluctuation theory, he made it one of his principal instruments for probing into the unknown. His work in his field after 1905 can be divided into two main categories. One series of papers clarifies the basis of the theory of fluctuations, confirming what Einstein had suspected from the beginning. One does not need any specific knowledge about the molecular structure of a system in order to get at the fluctuations of its macroscopic properties about their average or thermodynamic values. Boltzmann's principle, which relates entropy to probability, became the cornerstone of this development. Einstein inverted Boltzmann's equation, using it to express the probability of a state in which the variables deviate from their average values in terms of the entropy change involved in this deviation. This inversion of Boltzmann's principle, mathematically simple but physically profound, led him to a statistical thermodynamics freed entirely from its mechanical origins.[52]

In the second series of papers Einstein used the fluctuation theory based on Boltzmann's principle to explore the nature of radiation. With its help he demonstrated the unavoidability of a quantum theory and pointed to the need for a new theory that would fuse the wave and particle aspects of radiation. The Brownian motion of a mirror suspended in blackbody radiation became one of his favorite devices for thinking about the radiation problem, and it played a prominent part in a number of his works.[53]

Both these lines of development were peculiarly his own. When Einstein began following them, Smoluchowski was the only other physicist who took fluctuations seriously at all. Certainly no one else was ready to rely on arguments based on fluctuations as a guide to the obscurities associated with quanta.

At the First Solvay Conference in 1911, when it seemed "as if the ground had been pulled out from under one's feet, with no firm foundation on which to build to be seen anywhere,"[54] Einstein was ready to give the Boltzmann principle the same trust that he gave to the conservation of energy. "We should admit its validity without any reservations," he declared.[55] He had no doubts about the path he had chosen.

ACKNOWLEDGMENTS

This work was supported in part by a grant from the National Science Foundation.

NOTES

1. Lionel Trilling, *The Opposing Self*, New York, 1955, p. 5.

2. Albert Einstein, "Autobiographical Notes," in Paul Arthur Schilpp, ed., *Albert Einstein: Philosopher-Scientist*, Evanston, Ill., 1949, pp. 14-17. Translations from the German are mine unless a translator is indicated. Previously published translations have been reviewed and modified or corrected where appropriate.

3. Ernst G. Straus, "Memoir," in A. P. French, ed., *Einstein: A Centenary Volume*, Cambridge, Mass., 1979, pp. 31-32.

4. Einstein, "Autobiographical Notes," pp. 18-19.

5. Einstein to Marcel Grossmann, n.d. (summer 1901). I thank the Estate of Albert Einstein for permission to quote this unpublished letter.

6. A. Einstein, "Kinetische Theorie des Wärmegleichgewichtes und des zweiten Hauptsatzes der Thermodynamik," *Annalen der Physik* **9** (1902), pp. 417-433. The paper was received by the *Annalen* on 26 June 1902. According to a memorandum by Friedrich Haller, director of the Patent Office, Einstein was appointed to his position there on 16 June 1902. See Max Flückiger, *Albert Einstein in Bern*, Bern, 1974, p. 65, where the memorandum is reproduced.

7. See M. J. Klein, "Mechanical Explanation at the End of the Nineteenth Century," *Centaurus* **17** (1972), pp. 58-82.

8. Einstein, "Kinetische Theorie," p. 427.

9. Ibid., p. 433.

10. A. Einstein, "Eine Theorie der Grundlagen der Thermodynamik," *Annalen der Physik* **11** (1903), pp. 170-187.

11. Einstein to Besso, January 1903, in Albert Einstein and Michele Besso, *Correspondance 1903-1955*, ed. Pierre Speziali, Paris, 1972, p. 3.

12. Einstein to Besso, 17 March 1903, ibid., p. 14.

13. Heinrich Hertz, *The Principles of Mechanics Presented in a New Form* (1894), trans. D. E. Jones and J. T. Walley, New York, 1956, introduction, pp. 1-41. Also see Einstein, "Autobiographical Notes," pp. 30-31.

14. Einstein did not read Gibbs's *Elementary Principles* (see note 20 below) until some years later. See A. Einstein, "Bemerkungen zu den P. Hertzschen Arbeiten: 'Über die mechanischen Grundlagen der Thermodynamik,' " *Annalen der Physik* **34** (1911), pp. 175-176.

15. Einstein, "Grundlagen der Thermodynamik," p. 184.

16. A. Einstein, "Folgerungen aus den Capillaritätserscheinungen," *Annalen der Physik* **4** (1901), pp. 513-523; "Über die thermodynamische Theorie der Potentialdifferenz zwischen Metallen und vollständig dissociirten Lösungen ihrer Salze, und über eine elektrische Methode zur Erforschung der Molekularkräfte," ibid., **8** (1902), pp. 798-814. Einstein was soon referring to these papers as "my two worthless beginner's works." Einstein to Johannes Stark, 7 December 1907; quoted in A. Hermann, "Albert Einstein und Johannes Stark," *Sudhoffs Archiv* **50** (1966), p. 272.

17. A. Einstein, "Zur allgemeinen molekularen Theorie der Wärme," *Annalen der Physik* **14** (1904), pp. 354-362.

18. Einstein does not use k. His equations in this paper are expressed in terms of $\varkappa$, which is equal to $k/2$.

19. Einstein refers to Boltzmann's book, *Vorlesungen über Gastheorie*, 2 vols., Leipzig, 1896-1898, and in particular to section 42 of the second volume, in connection with his calculation of the average kinetic energy. Boltzmann does not even introduce a constant like k into his discussion.

20. J. Willard Gibbs, *Elementary Principles in Statistical Mechanics: Developed with Especial Reference to the Rational Foundation of Thermodynamics*, New Haven, 1902.

21. Ludwig Boltzmann, *Lectures on Gas Theory*, trans. S. G. Brush, 2 vols. in 1, Berkeley, 1964, p. 318.

22. Gibbs, *Elementary Principles*, p. 74.

23. Einstein, "Zur allgemeinen molekularen Theorie der Wärme," p. 354. On 14 April 1904 Einstein wrote to his friend Conrad Habicht, "I have now found the relationship between the size of the elementary units of matter and the wave lengths of radiation in an extremely simple way." Quoted in Carl Seelig, *Albert Einstein und die Schweiz*, Zurich, 1952, p. 64.

24. Boltzmann briefly summarizes Josef Loschmidt's method for estimating molecular diameters and numbers, including his numerical results, in *Lectures on Gas Theory*, pp. 94-95. On p. 264, however, he refrains from giving the results of a calculation following J. D. van der Waals, saying that it would be out of place in his book to analyze the experimental data in order to choose the most reliable numbers.

25. Walther Nernst, *Theoretische Chemie*, 2d ed., Stuttgart, 1898.

26. In *Lectures on Gas Theory*, Boltzmann wrote that the number of molecules in a cubic centimeter of nitrogen at atmospheric pressure and 25°C "falls between 2½ and 250 trillions" (p. 95). This corresponds to a value of Avogadro's number between 6.1×10^{22} and 6.1×10^{24}. Nernst's numbers correspond to a value of Avogadro's number equal to 1.2×10^{24}. See his *Theoretische Chemie*, p. 394. This is based on van der Waals's calculation.

27. J. C. Maxwell, "Molecules," *Nature* **8** (1873), p. 437. See also M. J. Klein, "The Historical Origins of the Van der Waals Equation," *Physica* **73** (1974), pp. 28-47.

28. M. Planck, "Über das Gesetz der Energieverteilung im Normalspektrum," *Annalen der Physik* **4** (1901), pp. 553-563. There is no reason to think that Einstein knew Planck's first publication on this subject, "Zur Theorie des Gesetzes der Energieverteilung im Normalspektrum," *Deutsche physikalische Gesellschaft, Verhandlungen* **2** (1900), pp. 237-245.

29. Planck was not too happy with the designation of k as Boltzmann's constant, "understandable" as it might have been. He pointed out that Boltzmann never used this constant or inquired into its numerical value. M. Planck, "Wissenschaftliche Selbstbiographie," in his *Physikalische Abhandlungen und Vorträge*, Braunschweig, 1958, vol. 3, p. 395.

30. See M. Planck, "Über irreversible Strahlungsvorgänge," *Annalen der Physik* **1** (1900), pp. 120-122. When Planck derived his new distribution law, he emphasized that the theory led to "other relationships" that would, in his opinion, be "of considerable importance for other fields of physics and also of chemistry." He underlined this assertion by setting off his calculation of the basic constants when he published the more detailed account in the *Annalen*: M. Planck, "Über die Elementarquanta der Materie und der Elektrizität," ibid., **4** (1901), pp. 564-566.

31. For a discussion of attempts between 1897 and 1910 to measure

the charge on an ion, see Robert Andrews Millikan, *The Electron*, Chicago, 1917, pp. 43-63, 155-158.

32. This remark is to be found at the conclusions of both "Energieverteilung im Normalspektrum" and "Über die Elementarquanta."

33. Einstein was evidently impressed by Planck's determination of k and also interested in seeing how much it depended on the assumptions of Planck's theory. He showed in 1905 that Planck's distribution law—and so also the assumptions that led to it—were not needed for obtaining Planck's value of k. See A. Einstein, "Über einen die Erzeugung und Verwandlung des Lichtes betreffenden heuristischen Gesichtspunkt," *Annalen der Physik* **17** (1905), pp. 132-148, esp. pp. 136-137.

34. A. Einstein, "Über die von der molekularkinetischen Theorie der Wärme geforderte Bewegung von in ruhenden Flüssigkeiten suspendierten Teilchen," *Annalen der Physik* **17** (1905), pp. 549-560. This and four other papers are collected in A. Einstein, *Investigations on the Theory of the Brownian Movement*, ed. R. Fürth, trans. A. D. Cowper, New York, 1956.

35. For the history of the work on Brownian motion see Mary Jo Nye, *Molecular Reality*, New York, 1972, esp. chaps. 1 and 2. Also see Stephen G. Brush, *The Kind of Motion We Call Heat*, Amsterdam, 1976, pp. 655-672. Jean Perrin describes his own classic experiments in *Atoms*, trans. D. L. Hammick, London, 1916.

36. Einstein to Habicht, n.d. (spring 1905), quoted in Seelig, *Einstein und die Schweiz*, p. 77. In a second paper on this subject, completed in December 1905, Einstein reports that soon after the appearance of his first work he was informed by Professor H. Siedentopf that Siedentopf and others had already been convinced by their observations that Brownian motion was caused by the thermal motion of the molecules in the liquid. Einstein also remarks that the "slender experimental material" now available to him confirms the correctness of his results at least in order of magnitude. He knows now that he is dealing with Brownian motion. See A. Einstein, "Zur Theorie der Brownschen Bewegung," *Annalen der Physik* **19** (1906), pp. 371-381.

37. Brush, *Motion We Call Heat*, p. 673, quotes the passage from the "Autobiographical Notes" in which Einstein describes his reason for studying the motion of suspended particles and says that he did this work "without knowing that observations concerning the Brownian motion were already long familiar" (pp. 46-47). Brush's skepticism about "the accuracy of these statements" seems quite unjustified to me in view of the evidence cited. See also Einstein to Besso, 6 January 1948, *Correspondance*, p. 391.

38. Marian von Smoluchowski, "Zur kinetischen Theorie der Brownschen Molekularbewegung und der Suspensionen," *Annalen der Physik* **21** (1906), pp. 756-780. For Smoluchowski, see also Armin Teske, *Marian Smoluchowski, Leben und Werk*, Warsaw, 1977.

39. See Smoluchowski's remarks at the beginning of his paper and the letter from Smoluchowski to Perrin, quoted by Teske, *Smoluchowski*, p. 161.

40. The clearest discussions of the relationship between the second law and fluctuation phenomena are Smoluchowski's. See his "Experimentell nachweisbare, der üblichen Thermodynamik widersprechende Molekularphänomene," *Physikalische Zeitschrift* **13** (1912), p. 1069, and "Gültigkeitsgrenzen des zweiten Hauptsatzes der Wärmetheorie," in *Vorträge über die kinetische Theorie der Materie und der Elektrizität* (The Wolfskehl Lectures, 1913), Leipzig, 1914.

41. Boltzmann, *Lectures on Gas Theory*, pp. 215-216.

42. See Einstein's reference to Ostwald in his first paper, "Folgerungen." See also the letters written in 1901 by Einstein and his father to Ostwald, quoted in Banesh Hoffmann and Helen Dukas, *Albert Einstein: Creator and Rebel*, New York, 1972, pp. 32-34.

43. Wilhelm Ostwald, *Grundriss der allgemeinen Chemie*, 3d ed., Leipzig, 1899, p. vi.

44. Einstein, "Autobiographical Notes," pp. 20-21.

45. Einstein to Besso, 6 January 1948, *Correspondance*, p. 391.

46. Mach, *Die Principien der Wärmelehre. Historische-Kritische Entwickelt*, Leipzig, 1896, pp. 428-429.

47. Ibid., p. 364.

48. Although Ostwald and others were completely convinced of the reality of atoms by the confirmation of the Brownian motion theory and other experimental results obtained between 1900 and 1910, Mach was not. See John T. Blackmore, *Ernst Mach: His Work, Life, and Influence*, Berkeley, 1972, pp. 319-323.

49. Perrin, *Atoms*, p. 109.

50. A. Einstein, "Theoretische Bemerkungen über die Brownsche Bewegung," *Zeitschrift für Elektrochemie* **13** (1907), pp. 41-42. Einstein's remarks were a criticism of recent work by Th. Svedberg. See Milton Kerker, "The Svedberg and Molecular Reality," *Isis* **67** (1976), pp. 190-216.

51. See Einstein, "Erzeugung und Verwandlung des Lichtes." Note also that in his second Brownian motion paper, "Zur Theorie der Brownschen Bewegung," Einstein derives the long wave-length limit of the radiation distribution from his general statistical theory. That this is "not the true law of radiation" he attributes to "a fundamental incompleteness in our physical conceptions." See Einstein, *Investigations*, pp. 26-27.

52. See, in particular, A. Einstein, "Über die Gültigkeitsgrenze des Satzes vom thermodynamischen Gleichgewicht und die Möglichkeit einer neuen Bestimmung der Elementarquanta," *Annalen der Physik* **22** (1907), pp. 569-572; "Theorie der Opaleszenz von homogenen Flüssigkeiten und Flüssigkeitsgemischen in der Nähe des kritischen Zustandes," ibid., **33** (1910), pp. 1275-1298.

53. See my paper, "Einstein and the Wave-Particle Duality," *Natural Philosopher* **3** (1964), pp. 1-49.

54. Einstein, "Autobiographical Notes," pp. 44-45.

55. A. Einstein, in *La théorie du rayonnement et les quanta*, ed. P. Langevin and M. de Broglie, Paris, 1912, p. 436.

Max Jammer

EINSTEIN AND QUANTUM PHYSICS

THE SUBJECT "Einstein and Quantum Physics" has a number of remarkable features. From the purely *human* point of view, it presents the heroic and tragic struggle of one of the greatest physicists of all time to proceed, unyieldingly and true to himself, along a lonely road, one aloof from the mainstream of physics but for him the only road to a deeper insight into the mysteries of the nature of matter. From the *historical* point of view, it offers the unique case of a scientist decisively furthering the rise and development of a scientific theory and subsequently rejecting, no less decisively, the very same theory once it gained general acceptance. From the *philosophical* point of view, it touches upon profound problems on the borderline between physics and philosophy. Treated superficially, the subject may easily be misunderstood—as indeed it has often been.

Modern physics is based on the theory of relativity and the quantum theory. Despite critics like E. T. Whittaker or G. Keswani, it is generally acknowledged that the theories of relativity are the result primarily of Einstein's work. However, it is not so well known that the development of quantum physics, too, owes a great debt to Einstein. Surely, every student of modern physics knows of Einstein's contributions to quantum physics. A biographer of Einstein even declares, "No physicist had more to do with the creation of quantum physics than Einstein. His work in this field would have been in and of itself a full scientific career for any other physicist." But a proof of this statement is given neither in this biography nor anywhere else.

It is my intention, first, to show how decisively Einstein's contributions promoted the development of the quantum theory and, second, to analyze the motives that led Einstein to reject the theory that owes to him so much. Many of the arguments on which my analysis is based can be found in the literature, especially in the writings of Martin Klein and Arthur Fine, and most of the historical facts to which I shall refer are mentioned in biographies of Einstein, such as that by Carl Seelig. However, by viewing these diverse items in a coherent perspective, we will be led to two conclusions that in my opinion are not yet sufficiently recognized: first, that Einstein was in fact one of the chief architects of the quantum theory; and second, that, contrary to widespread opinion, he rejected the theory not because *he*, Einstein—owing perhaps to intellectual inertia or senility—was too conservative to adapt himself to new and unconventional modes of thought, but, on the contrary, because *the theory* was in his view too conservative to cope with the newly discovered empirical data. The following analysis owes much to the historical investigations carried out by my friends Martin J. Klein, Arthur Fine, and Banesh Hoffmann.

∞

Einstein began his intellectual career with a profound study of thermodynamics. He also delved deeply into the foundations of statistical mechanics. His study of statistical fluctuations, in particular, proved remarkably fruitful, not only in his investigation of the Brownian motion, but also in his treatment of the radiation problem, which he transformed thereby into a problem of quanta. In fact, it was this inquiry that led him to the first of the three great papers of 1905. Although called merely "Einstein's paper on the photoelectric effect," it was of supreme importance for the whole of modern physics, for it introduced the concept of quanta of radiation.

Early in 1905, Einstein asked his friend Conrad Habicht for a copy of Habicht's recently completed dissertation and added, "In return I can promise you four papers, the first of which I shall soon be able to send to you as I am getting some free copies. It deals with radiation and energy characteristics of light and is very revolutionary as you will see. . . ." With these words Einstein described his paper "Über einen die Erzeugung und Verwandlung des Lichtes betreffenden heuristischen Gesichtspunkt" ("On a heuristic viewpoint concerning the production and transformation of light"), the first of his three papers published in volume

17 of the *Annalen der Physik.* About the third and most famous of these, his relativity paper, "Zur Elektrodynamik bewegter Körper" ("On the electrodynamics of moving bodies"), he wrote only this to Habicht: "Not yet fully worked out, it deals with the electrodynamics of moving bodies and is based on a modification of the theory of space and time; the purely kinematical part of this paper will certainly be of interest to you."

It was not the relativity paper but the one on light quanta that Einstein characterized as revolutionary—a judgment fully justified. For as we know today, special relativity differs from classical physics ultimately only in the replacement of the Leibniz postulate on intrasystemic simultaneity by the Einstein postulate. The latter, by depriving this relation of transitivity, opens the way to a straightforward construction of the causal topology (and metric) of Minkowski space (special relativity thus viewed, is, strictly speaking, even more classical than Newtonian physics), whereas the very notion of light quanta, basically a quantum field-theoretic notion, implies a profound change in the whole foundations of the conceptual framework of classical physics.

How, then, did Einstein arrive at this revolutionary conclusion? Confining himself to the limiting case of Planck's radiation law and hence to Wien's law, he calculated the entropy of such radiation contained in a volume V. Using the Boltzmann relation, he then showed that the probability of finding the entire radiation energy E at an arbitrary instant within a partial volume V' is given by the expression

$$W = \left(\frac{V'}{V} \right)^{\frac{E}{h\nu}}$$

In the case of n non-interacting particles, the probability for such a distribution is given by the same formula, with the exponent replaced by n. Einstein therefore concluded that "monochromatic radiation of low density behaves thermodynamically as if it consists of independent energy quanta of magnitude $h\nu$." Finally, he showed that this hypothesis explains in a simple and straightforward way a multitude of theretofore unexplained empirical rules, such as Stoke's law in photoluminescence, the laws of photoionization, and the law of the photoelectric effect.

The reasoning that led Einstein to the idea of "photons," as they were later called, demonstrated an independence of thinking and intellectual audacity that in its irreverence of accepted patterns of thought went far beyond the work of Max Planck. For although Planck's introduction of the quantum of action may rightly be said to have inaugurated the era of quantum physics, ultimately it was but the result of a compromise, of an interpolation between two expressions, involving entropy and energy, one corresponding to Wien's law and the other to the Rayleigh-Jeans formula. In fact, the introduction of quanta was for Planck, to use his own words, "an act of desperation," and he tried repeatedly, though unsuccessfully, to accommodate it "somehow" ("irgendwie") within the framework of classical physics.

Einstein, in contrast, was from the very start uninhibited and ready to discard concepts that had lost their validity in the light of new experience. He once wrote:

> Concepts which have proved useful for ordinary things easily assume so great an authority over us, that we forget their terrestrial origin and accept them as unalterable facts. They then become labelled as "conceptual necessities," *a priori* situations, etc. The road of scientific progress is frequently blocked for long periods by such errors. It is therefore not just an idle game to exercise our ability to analyse familiar concepts, and to demonstrate the conditions on which their justification and usefulness depend. In this way they are deprived of their excessive authority. Concepts which cannot be shown to be valid are removed. Those which had not been coordinated with the accepted order of things with sufficient care are corrected, or they are replaced by new concepts when a new system is produced which, for some reason or other, seems preferable.

For what reasons, then, did Einstein replace the notion of a continuous wave field of radiation by that of light quanta? True, he admitted, the wave theory of light accounts perfectly well for such purely optical phenomena as interference or diffraction. But, he added, optical observations refer to time averages rather than to instantaneous values. The continuous theory of light may therefore lead to contradictions with experience when applied to the phenomena of emission and transformation of light. These contradictions can be removed, he continued, if one assumes that the energy is discontinuously distributed in space so that light consists of a finite number of energy quanta, localized in space, which can be produced or absorbed only as complete units. Moreover, apart from Lord Rayleigh and independently of him, Einstein was the first to apprehend fully the difficulties resulting from the equipartition law of statistical mechanics when combined with Maxwell's theory, difficulties known, in Paul Ehrenfest's terminology, as "the ultraviolet catastrophe."

Thus, by exposing the limitations of the applicability of accepted concepts, Einstein recognized the need to modify or replace them with new concepts, no matter how daring, provided that within the former range of applicability they reduce to the older conceptions. This method of conceptual refinement, characteristic of the whole of Einstein's scientific work, had its first exemplification in his heuristic proposal of light quanta. The paper was also the first in which he applied the technique of using thought-experiments (*Gedanken-Experimente*) for the derivation of far-reaching conclusions, a device he subsequently applied with great skill and success on various occasions.

Reviewing the logical structure of Einstein's argument for light quanta, we note that the premises were theorems from thermodynamics (Wien's law, the Boltzmann relation) and, finally, an *argument by analogy*, based on comparing the two probability expressions mentioned above. This argument by analogy—the redundancy of which, by the way, was recognized only in 1971, when J. Dorling rederived Einstein's conclusion from the same premises but without the analogy argument—left Einstein's reasoning open, as is the case with every argument by analogy, to the objection that it lacked logical cogency. It was a free creation of the human mind, not a logic from experience. No wonder that Robert Millikan, who nine years later verified experimentally the linear dependence of the photoelectronic energy on the illuminating frequency, called Einstein's proposal a "bold, not to say reckless, hypothesis." In fact, prior to the discovery of the Compton effect in 1923, virtually no one shared Einstein's conviction in the reality of light quanta. For it was hard to imagine that a successful theory like Maxwell's should be discarded on the flimsy grounds of a fluctuation argument.

Einstein presented another argument in support of his claim in his 1909 paper "Zum gegenwärtigen Stand des Strahlungsproblems" ("On the present state of the radiation problem"). By weakening the conditions of the thought-experiment that had led him to this hypothesis, he strengthened the cogency of his argument. Instead of considering the highly improbable distribution of the total energy in a partial volume, he now simply calculated the mean-square fluctuation of energy and obtained an expression composed of two additive terms: the first corresponded to the classical theory of waves, whereas the second, predominating in the range of validity of Wien's law, could be understood only by accepting the hypothesis of light quanta.

At the end of September 1909, speaking for the first time at a scientific convention, Einstein read "Über die Entwicklung unserer Anschauungen über das Wesen und die Konstitution der Strahlung" ("On the development of our views on the nature and constitution of radiation") at the Salzburg Meeting of German Scientists. In this paper, which Wolfgang Pauli once called "a landmark in the development of theoretical physics," Einstein declared that the next phase in the development of theoretical physics would lead to a theory of light based on both wave and particle conceptions. To prove that these two structural properties need not be incompatible, Einstein constructed at the end of his paper a model along roughly the following lines: the energy of the electromagnetic field is assumed to be localized in singularities surrounded by fields of forces subject to a superposition principle in such a way that they produce in their totality a field of waves similar to that of Maxwell's theory.

Einstein's proposal to regard quanta of energy as singularities of continuous fields exerted a profound influence upon Louis de Broglie, Max Born, and others, and thereby on the development of quantum mechanics as a whole. But before I discuss this aspect of the development of the quantum theory, let me say a few words about Einstein's work on specific heats. For it was Einstein's study of specific heats that changed the *hypothesis* of quanta into the *theory* of quanta, as I shall now try to explain.

A scientific hypothesis becomes a scientific theory only if it also accounts for phenomena other than those for which it has been conceived to explain. It was therefore of fundamental importance that Einstein recognized, as early as 1907, that the application of the concept of quanta was not limited to radiation processes. In his 1907 paper "Die Plancksche Theorie der Strahlung und die Theorie der spezifischen Wärme" ("Planck's theory of radiation and the theory of specific heats") he wrote:

> If we cannot explain the elementary processes of energy exchange between radiation and matter on the accepted molecular-kinetic theory, do we not have to modify likewise our ideas about the periodic phenomena dealt with in the molecular theory of heat? I think, there can be no doubt about the answer. If Planck's theory of radiation strikes to the heart of the matter, then we must also expect to find contradictions between the present kinetic-molecular theory and experience in other areas of the theory of heat, contradictions that can be resolved by the route just traced.

The contradictions Einstein had in mind referred, of course, to the discrepancies between the classically derived Dulong-Petit theorem and the observed values of specific heats of monatomic solids at low temperatures, discrepancies known for some time, especially in the case of light elements such as beryllium, boron, carbon, and silicon. I need not explain how Einstein, using his new ideas, succeeded in accounting for these facts and how the Dulong-Petit formula, just like the Rayleigh-Jeans formula for radiation, turned out to be a limiting law, valid only for small values of $h\nu/kT$. It seems doubtful, however, whether in those early years the methodological importance of Einstein's work was fully recognized. Fortunately, external circumstances cooperated toward the same effect.

While investigating specific heats at low temperatures in connection with his discovery of the Third Law of Thermodynamics, Walther Nernst, one of the most influential personalities in the academic world at that time, realized that his measurements agreed fairly well with Einstein's theoretical predictions. Nernst became deeply interested in Einstein's ideas and soon declared himself an ardent advocate of the quantum theory. In an address before the Berlin Academy of Sciences in 1911 he observed, "At present, quantum theory is essentially only a rule for calculation, of apparently a very strange, one might even say grotesque, nature; but it has proven so fruitful by the work of Planck, as far as radiation is concerned, and by the work of Einstein, as far as molecular mechanics is concerned, that it is the duty of science to investigate it in as many ways as possible and to subject it to experimental examination."

With these ideas in mind, and supported by the Belgian industrialist Ernest Solvay, Nernst organized in 1911 what became known as the First Solvay Congress. It played an important role in the development of quantum physics. Suffice it to recall that Niels Bohr became interested in quantum physics as a result of hearing a report of the Brussels discussions from Lord Rutherford. Similarly, after reading the proceedings that his brother Maurice, the secretary of the meeting, had to prepare, de Broglie became so excited that "I decided to devote all my efforts to investigate the real nature of the mysterious quanta . . . whose profound meaning had not yet been understood."

But Einstein's influence was not merely circumstantial, as these examples may suggest. In addition to the fortuity of such propitious external factors, Einstein himself exerted a direct influence. Nernst, for example, was inspired by Einstein's extension of the quantum conceptions to the molecular theory of heat and was one of the first to investigate the possibilities of constructing a quantum theory, without recourse to radiation phenomena, on purely thermodynamical considerations. Since then, a number of scientists have attempted to establish quantum theory on thermodynamic foundations not connected directly with radiative processes, most recently Otto Stern in his paper "On a Proposal to Base Wave Mechanics on Nernst's Theorem," published in the *Helvetica Physica Acta* in 1962.

Furthermore, Einstein's thinking, both in form and substance, played an important role in the development of modern quantum mechanics, which was born in May 1925 when Werner Heisenberg replaced the classical concept of a continuous orbit of electrons with the notion of a totality of frequencies and amplitudes of the emitted radiation, which, in contrast to the former, are observable quantities. The methodological principle underlying Heisenberg's discovery of matrix mechanics was the idea to admit only observable quantities, an idea that Einstein's work on special relativity, with its rejection of absolute time, the ether, etc., had suggested to him, as Heisenberg repeatedly declared.

Wave mechanics, too, owes a great debt to Einstein. Here it was not so much the methodology but rather the substance of Einstein's work that was the influential factor. As Erwin Schrödinger himself admitted, "The whole theory would certainly not have originated yet, and perhaps never would have (I mean, not from me), had I not had the importance of de Broglie's ideas really brought home to me by your second paper on gas degeneracy."

The question "Why was it Schrödinger who discovered wave mechanics?" has been investigated by a number of historians of modern physics. Perhaps the answer lies in his 1922 paper, "Über eine bemerkenswerte Eigenschaft der Quantenbahnen eines einzelnen Elektrons" ("On a remarkable property of the quantum orbits of an individual electron"), a differential-geometric study of such orbits on the basis of Hermann Weyl's extension of Einstein's general relativity, a study in which Schrödinger "got hold of the resonance character of the [Bohr-Sommerfeld] quantum condition long before de Broglie. . . ." Or perhaps the answer lies in the impact upon Schrödinger of de Broglie's 1924 thesis "Recherches sur la théorie des quanta," whose point of departure for associating a periodic element to the corpuscular concept was Einstein's special theory of relativity. Whatever the answer, wave mechanics would not have originated, at least not in the way it did, without Einstein.

Interestingly, Einstein's ideas were a determining factor not only for the development of the *formalisms* of quantum mechanics but also for their *interpretations*. In fact, all major interpretations are rooted somehow in Einstein's ideas. Let me remind you that even the unconventional "many-worlds interpretation" originated in an attempt to reformulate quantum theory in a form suitable for application to general relativity and to provide a quantum description of a closed universe with which no external observer can be associated.

The point is even more obvious with the most widely accepted interpretation, Born's probabilistic interpretation of the wave function. Although persistently rejected by Einstein, it was, according to its originator, but an elaboration of Einstein's conceptions. When Born, as a result of his study of collision processes, advanced in 1926-1927 his thesis that ψ^2 represents a probability density, he claimed to have merely followed the footsteps of Einstein. And in accepting the Nobel Prize in 1954 "for his fundamental work in quantum mechanics and especially for his statistical interpretation of the wave function," Born stated:

> Einstein regards the statistical interpretation as unsatisfactory and has always tried to refute it. Yet the interpretation of the square of the wave function goes back to Einstein himself. He had declared that the average density of the photons in a beam of light must agree with the energy of the electromagnetic waves which describe this beam. It is the same idea which I proposed in 1927 as the interpretation of the Schrödinger wave function. . . .

Time does not allow me to show that other interpretations as well, even the stochastic interpretation, can be traced back to Einstein's ideas. Even if "quantum mechanics" stands for the formalism as well as the interpretation, I hope to have given sufficient evidence for the truth of the claim: "No physicist had more to do with the creation of quantum mechanics than Einstein."

Turning now to the second part of my talk, I shall refer to Einstein's philosophical writings, including his Herbert Spencer Lecture, "On the Methods of Theoretical Physics" (1933), his essay "Physics and Reality" (1934), his article "Quantenmechanik und Wirklichkeit" (1948), his talks collected in books such as *The World as I See It*, and, last but not least, his "Autobiographical Notes" (1949), in which he discussed the presuppositions that determined his scientific work and his anticipations as to the possibilities of future developments in theoretical physics. His view concerning the incomprehensible comprehensibility of the world of our sense experiences, his contention concerning the logical unbridgeability between sense experiences and scientific conceptions, his view of the latter as free creations of the human mind that are nonetheless subject to restrictive conditions, his insistence on an objectively given physical reality—all these are topics that undoubtedly will be discussed at this conference. Allow me therefore not to delve into the intricate problem of Einstein's position on the issue of the interaction between science and metaphysics. Of course, Einstein's philosophy of science was based on metaphysical presuppositions, just as was that of Heisenberg, of Bohr, or of any other of the great masters in quantum mechanics. And he was entitled to that. For as E. A. Burtt once put it: "there is no escape from metaphysics . . . the only way to avoid becoming a metaphysician is to say nothing."

Let me therefore refer very briefly to only those determining principles in Einstein's philosophy that had a direct bearing upon his attitude towards quantum physics and examine how they affected his thinking on this matter. The principles under discussion, I think, are essentially three. I shall call them briefly the maxims of *universality*, of *determinism*, and of *physical reality*.

The first maxim, permeating his whole theoretical work in physics, manifested itself in his unceasing search for a unifying theory, comprising, if possible, all of physics and explaining all known phenomena on a minimum of assumptions. His search for such a unification can be recognized already in his above-mentioned paper of 1909, "Zum gegenwärtigen Stand des Strahlungsproblems" ("On the present state of the radiation problem"). That paper concluded with a speculation on the possibility of establishing a relation between Planck's quantum of action h and the quantum of electric charge e, which he justifiably called "a stranger in the Maxwell-Lorentz electrodynamics."

It was this maxim that later, after his successful construction of general relativity, inspired him to search for a field-theoretical formulation of quantum physics. On March 3, 1920, he wrote to Born, "In my spare time I always brood about the problem of the quantum theory from the point of view of relativity. I do not think the theory can work without the continuum. But I do not seem to be able to give a tangible form to my pet idea, which is to understand the structure of the quanta by redundancy in determination, using differential equations."

How Einstein thought to incorporate the quantum into a general field theory is explained in his 1923 paper "Bietet die Feldtheorie Möglichkeiten für die Lösung des Quantenproblems?" ("Does the field theory provide possibilities for solving the problem of quanta?"). In ordinary mechanics, he pointed out, only the temporal evolution of the initial state of the system is determined by specific laws, namely, the differential equations of the laws of motion; the initial states, however, can be chosen at will. In quantum physics, in contrast, the initial state, too, seems to be subjected, as the quantum conditions indicate, to definite laws. These facts suggest a solution of the quantum problem by "overdetermining" the equations: the number of differential equations must exceed the number of field variables involved.

The very use of the apparatus of field-theoretical concepts at a time when quantum field theory was still a matter of the future and the emphasis on continuity and laws of motion in the form of differential equations gave additional support to Einstein's metaphysically rooted repugnance toward probabilism or indeterminism in physics. His persistent adherence to the second maxim, or, in other words, his outspoken opposition to the probabilistic interpretation of quantum mechanics, is well known.

Early in 1924 there appeared the then much discussed Bohr-Kramers-Slater paper on "The Quantum Theory of Radiation," which rejected even the principles of conservation of energy and momentum for individual processes and contended that "the occurrence of transition processes for the given atom itself, as well as for the atoms with which it is in mutual communication, is connected . . . by probability laws." Einstein wrote to Born: "Bohr's opinion about radiation interests me very much. But I should not want to be forced into abandoning strict causality without defending it more strongly than I have so far. I find the idea quite intolerable that an electron exposed to radiation should choose *of its own free will*, not only its moment to jump off, but also its direction. In that case, I would rather be a cobbler, or even an employee in a gaming-house, than a physicist."

Another often quoted example of Einstein's aversion to the spirit of Copenhagen and Göttingen is found in a letter written on 4 December 1926 to Born: "The quantum mechanics is very imposing. But an inner voice tells me that it is not yet the real thing. The theory says a lot, but does not really bring us any closer to the secret of the 'old one'. I, at any case, am convinced that *He* is not playing at dice."

How arduously Einstein was toiling to establish a connection between quantum mechanics and general relativity and how, in particular, he tried to reconcile Heisenberg's indeterminacy relations with a causal and continuous field theory can be seen in a paper from 1931. In spite of his reluctance to admit higher-dimensional generalizations of general relativity, Einstein published in collaboration with Walter Mayer "Einheitliche Theorie von Gravitation und Elektrizität." Here, contrary to Weyl's introduction of an additional gauge vector field and more in accordance with Kaluza's approach, Einstein made use of five vectors in a four-dimensional space-time. It seems to me not impossible that Einstein cherished the hope that such an approach could explain the Heisenberg indeterminacies as resulting from a loss of information, caused by suppressing one dimension when projecting a fully deterministic five-dimensional description into a four-dimensional space-time continuum. At the same time, it would provide a simple explanation why the quantum theory, in spite of its incompleteness, was so impressively successful. The failure to achieve his objective by such a generalization of the differential geometry of space-time did not lead Einstein to give up his conviction in the essential truth of his viewpoint.

In the following years, and especially in the course of the famous debate with Niels Bohr at the Fifth and Sixth Solvay Meetings in October 1927 and October 1930, respectively, Einstein fought desperately against being "forced into abandoning strict causality." By skillfully designed thought-experiments, he tried to refute the Heisenberg relations. Bohr's account of this debate, called "one of the great masterpieces of modern scientific reporting," is well known, and I shall not go into any details of this episode. Let me only remind you of one of its highlights, namely, how Einstein advanced his photon-box thought-experiment in order to refute the energy-time indeterminacy relation, and how Bohr, on the basis of Einstein's own theory that the rate of a clock depends on the gravitational potential, rebutted Einstein's argument. Einstein was defeated, but not convinced. He felt—as, by the way, did Bohr—that the whole story of the photon-box had not yet been told.

This episode was the climax of the Bohr-Einstein debate, not only because of the dramatic features involved, but also because it became a turning point in Einstein's attitude toward quantum mechanics. From that time on, he gave up any hope of refuting the orthodox quantum theory on grounds of internal inconsistency and concentrated instead on demonstrating not the *inconsistency* but the *incompleteness* of this theory.

As we know from his correspondence with Paul Ehrenfest and from his discussions with Leon Rosenfeld in the early thirties, Einstein conceived his incompleteness argument, which in its final version appeared in the well-known Einstein-Podolsky-Rosen (EPR) paper of 1935, by a gradual modification of the photon-box thought-experiment. Time does not allow me to describe, step by step, how this modification took place. Suffice it to point out that already on 4 November 1931, at a Berlin physics colloquium, Einstein presented a variation of the original thought-experiment that enabled him to show that *either* the energy *or* the time of emission of the photon, but not both at the same time, can be predicted precisely, even after the photon has left the box. If one regards the box and the photon as two entangled systems, one must conclude that whether one or the other of two quantum-mechanically incompatible properties could be ascribed to the photon depends on what measurement one decides to perform on the box, even if it is separated from the photon by an arbitrarily large, spacelike interval. Clearly, almost all logico-physical ingredients of the EPR argument were already at hand, with the difference that in the EPR paper the box was replaced by a second particle that was supposed to have interacted with the original particle, and the whole process was described in the mathematical language of quantum mechanics.

However, one essential point of the EPR argument had not yet found its explicit expression in Einstein's Berlin colloquium talk, which brings us to the core of the conflict between Einstein and the proponents of the orthodox theory. This point was Einstein's explicit reference and use of the third of the three maxims mentioned above. The notion of physical reality, as you recall, played an important role in the EPR paper. Einstein defined it by the sufficient condition: "If, without in any way disturbing a system, we can predict with certainty the value of a physical quantity, then there exists an element of physical reality corresponding to this physical quantity." Defining a theory as complete only if each element of physical reality has a counterpart in the theory made it easy to prove the incompleteness of the orthodox theory. From the thought-experiment it could be inferred that two quantities—even incompatible ones, such as position and momentum—may satisfy the reality condition; but incompatible quantities, associated with noncommutative operators, cannot simultaneously be represented in the theory.

I cannot discuss here the various reactions, not even Bohr's answer, that this paper evoked. There is one kind of reaction, however, that I would like to talk about, because it caused, I think, a widespread misconception concerning Einstein's ideas of how to solve the quantum problem.

It is often claimed that Einstein intended in the EPR paper to recommend the introduction of hidden variables. Thus John Stewart Bell's influential 1964 paper "On the Einstein Podolsky Rosen paradox" opens with this statement: "The EPR paradox . . . was advanced as an argument that quantum mechanics . . . should be supplemented by additional variables." Clauser, Horne, Shimony, and Holt declared in 1969 that Einstein "concluded that the quantum mechanical description of a physical system should be supplemented by postulating the existence of 'hidden variables', the specification of which would predetermine the result of measuring any observable of the system."

Einstein made very few explicit references to hidden-variable theories—naturally so, since the latter began to gain prominence only with the appearance of David Bohm's first paper on this subject in 1952, three years before Einstein's death. But documentary evidence, scanty as it is, discloses incontestably Einstein's opposition to these ideas. In a letter to Mauritius Renninger, dated 3 May 1953 Einstein was referring to Bohm's paper when he wrote: "I do not believe that such a theory is tenable." A few weeks later, in a letter to the physicist Aron Kuppermann, Einstein was more specific: "I think, it is not possible at all to obtain a description of individual systems by merely complementing the present statistical quantum theory." Thus, it would be unwarranted to claim, as has Michael R. Gardner, that the 1967 Kochen and Specker proof of the logical impossibility of hidden variable theories may be thought to have settled the issue between Bohr and Einstein in Bohr's favor.

Einstein never proposed a hidden variable theory. In his view a more radical step would be required to solve the quantum problem. Particular evidence comes from Einstein's last publication on the quantum problem, "Elementare Überlegungen zur Interpretation der Grundlagen der Quanten-Mechanik" ["Elementary Consideration concerning the Interpretation of the Foundations of Quantum Mechanics"], which he contributed to the collection of *Scientific Papers Presented to Max Born* (1953). Again using his technique of thought-experiments, Einstein considered the case of a small sphere or particle that moves freely between two perfectly reflecting walls separated from each other by a distance L. Solving the appropriate Schrödinger equation and applying Born's interpretation, he obtained for the momentum the well-known values $+hn/2L$ and $-hn/2L$, n being an integer. Einstein pointed out that, according to Born, it is not correct to imagine that this result makes sense for an individual particle; rather, it is meaningful only if we measure this quantity in a statistical ensemble of such particles. But since the result remains valid no matter how large the dimensions involved, it follows that, even for a "macrosystem," physical properties such as momentum have no objective observer-independent reality. Einstein then applied Bohm's approach to the same problem and showed that the resulting value of the momentum—in this case, zero—is equally unsatisfactory.

Neither theory satisfied Einstein's requirement that in the range of the applicability of the older conceptions the new concepts reduce—if not conceptually, at least operationally—to the older ones, or, applied to the case under discussion, that the motion of a "macrosystem" should approach the motion described by classical mechanics. Moreover, the question "Does quantum mechanics, in the case of macrosystems, yield, in reasonable approximation, the reality-description provided by classical mechanics?" must be answered in the negative, whether we use the orthodox theory or Bohm's so-called "causal theory" of quantum mechanics. Both approaches, Einstein concluded, are unsatisfactory, for "physics has to strive for a reality-description of the individual system."

Obviously, the key to an understanding of Einstein's attitude toward quantum mechanics lies in his emphasis on the third maxim, his conviction in the validity of the conception of an objectively given, observer-independent physical reality. On 15 April 1950 he wrote to Besso: "Strictly speaking, it is not so much the question of 'causality' which lies at the center of the whole matter, but rather the question of real existence and the question whether there exist rigorously valid laws for the theoretically described reality." Two years later, in another letter to Besso, Einstein most emphatically explained his opposition to the orthodox theory: "The present quantum theory is unable to provide the description of a real state of physical facts, but only of an (incomplete) knowledge of such. Moreover, the very concept of a real factual state is debarred by the 'orthodox' quantum theoreticians. The situation arrived at corresponds almost exactly to that of the good old Bishop Berkeley."

The orthodox point of view has as its starting point an idea with which Einstein fully agreed. It was the acknowledgment made by Bohr in 1927 in the opening statement of his famous Como Lecture: "The quantum theory is characterised by the acknowledgment of a fundamental limitation in the classical physical ideas when applied to atomic phenomena." However, the orthodox theory tries to overcome the difficulties emerging from this acknowledgment by two steps: on the one hand, it does *not* discard the concepts of classical physics but retains them in full; on the other hand, it imposes restrictions upon them by prohibiting, to say it in simple words, their full use at one and the same time, or, more precisely, by segregating them into two disjoint categories, called complementary, with the proviso that only concepts belonging to one and the same category can be ascribed simultaneously to microphysical objects or phenomena. The conservative retention of classical notions was justified on the grounds that, owing to the limitations of the human capacity of conceptualization, an unambiguous intersubjective communication about empirical data, such as laboratory experience, cannot be carried out without them. The second step, combined with the quantum postulate according to which "any observation of atomic phenomena will involve an interaction with the agency of observation not to be neglected," as Bohr put it, implies that no microphysical phenomenon or state of affairs can be accorded physical reality on its own.

Einstein found both these steps unsatisfactory. A theory that programmatically denies the existence of an observer-independent reality or is incapable of providing a theoretical presentation of it was for him unacceptable, or at least "incomplete." This was precisely the contention expressed in the EPR argument. Moreover, the same argument invalidated the implication of the quantum postulate, for in the case of entangled systems it is possible, as the argument showed, to obtain a precise measurement result for a system without in any way disturbing it.

As mentioned at the beginning of this discussion, Einstein was always of the opinion that "concepts which cannot be shown to be valid have to be removed." If newly discovered empirical knowledge shows that a concept hitherto successfully used has lost its validity or justification—*Berechtigung* was the term used by Einstein—it must also lose its applicability, its *Brauchbarkeit*. *Berechtigung* and *Brauchbarkeit* were for Einstein inseparable.

In May 1928, Einstein received a letter from Schrödinger that enclosed a copy of a letter from Bohr in which Bohr defended his retention of the concepts of classical physics. Einstein replied to Schrödinger: "I think that you have hit the nail on the head. . . . Your claim that the concepts p, q [momentum, position] will have to be given up, if they can only claim such a 'shaky' meaning, seems to me to be fully justified. The Heisenberg-Bohr tranquilizing philosophy—or religion?—is so delicately contrived that, for the time being, it provides a gentle pillow for the true believer from which he cannot very easily be aroused. So let him lie there."

Clearly, in Einstein's view concepts of classical physics to which Bohr clung tenaciously, despite their demonstrated inapplicability, were, as he called them, but "a gentle pillow" or a palliative duping us to ignore, if I may say so, reality in both the metaphorical and the literal sense of the expression. Einstein, in contrast, was not only ready to face the difficulties of giving up altogether the classical concepts but thought it imperative to replace them by fundamentally novel conceptions, however unconventional, provided they enable us to cope with the newly acquired empirical data without doing violence to our conception of physical reality. That Einstein remained within the confines of field-theoretical considerations in this search should not be interpreted as a conservative or even a reactionary refusal to admit the legitimacy of radically new conceptions. Late in his life he declared: "Adhering to the continuum originates with me not in a prejudice, but arises out of the fact that I have been unable to think up anything organic to take its place." We should be careful not to confuse lack of success with lack of intention.

In his introduction to the *Born-Einstein Letters*, Heisenberg wrote:

> Most scientists are willing to accept new empirical data and to recognize new results, provided they fit into their philosophical framework. But in the course of scientific progress it can happen that a new range of empirical data can be completely understood only when the enormous effort is made to enlarge this framework and to change the very structure of the thought processes. In the case of quantum mechanics, Einstein was apparently no longer willing to take this step, or perhaps no longer able to do so.

True, Einstein failed in his lifelong efforts to solve the quantum problem, not *because* of his unwillingness to take the step mentioned by Heisenberg, but *in spite of* his willingness. Einstein acknowledged this failure. At the age of thirty-eight, in a 19 March 1917 letter to Besso, Einstein asked, "I feel that the real problem which the eternal poser of riddles has put before us is not yet understood at all. Will we ever in our life-time get hold of the redeeming idea?" Thirty-eight years later, on 3 February 1955, in a letter written to Laue ten weeks before his death, Einstein gave the answer: "If I have learnt anything by racking my brains throughout my long life, it is this, that we are still farther away from a profound understanding of the elementary processes than most of our contemporaries assume. . . ."

Einstein's lifelong labor, though not crowned with success, was not in vain. His great adversary, Niels Bohr, declared in 1961 that "were it not for Einstein's challenge, the development of quantum physics would have been much slower." Einstein's was one of those failures of which the author of the ancient book *On the Sublime* said, "In great attempts it is glorious even to fail."

Reception of Einstein's Scientific Ideas

P. A. M. Dirac

THE EARLY YEARS OF RELATIVITY

I AM VERY HAPPY to have this opportunity of talking at the Einstein Symposium because I am a very great admirer of Einstein.

In this Symposium we have heard many historical talks. Historians collect all the documents they can find referring to their subject, assess those documents, collate them, and then give us a detailed account of what happened. I am not going to give you a talk of that nature, because I am not a historian. What I would like to talk to you about is the arrival of relativity as it appeared to someone who lived through it.

I was in England at the time, and for the most part I shall be talking about the coming of relativity to England, although I think it happened very much the same in other Western countries. What I want to emphasize is the tremendous impact that relativity had on the general public. I think that the historians have not emphasized this sufficiently in their talks.

First, I must describe the background. The time I am speaking of is the end of the First World War. That war had been long and terrible. It had been a war in which not very much had happened, from the military point of view. We had battle lines that remained almost static year after year, but we also had tremendous casualties, ever increasing casualties, and nearly everyone had lost close relatives or friends. Then the end of this war came, rather suddenly and unexpectedly, in November 1918. There was immediately an intense feeling of relaxation. It was something dreadful that was now finished. People wanted to get away from thinking about the awful war that had passed. They wanted something new. And that is when relativity burst upon us.

I can't describe it by other words than by saying that it just *burst* upon us. It was a new idea, a new kind of philosophy, and it aroused interest and excitement in everyone. The newspapers, as well as the magazines, both popular and technical, were continually carrying articles about it. These articles were mainly written from the "philosophical" point of view. Everything had to be considered relatively to something else. Absolutism was just a bad idea that one had to get away from. There was no public radio in those days, so we did not have that kind of propaganda; but all the written material that we had was devoted to bringing out this new idea of relativity. In most of what people wrote there was no real physics involved. The people writing the articles did not understand the physics.

At that time I was sixteen years old and a student of engineering at Bristol University. You might think it was rather unusual for someone only sixteen years old to be a student, but it was not unusual when you consider the times. All the young men had been taken away from the universities to serve in the army. There were some professors left, those who were too old to serve in the army and those who were not physically fit; but they had empty classrooms. So the younger boys were pushed on, as far as they were able to absorb the knowledge, to fill up these empty classrooms. That is how I came to be an engineering student at the time.

I was caught up in the excitement of relativity along with my fellow students. We were studying engineering, and all our work was based on Newton. We had absolute faith in Newton, and now we learned that Newton was wrong in some mysterious way. This was a very puzzling situation. Our professors were not able to help us, because no one really had the precise information needed to explain things properly, except for one man, Arthur Eddington.

Eddington was an astronomer. He had kept in touch with his friend Willem de Sitter, an astronomer in Holland. Holland was neutral at that time, so they could write to each other. De Sitter had kept in contact with Einstein, and in that roundabout way Eddington was in contact with Einstein and had heard all about the development of relativity theory.

At the end of 1918, the end of the war, relativity was not really a new idea. It had dated from 1905, a good many years before the war started. But no one had heard about relativity or about Einstein previously, except for a few specialists at the universities. No one in the engineering faculty at Bristol had heard anything about these things, and it was all completely new to us. Then we had Eddington to explain things.

Eddington was very good at popular exposition. He had a great talent for it, and he applied his talent to explaining the foundations of relativity theory to the general public. He told us about the need to consider that we cannot communicate instantaneously with people at a distance. We could communicate only with the help of light signals and make a note of the time the light signals were sent out and the time we received an answer, and we had to work on that. Eddington also told us about the Michelson-Morley experiment. It was quite an old experiment, but one whose importance had not been previously appreciated. Albert A. Michelson and Edward W. Morley had attempted to determine the velocity of the earth through the ether, but, surprisingly, their experiment failed to give any definite answer. The only way to account for this failure was to suppose a very peculiar behavior of measuring rods and clocks. Moving measuring rods had to be subject to a kind of contraction, called the Lorentz-Fitzgerald contraction. Clocks had to have their rates slowed up when they were moving.

All this was very hard to explain to the general public, but still Eddington and other people wrote innumerable articles about it, doing the best they could. The engineering students were not very much better off. We were told that in some way the absolute scheme of things that we had been using in our engineering studies had to be modified, but there was no very definite way to make the modification. We were not given any definite equations.

Now, Eddington was an astronomer and was very interested in the general theory of relativity, and especially in testing its astronomical consequences. The theory predicts a motion of the perihelion of Mercury. It was easy to work this out and to check that the observations agreed with the Einstein theory. This was the first big triumph of the Einstein theory.

Then there was the question of observing whether light would be deflected when it passed close by the sun, another requirement of the Einstein theory. This is something that can be tested observationally only at a time of total eclipse. Eddington had heard about these things already in 1916, and he immediately set to work to find out if there was a favorable total eclipse coming up soon. He found that there would be a very favorable one occurring in May 1919—very favorable because the time of totality was long and also because the sun was then in a very rich field of stars, so that there would be many stars to observe. Eddington immediately set to work to plan expeditions to make the needed observations. Of course, he knew very well that there would be no hope of making such observations while the war was on, but he was hoping that the war would be over in time. The war was over in November 1918, which was in time, and the eclipse expeditions were sent out. They brought back their photographic plates and measured them up, and they got results in agreement with the Einstein theory. Then the public really went wild. The new theory was proved, and everyone was so excited about it.

I shall mention just one example to illustrate the sort of enthusiasm that prevailed. In a detective story called *The Bishop Murder Case*, an important clue was provided by a piece of paper on which some of the Einstein equations were written down. The theory of relativity was woven into the plot of the story, with the result that the book had a big sale. All the young people were reading it. That shows you the tremendous excitement that pervaded all fields of thought. It has never happened before or since in the history of science that a scientific idea has been so much caught up by the public and has produced so much enthusiasm and excitement.

Yet with all that was written about relativity, we still had very little accurate information. Professor Broad at Bristol gave a series of lectures about relativity, which I attended, but they dealt mostly with philosophical aspects. He did, however, give some information about the geometry of special relativity, in particular about Minkowsky's space, and I began to get some definite information about the theory from him.

We really had no chance to understand relativity properly until 1923, when Eddington published his book, *The Mathematical Theory of Relativity*, which contained all the information needed for a proper understanding of the basis of the theory. This mathematical information was interspersed with a lot of philosophy. Eddington had his own philosophical views, which, I believe, were somewhat different from Einstein's, but developed from them.

But there it was, and it was possible for people who had a knowledge of the calculus, people such as engineering students, to check the work and study it in detail. The going was pretty tough. It was a harder kind of mathematics than we had been used to in our engineering training, but still it was possible to master the theory. That was how I got to know about relativity in an accurate way.

Eddington's book did not give any information about the struggles that Einstein had gone through in order to set up his theory. It just gave us the completed result. I have been very interested in the lectures given by historians at the various Einstein symposia, which enable me to understand better both Einstein's struggles and also his appreciation of the need for beauty in the mathematical foundation. Einstein seemed to feel that beauty in the mathematical foundation was more important, in a very fundamental way, than getting agreement with observation.

This was brought out very clearly in the early work about the theory of the electron. Hendrik Lorentz had set up a theory for the motion of electrons that was in agreement with Einstein's principles, and experiments were made by Walter Kaufmann to see whether this theory was in agreement with observation. The resulting observations did not support the theory of Lorentz and Einstein. Instead, they supported an older theory of the electron, given by Max Abraham. Lorentz was completely knocked out by this result. He bewailed that all his work had gone for nothing. Einstein seems not to have reacted very much to it. I do not know just what he said—that's a question for the historians to decide. I imagine that he said, "Well, I have this beautiful theory, and I'm not going to give it up, whatever the experimenters find; let us just wait and see."

Well, Einstein proved to be right. Three years later the experiments were done again by someone else, and the new experiment supported the Lorentz-Einstein view of the electron. And some years after that, a fault was found in the apparatus of Kaufmann. So it seems that one is very well justified in attaching more importance to the beauty of a theory and not allowing oneself to be too much disturbed by experimenters, who might very well be using faulty apparatus.

Let us return to the general theory of relativity. The observations of the eclipse expeditions supported the theory. Right from the beginning there was agreement. We then had a satisfactory basis for the development of relativity. At that time I was a research student and very much enjoyed the new field of work that was opened up by relativity. One could take some previous piece of work that had been expressed in nonrelativistic language and turn it into the relativistic formalism, get a better understanding of it, and perhaps find enough material to publish a paper.

It was about that time, I think in 1924, that A. H. Compton visited Cambridge and spoke at the Cavendish Laboratory about his experiments on the Compton effect. These experiments involved both the light quantum hypothesis and some of the fundamental relations of special relativity. I can remember very well this colloquium. I found Compton's results very convincing, and I think most of the members of the audience were also convinced by Compton, although there were a few people who held out against the light quantum hypothesis.

The situation was completely changed in 1925, when Werner Heisenberg introduced his new quantum mechanics. This was a theory in which it was soon found out that the basic idea was to have dynamic variables that do not satisfy the commutative law of multiplication. That is to say, $a \times b$ is not equal to $b \times a$. Quite a revolutionary idea, but an idea that rapidly proved to have great success.

Now, in spite of that success, Einstein was always rather hostile to quantum mechanics. How can one understand this? I think it is very easy to understand, because Einstein had been proceeding on different lines, lines of pure geometry. He had been developing geometrical theories and had achieved enormous success. It is only natural that he should think that further problems of physics should be solved by further developments of geometrical ideas. Now, to have $a \times b$ not equal to $b \times a$ is something that does not fit in very well with geometrical ideas; hence his hostility to it.

I first met Einstein at the 1927 Solvay Conference. This was the beginning of the big discussion between Niels Bohr and Einstein, which centered on the interpretation of quantum mechanics. Bohr, backed up by a good many other physicists, insisted that one can use only a statistical interpretation, getting probabilities from the theory and then comparing these probabilities with observation. Einstein insisted that nature does not work in this way, that there should be some underlying determinism. Who was right?

At the present time, one must say that, according to Heisenberg's quantum mechanics, we must accept the Bohr interpretation. Any student who is working for an exam must adopt this interpretation if he is to be successful in his exams. Once he has passed his exams, he may think more freely about it, and then he may be inclined to feel the force of Einstein's argument.

In this discussion at the Solvay Conference between Einstein and Bohr, I did not take much part. I listened to their arguments, but I did not join in them, essentially because I was not very much interested. I was more interested in getting the correct equations. It seemed to me that the foundation of the work of a mathematical physicist is to get the correct equations, that the interpretation of those equations was only of secondary importance.

Right from the beginning of quantum mechanics, I was very much concerned with the problem of fitting it in with relativity. This turned out to be very difficult, except in the case of a single particle, where it was possible to make some progress. One could find equations for describing a single particle in accordance with quantum mechanics, in agreement with the principle of special relativity. It turned out that this theory provided an explanation of the spin of the electron.

Also, one could develop the theory a little further and get to the idea of antimatter. The idea of antimatter really follows directly from Einstein's special theory of relativity when it is combined with the quantum mechanics of Heisenberg. There is no escape from it. With just a single, uniform line of argument one goes right up from special relativity to antimatter.

This was all very satisfactory so long as one considered only a single particle. There remained, of course, the problem of two or more particles interacting with each other. Then one soon found that there were serious difficulties. Applying the standard rules, all one could say was that the theory did not work. The theory allowed one to set up definite equations. When one tried to interpret those equations, one found that certain quantities were infinite according to the theory, when according to common sense they should be finite. That was a very serious difficulty in the theory, a difficulty that still has not been completely resolved.

Physicists have been very clever in finding ways of turning a blind eye to terms they prefer not to see in an equation. They may go on to get useful results, but this procedure is of course very far from the way in which Einstein thought that nature should work.

It seems clear that the present quantum mechanics is not in its final form. Some further changes will be needed, just about as drastic as the changes made in passing from Bohr's orbit theory to quantum mechanics. Some day a new quantum mechanics, a relativistic one, will be discovered, in which we will not have these infinities occurring at all. It might very well be that the new quantum mechanics will have determinism in the way that Einstein wanted. This determinism will be introduced only at the expense of abandoning some other preconceptions that physicists now hold. So, under these conditions I think it is very likely, or at any rate quite possible, that in the long run Einstein will turn out to be correct, even though for the time being physicists have to accept the Bohr probability interpretation, especially if they have examinations in front of them.

There are two other subjects that I would like to talk about. The first concerns an incompleteness in the Einstein theory of gravitation. This Einstein theory gives us field equations, but one cannot solve them and obtain precise results without having some boundary conditions, that is, some conditions that one can use to refer to what space is like at a very great distance. To establish the boundary conditions, we need to have a theory of cosmology, a theory that tells us what the universe as a whole looks like when we smooth out the irregularities caused by stars and galaxies, and so on. Einstein himself realized from the beginning the need for such boundary conditions, and he proposed a model of the universe, a model in which space is limited, although unbounded. That was a new idea for people to get used to, one on which Eddington brought to bear his remarkable powers of explaining to try to get it across to the general public. The idea is not that difficult to comprehend; you can think of the surface of the earth, a region of space in two dimensions that is finite but still unbounded.

Einstein set up a cylindrical model of the universe, but it was soon found that it had to be abandoned because of its static character. Observations showed us that things that are very far away are receding from us with velocities that increase as the distance increases. Einstein's model did not account for that phenomenon. Therefore, it had to be abandoned.

Soon afterward, another model was proposed by de Sitter. De Sitter's model did give correctly the distant matter receding from us, but it gave zero for the average density of matter, which of course is in disagreement with observation. Therefore, this model also had to be abandoned.

Many other models were worked out by Alexander Friedman, Georges Lemaitre, and others, all consistent with Einstein's basic field equations. Among all these other models, I want to call your attention to one that was brought out jointly by Einstein and de Sitter. This model was satisfactory in that it gives distant matter receding from us and also gives a non-zero average density of matter about in agreement with observation. In all its basic features it appears to be satisfactory.

This Einstein–de Sitter joint model I would like to propose as the one that should be generally accepted, because it is the simplest of all the models that are not in some elementary way in disagreement with the facts. We ought to keep to the simplest model until something turns up that causes us to depart from it.

The Einstein–de Sitter model proposes that the universe started with a big bang and will go on expanding forever. Many models claim that the universe will expand to a certain point and then contract again. This is an unneeded complication, and there is no observational evidence supporting it at the present. I feel that such models should not be taken as seriously as the simple Einstein–de Sitter model. The latter has the character that three-dimensional space is flat. It also has the character that the pressure in the smoothed-out universe is very small.

I should mention that all these different models will have their influence on the solutions of the Einstein equations we need in order to account for phenomena occurring in the solar system. But the differences are exceedingly small. They would not affect the wonderful agreement that we have between observation and Einstein's theory in the description of the solar system.

There is one other topic that I would like to talk about: the problem of the unification of the gravitational field and the electromagnetic field. Einstein was very much concerned with this unification. Those were the only two fields that he had to consider. Nowadays, physicists have other fields, but these other fields all involve short-range forces, forces that are significant only for particles extremely close together, lying inside an atomic nucleus. The gravitational force and the electromagnetic force are long-range forces. These forces fall off inversely proportional to the square of the distance. In some ways, they are more important, I think, than the other forces. One feels that there should be a close connection between them.

Einstein had achieved his very great success in accounting for the gravitational force in terms of geometry, and he thought that some generalization of the geometry would bring in the electromagnetic force also. A method of doing this was very soon discovered by Hermann Weyl. Weyl made a rather simple generalization in the geometry of Riemann, which Einstein was using. He supposed that the distance between two neighboring points does not have an absolute value. There is no natural absolute unit to which one could refer it. But one can transport it from one location in space to another, and the equations that govern this transport are such that, if you go around a closed loop and get back to your starting point, the final distance does not agree with the starting value. That led to a generalization of the geometry, which was soon found to provide just what was needed in order to bring in the electromagnetic field.

This seemed to be a very wonderful solution of the problem, but then there was a difficulty. Atomic events do provide a natural scale for measuring distances. You could refer all your distances to this atomic scale, and then there would be no point at all in having the uncertainty in distances introduced by Weyl's geometry. So this very beautiful theory of Weyl was reluctantly abandoned.

Einstein worked for the rest of his life on trying to solve this problem of unifying the gravitational and the electromagnetic fields. He tried one scheme after another. All were unsatisfactory. Other people have joined in this work without achieving any greater success than Einstein did. I have been wondering whether Einstein was limiting his ideas too much in these attempts to unify the gravitational and the electromagnetic fields. It seems to me that it is quite possible that one will have to bring in cosmological effects to arrive at a satisfactory solution of this problem. The way cosmological effects would show up would be this.

Let us accept the Einstein theory as it stands for all problems involving just classical theory, and only when we go over to atomic problems, let us require that some modification is needed. This modification can be expressed by saying that Planck's constant is not really a constant in the cosmological sense but must be considered as varying with the epoch, that is, with the time since the origin of the universe, the "Big Bang." If we have Planck's constant h varying, then we must also have the charge on the electron e varying, because e^2/hc is a dimensionless constant that plays an important role in physics and is observed to have the value 1/137, and it seems to be really constant. Thus, if h is varying, e must vary according to its square root. On this basis one could set up a new theory.

Before introducing such a drastic revision in basic ideas, it is desirable to have some confirmation of it by observation. If you have these atomic constants varying when referred to the Einstein picture, it will mean that atomic clocks will not keep the same time as the time of the Einstein theory. The time of the Einstein theory is the time that governs the motion of the planets around the sun, what astronomers call ephemeris time. So one could look to see whether there is any difference between atomic clocks and clocks based on ephemeris time.

Astronomers have been studying this question. In particular, T. C. Van Flandern, working at the Naval Research Observatory in Washington, has spent many years studying lunar motion, referred to both ephemeris time and atomic time. Lunar motion has been observed with atomic clocks since 1955. However, Van Flandern's results up to the present are still not conclusive. One must wait a little longer to see whether a difference really exists between the two ways of measuring time.

There is another possibility of checking these new ideas about variation of constants that has been followed up by I. I. Shapiro. His method consists in sending radar signals to one of the planets and observing the reflected radar signals, and then timing the to and fro journey with an atomic clock. In effect, he is observing distances in the solar system with atomic clocks, and there should be some discrepancy showing up if the atomic clocks do not keep the same time as ephemeris time.

I was talking to Shapiro just before I came to this Symposium and asked him about the latest information concerning his work. There is a very good chance of observing radar waves reflected from Mars because of the Viking Lander that landed on Mars in 1976, which can be used to send back reflected radar waves. Shapiro said that the time base that he has for the observations with the Viking expedition (just about two years) is not long enough for him to give a definite answer to this question. There were also observations of Mars made previously with a Mariner expedition, which give him a considerably longer time base, eight years instead of two. But he told me that another year or two of work would be needed to evaluate his results before he could answer the question of variation of the constants.

That is the situation at the present time. I am sorry that I cannot offer anything more definite than that. There are hopes that Shapiro will come out with a definite result, and if it is a positive result, we shall have a new basis for looking at the question of the unification of the gravitational and the electromagnetic fields. It might be possible to revive Weyl's geometry, which was abandoned only because it clashed with the distances provided by atomic events. So it seems that if this approach turns out to be correct, the whole theory of unifying these distances with gravitational theory would have to be gone into again. There may very well be greater chances of success.

Banesh Hoffmann

SOME EINSTEIN ANOMALIES

GEOMETRY

THE OFFICIAL TOPIC of this session is the reception of Einstein's scientific ideas. My contribution is not on that subject. But, in a sense, it is closely related: its topic could indeed be described as the reception of Einstein's scientific ideas—but by Einstein himself. I want to consider some puzzling aspects of his writings and to venture to draw from them hints as to some of his possible thought processes.

In his "Autobiographical Notes," written at age sixty-seven, Einstein tells of the sense of awe and wonder that seized him at the age of four or five when he first saw a magnetic compass. He continues:

> At the age of 12 I experienced a second wonder of a totally different nature: in a little book on Euclidean plane geometry that came into my hands at the beginning of a school year. Here were assertions, as for example the intersection of the three altitudes of a triangle in one point, which—though by no means evident—could nevertheless be proved with such certainty that any doubt appeared to be out of the question. This lucidity and certainty made an indescribable impression on me.[1]

The above seems clear enough. But something unexpected enters when we look at its context. Here is what immediately follows the words just quoted:

> That the axioms had to be accepted unproved did not disturb me. In any case it was quite sufficient for me if I could peg proofs upon propositions the validity of which did not seem to me to be dubious. For example, I remember that an uncle told me the Pythagorean theorem before the holy geometry booklet had come into my hands. After much effort I succeeded in "proving" this theorem on the basis of the similarity

of triangles; in doing so it seemed to me "evident" that the ratios of the sides of a right triangle would have to be completely determined by one of the acute angles. Only something that did not similarly seem to be "evident" appeared to me to be in need of any proof at all. (9-11)

There is an anomaly here. Before discussing it, however, let us try to reconstruct young Einstein's proof of the Pythagorean theorem. From his remarks about the ratios of the sides of similar right triangles, we can infer that it probably went along the following lines. In the right triangle ABC, with C the right angle, drop the perpendicular CD onto the hypotenuse AB. Since CD and CB are perpendicular, respectively, to AB and AC, $\angle CAB = \angle DCB$. Then, by similar right triangles, we have

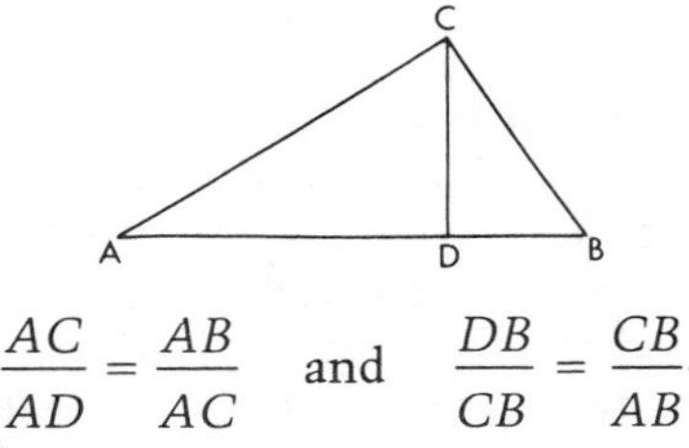

$$\frac{AC}{AD} = \frac{AB}{AC} \quad \text{and} \quad \frac{DB}{CB} = \frac{CB}{AB}.$$

These can be written

$$AC^2 = AB \cdot AD \quad \text{and} \quad CB^2 = AB \cdot DB.$$

By addition we obtain

$$AC^2 + CB^2 = AB(AD + DB) = AB \cdot AB = AB^2,$$

and we can well imagine young Einstein crying out Q.E.D. in delight—except that he may not have heard of that abbreviation at that pre-geometry-book time.

This brings us directly up against the anomaly. To see it, consider the situation chronologically. The young Einstein, wholly untutored in geometry, creates a striking proof of the Pythagorean theorem. This is no mean feat, and his proof is certainly an instance of the power of geometrical reasoning. Yet he feels no sense of wonder—or if he does, he fails to mention it when writing his "Autobiographical Notes." Some time later, young Einstein comes across a geometry text, and this time he is overwhelmed by a feeling of wonder and admiration that he recalls for the rest of his life.

Stated thus, the story seems incredible. Certainly, we owe it to Einstein to take the story seriously. Perhaps by accepting it we can gain a valuable insight. For there are significant differences between the two situations. The statement of the Pythagorean theorem had come to young Einstein *ex cathedra* as a fact, and any initial wonder could well have been dissipated in the course of his struggle to construct a proof. Moreover, there was a considerable component of algebra in his proof—if our reconstruction of it is to be trusted. Also, there was little that was *visually* striking in the Pythagorean theorem—nothing to match the picture of the concurrent altitudes. And above all, the theorem was presented to him as an isolated result.

In contrast, the holy geometry booklet opened up a whole new way of mental life. It began by listing axioms that, apparently, no reasonable person could doubt. And then, by simple, transparent, graphic, nonalgebraic reasoning, it generated from these self-evident truths a wealth of interrelated theorems that were not only general and unchallengeable but also, on occasion, beautiful, visually striking, and totally unexpected, like the theorem of the altitudes of any triangle. The spectacular results seemed out of all proportion to the meager means by which they were obtained. Perhaps it was this economy of means, this magical creation of seemingly irrefutable theorems out of practically nothing, that captivated young Einstein. If so, it is easy to understand why the effect of the holy geometry booklet transcended that of the proof of the Pythagorean theorem. Indeed, the fact that the undoubted Pythagorean elation went unmentioned in the "Autobiographical Notes" shows vividly how thoroughly it was transcended by the sense of wonder and revelation engendered by the holy geometry booklet.

Certainly, as we look at Einstein's major achievements it is hard to escape the conclusion that he was profoundly influenced by the beauty he found in the holy geometry booklet. We find in his work frequent echoes of both its economy of means and its deductions from basic principles.

KEEPING ABREAST OF LIGHT WAVES

One echo of the holy geometry booklet—and a powerful one—comes later in the "Autobiographical Notes":

> By and by I despaired of the possibility of discovering the true laws [of physics] by means of constructive efforts based on known facts. The

longer and the more despairingly I tried, the more I came to the conviction that only the discovery of a universal formal principle could lead us to assured results. The example I saw before me was thermodynamics. The general principle was there given in the assertion [Satze] that the laws of nature are such that it is impossible to construct a *perpetuum mobile* (of the first or second kind). How, then, could such a universal principle be found? (53)

Immediately following this declaration of the importance of basic principles—they remind us of the powerful axioms in the holy geometry booklet—there comes a famous passage that merits attention here because it will serve two purposes. The first is to rectify a widely held misconception, and the second is to discuss an anomaly that seems to have gone largely unnoticed.[2]

After ten years of reflection such a principle resulted from a paradox upon which I had already hit at the age of sixteen: If I pursue a beam of light with the speed c (speed of light in a vacuum), I should observe such a beam of light as a spatially oscillatory electromagnetic field at rest. However, there seems to be no such thing, whether on the basis of experience or according to Maxwell's equations. From the very beginning it appeared to me intuitively clear that, judged from the standpoint of such an observer, everything would have to happen according to the same laws as for an observer at rest relative to the earth. For how should the first observer know, or be able to determine, that he is in a state of fast uniform motion?

One sees that the germ of the special theory of relativity is already contained in this paradox. . . . (53)

First, with regard to the widely held misconception mentioned above, one can satisfactorily dismiss Einstein's paradox by pointing out that, according to the special theory of relativity, no observer can move as fast as light and thereby keep abreast of it. However, the situation is not that simple, as the following considerations will show. The Maxwell equations for light in a vacuum are

$$\nabla \times E + \frac{1}{c}\frac{\partial B}{\partial t} = 0, \quad \nabla \times B - \frac{1}{c}\frac{\partial E}{\partial t} = 0, \quad \nabla . E = 0, \quad \nabla . B = 0. \tag{1}$$

It is easy to verify that if f is any well-behaved function of one variable, the vectors

$$E = \Big(0, f(x - ct), 0\Big) \text{ and } B = \Big(0, 0, f(x - ct)\Big) \tag{2}$$

constitute a solution of these equations. The vectors represent a plane wave of light traveling with speed c in the x-direction, the vector E being parallel to the y-axis and the vector B parallel to the z-axis.

Now replace $x - ct$ in (2) by x. This is what one would naturally do in those prerelativity days of the sixteen-year-old Einstein in order to represent the light wave (2) as seen by the observer keeping abreast of it. The result is

$$E = \Big(0, f(x), 0\Big) \text{ and } B = \Big(0, 0, f(x)\Big). \qquad (3)$$

This represents a spatial electromagnetic field at rest, as envisioned by the young Einstein, and one easily confirms that (3) does *not* satisfy the Maxwell equations (1). We thus seem not only to have borne out young Einstein's insight but also to have done so *without needing to appeal to the relativistic limitation on the speed of an observer.*

This last is remarkable in itself. But let us not stop as if the matter were settled by it. Instead, let us similarly consider what the plane light wave (2) would look like if we moved in the positive x-direction with a speed v that, for vividness, we can think of as quite small—say, 5cm/sec—even though it can have any value whatever. Replacing x in (2) by $x + vt$, we obtain

$$E = \Big(0, f\big(x - (c - v)t\big), 0\Big) \text{ and } B = \Big(0, 0, f\big(x - (c - v)t\big)\Big). \quad (4)$$

As expected, this represents a plane wave of light moving relative to us with speed $c - v$ along the x-axis. But when we substitute it into the Maxwell equations (1), we find, perhaps to our initial astonishment, that, like (3), it fails to satisfy them—except when $v = 0$. Thus, if we were content to explain away the failure of (3) to satisfy the Maxwell equations by arguing that we cannot move with speed c, surely we should conclude from the similar failure of (4) that *we cannot move at all!*

Actually, this conclusion is not as absurd as it may seem. Indeed, we should have expected it, because Maxwell's theory was an ether theory and, as such, implied—or, in the prerelativistic era, seemed to imply—the existence of well-defined rest frames, namely, those at rest relative to the ether. According to *Newtonian* kinematics, the Maxwell equations are valid only in such rest frames. If we replace x in them by $x + vt$, as was done in going from (2) to (4), the Maxwell equations change their form, with v entering explicitly, but the waves (4) moving with speed $c - v$ satisfy the transformed Maxwell equations. From this point of view, the mathematical conflict, and with it the paradox, can be thought of as arising from making a nonrelativistic transformation on the solution, but either no transformation on the Maxwell equations or else a relativistic one (which leaves their form unchanged).

This brings us to the anomaly mentioned in introducing this topic. Obviously, at age sixteen Einstein did not think in terms of Newtonian versus Lorentz transformations. But it is not easy to say what he did have in mind at that time. Recall, first, that the principle of relativity states that the laws of physics are the same for all observers in uniform relative motion, and now recall these already-quoted words of the aged Einstein telling of his thoughts as a youth. Talking of the observer who is keeping abreast of a light wave, he says: "From the very beginning it appeared to me intuitively clear that, judged from the standpoint of such an observer, everything would have to happen according to the same laws as for an observer at rest relative to the earth" (53).[3] From this it seems abundantly clear that at the age of sixteen Einstein had formulated the principle of relativity and applied it to an optical problem. But let us now recall other words of Einstein that we have already cited: "How, then, could such a universal principle be found? *After ten years of reflection* such a principle resulted from a paradox upon which I had already hit at the age of sixteen" (53; emphasis added). Here the implication is clear that he did *not* have such a principle at age sixteen. Where lies the truth? Let me offer a suggestion.

Suppose first that at age sixteen Einstein believed that the speed of light is c relative to all observers. Then he would have had either the principle of relativity or the other of the two principles on which he later built the special theory of relativity. In either case, his statement about finding a principle "after ten years of reflection" would be utterly baffling.

The alternative is to assume that at age sixteen Einstein had no objection to light having a relative speed of $c - v$ except for the case $v = c$, which pertains to frozen light. But this, too, gives rise to a problem, one that becomes particularly striking if we consider the case of an observer keeping almost, but not quite, abreast of a light wave. According to classical ideas, this observer would see a wave such as that represented by (4), with $c - v = \varepsilon$, where ε is small. This is a sluggish light wave moving relative to the observer with speed ε. Did the young Einstein seriously consider this case? One is inclined to think that he did not, however unlikely that may seem. For, suppose he had done so. Would he have felt that light waves traveling relative to him with speeds of the order of a centimeter a second, or a millimeter a year, or a micron a century were acceptable, but frozen light was not? It seems much more likely that he would have felt that the sluggish light and its limiting case, the frozen light, were on a par, the former merging smoothly into the latter. But in that case we have no logical explanation of his finding the frozen light unacceptable.

I suggest that the young Einstein did not consider sluggish light and that the reason for his failure to do so was not logical but psychological. In the name of a principle of relativity whose name he could not have known at the time, he rejected frozen light. Having sensed the existence of a profound paradox, he may have experienced a psychological blocking that prevented him from giving serious consideration to sluggish light. For, subconsciously, he may have known that such consideration could tempt him to find sluggish light acceptable, and then frozen light too, in which case paradox would be lost. This psychological possibility is not as farfetched as some may think. The workings of intuition transcend those of the intellect, and, as is well known, innovation in science is often a triumph of intuition over logic.

Besides, did not most of us ignore sluggish light too?

PARTICLES OF LIGHT

Let us proceed now to 1905, the fabulous year in which Einstein's genius came to dazzling flower. The first of his research papers in that year sets forth the revolutionary idea that, somehow, light should be regarded as consisting of particles despite the overwhelming evidence that it consists of waves. It is sometimes said that the opening sections of papers by Einstein often go to the heart of the matter. In his paper on light quanta he begins with these words:

> A profound formal distinction exists between the theoretical concepts that physicists have formed regarding gases and other ponderable bodies and the Maxwellian theory of electromagnetic processes in so-called empty space. While we consider the state of a body to be completely determined by the positions and velocities of a very large, but finite, number of atoms and electrons, we make use of continuous spatial functions to describe the electromagnetic state of a given volume, and a finite number of parameters cannot be regarded as sufficient for the complete determination of such a state. According to the Maxwellian theory, energy is to be considered a continuous spatial function in the case of all purely electromagnetic phenomena, including light, while the energy of

a ponderable object should, according to the present conceptions of physicists, be represented as a sum taken over atoms and electrons. The energy of a ponderable body cannot be subdivided into arbitrarily small parts, while the energy of a beam of light from a point source (according to the Maxwellian theory of light or, more generally, according to any wave theory) is continuously spread over an ever increasing volume.[4]

These words do indeed go to the heart of the matter. And yet, in a revealing way, they do not. In his introductory remarks, Einstein tells us of a fundamental conflict of smooth versus granular when light is treated as a wave and matter as made up of atoms, this conflict being apt to obtrude itself when one considers the interaction of light and matter. Einstein avoids that conflict by treating light, like matter, as granular, and to this extent his opening remarks go to the heart of the matter. But in his very next words he concedes that "the wave theory of light, which operates with continuous spatial functions, has worked well in the representation of purely optical phenomena and will probably never be replaced by another theory." To explain away the smoothness, he says, "It should be kept in mind . . . that the purely optical observations refer to time averages." Nevertheless, the fact remains that, as he well knows, he cannot offer a granular explanation of interference effects, and this means that he cannot be rid of the smooth light waves.

We may therefore sum up as follows. Einstein, in his introductory remarks, tells of the conflict of smooth versus granular—wave versus particle—when waves of light confront atomic matter. But he ends up in far deeper conflict than before. For instead of the initial conflict of smooth versus granular—wave versus particle—for light versus matter, he still has the conflict of smooth versus granular—wave versus particle—but now not for light versus matter but for light versus light itself. And *that* is the true heart of the matter. He starts by pointing out a conflict and then dares to try to resolve it by introducing a bigger conflict. Looked at in this way, Einstein's words reveal an extraordinary intuition, for in this 1905 paper, in an era when the particle theory had long been regarded as dead, Einstein was sensing what we now speak of as the wave-particle duality—and sensing it two eventful decades before anyone dared to formulate it. (Einstein never felt happy about that formulation, however.)

There is more to tell about the sureness of Einstein's intuition in connection with his idea of particles of light. In 1905, a mere thirteen weeks after presenting the idea, he sent in the paper "On the Electrodynamics of Moving Bodies,"[5] in which he presented what we now call the special theory of relativity. In it he makes no specific mention of particles of light. Rather, he builds on two postulates: the principle of relativity, and the principle that the motion of light is independent of that of its source. Each by itself is highly plausible. The principle of relativity holds in Newtonian mechanics, and Einstein is extending it to hold also for electromagnetic and other phenomena. The second principle is particularly interesting here because it seems trivially obvious if we believe that light consists of waves in an ether, for waves, once launched, are free of their source and move at the dictates of the medium in which they are propagated. We note in passing that, having stated the principle, Einstein boldly undercuts its foundations by remarking that in the theory to be presented "the introduction of a 'luminiferous ether' will prove to be superfluous." As is well known, he goes on to develop from his two principles (recall the axioms of the holy geometry booklet) a theory in which, for example, two observers in uniform relative motion will each legitimately claim that the other's clocks go at a slower rate than his own.

What, though, if one starts out with the belief that light consists of particles? Then the second postulate makes no sense at all,[6] and although that fact may seem decidedly discouraging, it actually gives rise to a powerful temptation. Consider Einstein's situation at the time. He had just proposed the idea of particles of light, which he knew was far too revolutionary to be readily accepted. He must have sought far and wide for applications that would help to give credence to the idea. He had included three such applications in the paper in which he presented it, and now there loomed a striking possibility of a fourth application: throw away the wave-related second postulate and merely let the particles of light behave like other Newtonian particles. Then the Newtonian principle of relativity will apply, and at once one will have a dazzlingly simple explanation of the repeated failures of optical experiments to measure the velocity of the earth—an explanation using familiar space and time, moreover. Think what a tremendous triumph this would have seemed for Einstein's idea of particles of light. The temptation must have been enormous—not to physicists in general, of course, but uniquely to Einstein, since he alone at that time believed in particles of light. Yet Einstein resisted the temptation, opting instead for the wave-related second postulate that led to the strange new behavior of time and space. Here we see once more the extraordinary sureness of his intuition.[7]

ACCEPTING GENERAL RELATIVITY

Einstein's ten-year journey from the special to the general theory of relativity offers further evidence of the sureness of his intuition, but here, as will be seen, an irony enters. His initial motivation and constant guide on the journey was an aesthetic striving after beauty and simplicity. He felt that having only one type of motion relative was inartistic and therefore probably poor physics. If uniform motion was relative, as was the case in the special theory of relativity, then *all* motion should be. Aesthetics demanded it. But he needed courage as well as a sure intuition to believe this or even to contemplate it, since there was striking evidence in everyday life that acceleration is absolute. Recall, for example, the plight of the straphanger in a crowded train. He can relax while the train is moving uniformly but must hold on tightly when it speeds up or slows down. The uniform motion of the train has no effect inside and is thus relative, but nonuniform motion of the train does have an effect inside and is therefore absolute. Or so it would seem.

Nevertheless, emboldened by his aesthetic belief, Einstein looked at the evidence anew and saw that it could be interpreted as supporting, instead of opposing, his view that all motion should be relative. As is well known, he realized that the proportionality of gravitational and inertial mass meant that all *mechanical* experiments in a uniformly accelerated laboratory far from massive gravitating bodies would be duplicated in an unaccelerated laboratory in a uniform gravitational field. For all its profundity, this realization is trifling when compared with Einstein's next step. In stating his special principle of relativity in 1905, he had taken a principle belonging to Newtonian mechanics and declared it valid for all physics—a typical Einsteinian unification. Here, similarly, noting the Newtonian equivalence of uniform acceleration and uniform gravitation for mechanics, he enunciated his principle of equivalence, which declared the equivalence valid not just for mechanics but for all physics. By this bold act of unification he could declare all motion relative; effects inside a closed vehicle that might be thought of as due to acceleration could be due to gravitation instead. This step was so bold and so beautiful, and its consequences so momentous, that I rank it among the most remarkable insights in the history of physics.

It is well known that the principle of equivalence led Einstein to his masterpiece, the general theory of relativity. But it did not lead him there directly, and it is here that the irony enters. For Einstein made mistakes, some of them surprisingly elementary. Let me tell of one that is particularly relevant to my topic.

For reasons that need not be discussed, Einstein proposed what he called *the principle of general covariance.* It required that the laws of physics be expressible in a form that is the same in all space-time coordinate systems. In practice, it meant that the laws were to be expressed in tensor form. However, after reaching the momentous conclusion that gravitation should be represented solely by the metrical tensor g_{ab}, he began to doubt the principle of general covariance because of problems with the conservation of momentum and energy, and he bolstered his doubts by means of the following erroneous argument.[8] If the metrical tensor g_{ab} represents the gravitational field, it should be uniquely determined by the tensor T_{ab}, which represents the sources of the gravitational field. In space-time, consider a region S in which there are no gravitational sources, so that $T_{ab} = 0$ there. Make a change of coordinates from x^a to $\bar{x}^a$, such that $\bar{x}^a = x^a$ outside S but $\bar{x}^a \neq x^a$ inside S. Then outside S we shall have $\bar{T}_{ab} = T_{ab}$, and inside S we shall have $\bar{T}_{ab}$ and T_{ab} both zero. Therefore the sources will be everywhere the same as before. But inside S we shall, in general, have $\bar{g}_{ab} \neq g_{ab}$, which shows that for the same overall distribution of sources there will be different gravitational fields, g_{ab} and $\bar{g}_{ab}$, inside S. This would contradict the requirement that the sources completely determine the field, which is essentially the requirement of causality.

It is clear from both the argument pertaining to the unchanging T_{ab} and the argument pertaining to the changing g_{ab} that Einstein was saying that if the components of a tensor change under a coordinate transformation, the tensor itself changes. But in fact, just the opposite is the case. Precisely because g_{ab} is a tensor, the g_{ab} and $\bar{g}_{ab}$ actually represent the same field even though $\bar{g}_{ab} \neq g_{ab}$. In general, we should *expect* the components of a tensor to change when the coordinates are changed. That is the nature of tensors. Indeed, it is almost the first thing one learns in studying the tensor calculus.

The error led Einstein seriously astray for two years. With profound misgivings, he gave up the principle of general covariance and allowed only linear coordinate transformations, because he thought this would save the conservation laws and also because with such transformations one could not have $\bar{x}^a = x^a$ in some regions and $\bar{x}^a \neq x^a$ in others. Later, after much labor, he allowed somewhat more general transformations while still shunning the principle of general covariance.

How could Einstein have made so elementary an error? We cannot simply argue that he was a physicist rather than a mathematician and thus did not know any better. He did know better. Besides, he was not alone. The erroneous argument appeared in a paper written with his good friend Marcel Grossmann, who was a first-rate mathematician and an expert on the tensor calculus. Indeed, he taught Einstein the essentials of the subject. So how could Einstein and Grossmann together have made so elementary and long-lasting an error?

I suggest that it was because Einstein had long been profoundly concerned about causality and determinism. His concern, as we know, was to manifest itself strikingly in his dislike of the probabilistic interpretation of quantum mechanics a decade later. Before that, the question had arisen in a different form in the special theory of relativity, in which an event *A* could occur earlier than an event *B* for one observer but later for another, in which case event *A* could not cause event *B*. In developing the general theory of relativity, an intense preoccupation with the question of causality could well have led Einstein to sense danger where none was present. We can picture him coming upon the seeming tensorial violation of causality, reacting with an instinctive alarm that overwhelmed reason, and seizing upon the makeshift and aesthetically displeasing pseudo-remedy of linear transformations. With the imagined danger thus contained, Einstein may well have experienced a psychological blocking that for a long time prevented him from reentering the region of alarm.

What of Grossmann? He had long held Einstein in the highest esteem. The intensity of Einstein's feelings about causality and the assurance with which he spoke of and dealt with the imagined danger must have been overwhelming—so much so as to lead Grossmann, the tensor expert, to acquiesce in the error. This manifestation of intense preoccupation with causality is not without interest in connection with Einstein's later dislike of the Copenhagen interpretation of quantum mechanics.

Ultimately, Einstein realized that his tensorial fears for causality had been groundless, and he returned to his principle of general covariance, which he had abandoned "only with a heavy heart." Progress was now swift. In 1915 the long-sought theory suddenly crystallized into a thing of compelling beauty. Applying it to the motion of the planet Mercury, Einstein found that, in the most natural manner and without any forcing, it accounted for the residual advance of 43″ of arc per century that was unexplained in the Newtonian theory. All this and more occurred in Berlin during the First World War.

Because of the war, there was no direct communication between Germany and Great Britain. However, the British astronomer and physicist Arthur Stanley Eddington, who was a Quaker, obtained details of the new theory through fellow scientists in neutral Holland, and he was immediately captivated by its beauty. In the midst of war, he and the Astronomer Royal began preparing for an eclipse expedition in 1919 to test Einstein's prediction of the gravitational bending of light rays. Although Einstein had developed his theory in enemy Berlin and although anti-German feeling was strong in Britain, Eddington's hope was not to refute the theory but to confirm it. To his delight, his hope was realized.

Aware of the historical importance of what had happened, the British put aside their war-engendered anti-German feelings and announced the eclipse results enthusiastically at a specially convened joint meeting of the Royal Society of London and the Royal Astronomical Society. It was clear to the scientists in the crowded hall that Newton's theory, which had held sway for 250 years, must yield to the theory perfected in enemy Berlin in time of war. The newspapers eagerly reported the event. To a war-weary world it came as a striking reminder that pencil and paper can transcend guns and high explosives. In these dramatic circumstances Einstein became world famous almost overnight.

Of particular interest is Einstein's own acceptance of his general theory of relativity. He had worked on it for ten years, sustained by an aesthetic intuition that showed him the way through complicated regions of thought and repeatedly brought him back to the right path when he strayed. When the theory suddenly crystallized, he was spellbound by its beauty and intrinsic simplicity. In 1916, three years before the eclipse verification, in a book explaining relativity for the layman, he told of the perihelion calculation. Then, speaking of the bending of light rays and another prediction, the gravitational red shift, he said, "I do not doubt that these deductions from the theory will be confirmed also." They were, of course.

Perhaps we think that the source of this confidence was the perihelion result. If so, we underrate Einstein. The main source was the profound Mozartean simplicity of the theory—its utter naturalness. To see this, let us look at the pace and the manner of its crystallization. The physics section of the Prussian Academy of Sciences met once a week. In a paper presented on 4 November 1915, Einstein was still slightly restricting the allowable coordinate transformations (the Jacobian of the transformation had to be unity).[9] A week later, on 11 November, he removed the restriction. His equations now conformed to the principle of general covariance.[10] The next week, on 18 November, he presented the perihelion calculations,[11] and on 25 November he added finishing touches to the field equations.[12] Naturally, he was jubilant. In a letter to his friend Paul Ehrenfest the following January, he wrote, "Imagine my joy at the feasibility of the general covariance and at the result that the equations yield the correct perihelion motion of Mercury. I was beside myself with ecstasy for days."

But now consider in more detail the paper that Einstein presented on 4 November 1915. The perihelion calculation was yet to be made. The principle of general covariance was yet to be fully incorporated: the equations were tensor equations only for certain types of transformations. But in the first approximation his equations did yield Newtonian results. At this stage Einstein could already see the beauty inherent in his fast-crystallizing theory. In this highly technical scientific paper of 4 November we find these remarkable and revealing words: "Dem Zauber dieser Theorie wird sich kaum jemand entziehen können der sie wirklich erfasst hat [Hardly anyone who has truly understood this theory will be able to resist being captivated by its magic]." Here speaks the quintessential Einstein—the supreme artist of science. Is it any wonder that, despite what he would have wished, we are celebrating the centennial of his birth so extravagantly?

NOTES

1. A. Einstein, "Autobiographical Notes," trans. P. A. Schilpp, in Schilpp, ed., *Albert Einstein: Philosopher-Scientist*, Evanston, Ill., Library of Living Philosophers, 1949, p. 9. Page references in the text will be to this edition. The English translations have occasionally been modified.

2. Banesh Hoffmann, "An Einstein Paradox," *Proceedings of the New York Academy of Sciences* **36** (1974), pp. 730-737.

3. Einstein's next sentence is garbled. He evidently meant to say that otherwise the first observer would be able to determine that he was in a state of fast uniform motion. That, of course, would be incompatible with the principle of relativity.

4. A. Einstein, "Über einen die Erzeugung und Verwandlung des Lichtes betreffenden heuristischen Gesichtspunkt," *Annalen der Physik* **17** (1905), pp. 132-148. The translation is by A. B. Aarons and M. B. Shepard, in *American Journal of Physics* **33** (1965), pp. 367-374.

5. A. Einstein, "Zur Elektrodynamik bewegter Körper," *Annalen der Physik* **17** (1905), pp. 891-921; reprinted in *The Principle of Relativity: A Collection of Original Memoirs on the Special and General Theory of Relativity by H. A. Lorentz, A. Einstein, H. Minkowski and H. Weyl,* trans. W. Perrett and G. B. Jeffrey, New York, Dover, n.d., pp. 37-65.

6. Once we have the fully developed special theory of relativity, of course, it does turn out to make sense, provided we assume that light quanta have zero rest mass. But we are dealing here with basic postulates at the initial stages of the theory.

7. *Note added in proof.* I am indebted to John Stachel for bringing to my attention a 1912 letter from Einstein to Paul Ehrenfest and a 1922 letter from Einstein to Mario Visciglini, which show that Einstein had indeed considered and rejected the idea of applying the concept of particles of light to the problem of relativity.

8. A. Einstein and M. Grossmann, "Entwurf einer Verallgemeinerten Relativitätstheorie und eine Theorie der Gravitation," *Zeitschrift für Mathematik und Physik* **62** (1913), pp. 225-244.

9. A. Einstein, "Zur allgemeinen Relativitätstheorie," *Preussische Akademie der Wissenschaften, Sitzungsberichte* (1915), pt. 2, pp. 778-786.

10. Ibid., pp. 799-801.

11. A. Einstein, "Erklärung der Perihelbewegung des Merkur aus der allgemeinen Relativitätstheorie," ibid., pp. 831-839.

12. A. Einstein, "Feldgleichungen der Gravitation," ibid., pp. 844-847.

Loren R. Graham

THE RECEPTION OF EINSTEIN'S IDEAS: TWO EXAMPLES FROM CONTRASTING POLITICAL CULTURES

THE ADVENT of the theory of relativity was a brilliant and startling event that attracted the attention of a wide public. Laypeople often found it irresistible, yet baffling. In response to the challenge, writers in many countries attempted to make the theory accessible to nonscientists. In the process of interpreting relativity in a way that eased comprehension, these authors often attached philosophical or ideological messages to the theory that were rooted in their own preferences about the meaning of science for social values.

Einstein himself was opposed to attempts to use the theory of relativity as a vehicle for a message with general social, philosophical, or political overtones. In a 1918 article defending relativity from attacks in Germany, Einstein emphasized that when he used the term "relativist," he meant a "physical relativist" and certainly not a "philosophical relativist," that is, one who attacks all absolute principles.[1] He later commented, "I believe that the present fashion of applying the axioms of physical science to human life is not only a mistake but has something reprehensible to it."[2] He was particularly upset by the opinion of some writers, physicists among them, that physics could shed light on questions of human will and values. He chided his fellow physicist Niels Bohr for writing popular articles in which he extended quantum mechanics to the social sciences, such as anthropology.[3] When the usually "clear-thinking Bohr" begins to write on topics like complementarity, Einstein once caustically observed, "he thinks of himself as a prophet."[4]

But once relativity theory had been launched by its author, it assumed a life of its own, and Einstein's admonitions often went unheeded. The public demanded an explanation of the broader meaning of this revolution in the physicists' view of the universe, and many physicists were not immune to the temptation to provide such interpretations. A host of popular discussions of the "meaning" of relativity soon appeared, greatly varying in sophistication.

In this paper I shall compare examples of interpreters and popularizers of Einstein's theory of relativity in virtually antithetical political cultures, England and the Soviet Union. I shall then propose that these interpreters of Einstein are best understood by viewing them from two standpoints: first, from the standpoint of the characteristics of their own personalities, and, second, from the standpoint of the political cultures in which they lived.

My initial task is to choose a prominent British interpreter of relativity and a prominent Soviet one. A number of candidates are available in both countries, and I should be explicit about the criteria for my choices. First of all, I think it will be more intellectually interesting—although perhaps less entertaining—if I consider only people who had strong claims for authoritative knowledge about the subject. At the same time, I shall seek candidates who had genuine influence outside the scientific community. In summary, then, my criteria of choice are: international rank as a specialist in some aspect of relativity theory, and unusual influence as an interpreter and defender of relativity physics to educated nonphysicists.

If one looks at the generation of British scientists roughly contemporaneous with Einstein, the two who seem to qualify best in terms of my criteria are James Jeans and Arthur S. Eddington. Both were highly qualified and honored scientists, recognized for their achievements in astronomy and theoretical physics, and both were prolific popular authors who attempted to translate the new physics into a language understandable to their nonscientist colleagues. In my opinion, a consideration of the views of either man would be enlightening for the purposes I have set forth. I think, however, that a case can be made for Eddington as the more appropriate. His name is more closely linked to relativity theory than is that of Jeans, both because he participated in the famous expeditions of 1919 that supported the theory and because he assumed more directly than Jeans the responsibility for making relativity familiar to the English-speaking world. Eddington was a man who, as the British physicist J. J. Thomson commented in 1930, "by his eloquence, clearness and literary power persuaded multitudes in this country and America that they understand what relativity means."[5] At least one generation of American and British citizens was more familiar with Eddington's interpretation of relativity than with any other on the subject. His particular viewpoints have significance, then, for the reception that relativity theory gained among educated nonscientists in England and America.

If one turns to the Soviet Union with the same criteria in mind, no physicist stands out with quite the clarity of Jeans and Eddington in England. A number of candidates could be mentioned: A. A. Fridman (Friedmann), O. Iu. Shmidt, S. I. Vavilov, and V. A. Fok (Fock). All were physicists or mathematicians who defended relativity theory against crude attacks. A. A. Friedmann was a relativist of the first rank whose 1922 paper in *Zeitschrift für Physik* won Einstein's belated admiration and laid the foundation for expanding theories of the universe.[6] His early death in 1925, however, precluded the possibility that he might become a major influence in the shaping of subsequent popular interpretations of relativity theory. O. Iu. Shmidt, mathematician, polar explorer, and vocal Marxist, never undertook an extensive interpretation of relativity theory, except to affirm that it was in thorough accordance with dialectical materialism, thus dismissing those militant Marxists who said otherwise.[7] S. I. Vavilov, future president of the Academy of Sciences, defended relativity scientifically and criticized it philosophically; his work on this topic was only one of his many activities, however, and never achieved a pervasive influence.[8] Many other prominent Soviet physicists—Igor Tamm, Lev Landau, Peter Kapitsa—avoided philosophical discussions of physics in print whenever possible, although their support for relativity physics was never in doubt.

The person who emerges from this consideration as the most likely comparison with Eddington is Vladimir A. Fock, a scholar known to all Soviet citizens interested in the philosophy of physics. Although sixteen years younger than Eddington, he performed the same task of interpreting the theory of relativity in a fashion that was, during a certain period of time, particularly appealing to his culture. Furthermore, Fock meets our criterion of scientific quality. His works in quantum mechanics, the theory of electromagnetic diffraction, and general relativity are well known. His name is carried in the title of the Klein-Fock relativistic wave equation for a particle with no spin in an electromagnetic field. His most outstanding characteristic was his mathematical talent. The usually reserved Paul Ehrenfest once remarked that there was no mathematical problem with which Fock could not deal; he added that "Fock kann einen Stiefel ausrechnen."[9]

Fock, like Eddington, won acclaim with nonscientists in his society because his interpretation of relativity physics accorded with the nonscientific predispositions (some would say prejudices) of many members of that society. Fock's articles were featured in philosophical as well as physics journals, and he wrote on relativity in the main Soviet newspaper, *Pravda.*[10] He was the author of the article "The Theory of Relativity" for the second edition of the *Large Soviet Encyclopedia.* He won a Stalin Prize, a Lenin Prize, and was given the order "Hero of Socialist Labor." In the late forties and fifties Fock was the Soviet Union's leading authoritative interpreter of relativity physics to the wider Soviet public. At certain moments he had to fend off attacks from Stalinist philosophers who thought his interpretation too partial to Einstein and not sufficiently Marxist, and at other times he faced the opposition of many Soviet physicists who considered him too critical of Einstein and too dogmatic in his own Marxism.

For different reasons, Eddington also faced criticism from philosophers and physicists in his country—from philosophers who thought he was writing bad philosophy, and from physicists who disliked the religious mysticism embedded in his writings about physics. Two additional similarities between Fock and Eddington deserve mention. First, they wrote well-known books with almost identical titles: Eddington's was *Space, Time and Gravitation*; Fock's, *The Theory of Space, Time and Gravitation*. Second, the philosophic interpretations of relativity theory advanced by both men no longer enjoy their earlier popularity; indeed, Eddington's effort to find religious meaning in relativity theory appears idiosyncratic to many English-language readers today, and there are signs that something roughly similar has happened in the Soviet Union to Fock's Marxist interpretation of relativity. Yet both men, at specific moments of history, were the best-known interpreters of relativity theory in their respective societies, and these contrasting confluences of interpretations of science and cultural receptivity deserve the attention of intellectual historians.

ARTHUR STANLEY EDDINGTON

Arthur Stanley Eddington was one of the most distinguished astronomers and astrophysicists of the twentieth century. His mathematical genius was widely recognized from the moment he won the position of "first wrangler" in the mathematics tripos examination at Cambridge University, the first second-year student to do so in the history of the university. In later years his pioneering work on astrophysics included a wide range of problems—stellar motions and constitution, radiation pressure, interstellar matter, the mass-luminosity relationship, and relativistic cosmology. Einstein once commented that Eddington's technical treatise *Mathematical Theory of Relativity* was the finest presentation of the subject in any language. Harlow Shapley of Harvard wrote upon Eddington's death: "It is probably correct to say that during the past 25 years Eddington's influence on fellow astronomers has been greater than that of any other man."[11] If Eddington's more popular writings on relativity theory sometimes caused his scientific colleagues to wince, they could hardly blame the public for believing that Eddington should know what he was talking about.

Eddington wrote four or five books that enjoyed large circulation, especially *The Nature of the Physical World* (1928) and *Science and the Unseen World* (1929). One of his goals in these books was popularization—to make the new physics comprehensible to a large audience. A master of metaphor and literary craftsmanship, he embellished his discussions of science with citations from poetry and the classics of English literature, which he knew well. Beyond popularization, however, Eddington had another goal: the exploration of the significance of the new science for social values. His emphasis upon religious mysticism is the aspect of Eddington's writings that distinguishes them most clearly from the works of the host of other popularizers of relativity physics. And it was this emphasis that attracted the attention of the public.

Eddington constructed an interpretation of the place of science in modern thought that underscored the methodological importance of relativity theory. Einstein had insisted that each physical quantity be defined as the result of certain operations of measurement, and he showed that by examining these operations more closely than anyone before had done, a logical opening appeared through which a new concept of time, or simultaneity, could be drawn. The consequence for physics was a revolution in thought.

Eddington believed that Einstein's method had consequences outside of physics as well, for to him it pointed toward a sharper delineation of the realm of science's authority. Einstein had demonstrated, Eddington argued, that the world of science is built upon a much smaller part of the total experience of human beings than earlier believed. Science, Eddington maintained, is limited to the metrical realm, even though many of our experiences are nonmetrical: "The cleavage between the scientific and extra-scientific domain of experience is, I believe, not a cleavage between the concrete and transcendental, but between the metrical and non-metrical."[12]

Einstein, in one of his earliest attempts to explain relativity theory to nonscientists, described hypothetical scenarios in which trains ran down railroad tracks at constant speed while lights were emitted from the trains and lightning flashes struck the track ahead and behind the trains.[13] Max Born, in a similar early explanation of relativity, used the example of ships and barges.[14] Both Einstein and Born spoke of hypothetical observers who informed the reader about the coincidences of the light flashes with other physical events, such as the position of the hands of a clock. They then drew the conclusion that the concept of simultaneity has meaning only when directly linked to the velocity of light. The reader was admonished to pay attention only to the measurements of physical events and to abandon common-sense notions such as absolute simultaneity or light-bearing ethers.

Eddington thought that these scenarios had a broader meaning: they showed not only that many common-sense notions are scientifically incorrect but also that many ordinary human perceptions are superfluous to the scientist. Even such sensations as color have no place in the scientific world as interpreted by Eddington. "There is no colour in the physical world," he wrote. "Everything that we [scientists] assert can be verified by a colour-blind person."[15]

To Eddington a scientist is a person who has "eliminated all superfluous senses." What senses would be left to such a person? Only enough to read a pointer on a scientific instrument: "When we stripped our ideal observer of most of his sense organs we left him part of an eye that he might observe coincidences."[16] All physical observations are reduced to coincidences between an indicator and a graduated scale.

Of course, Eddington granted that physical science goes far beyond its initial information from pointer readings, but he insisted that this starting point was the whole of the *empirical* part of physics. The rest was a creation of the human mind, which he sharply differentiated from the physical world. Man builds his world out of *relations* and *relata*, but "ultimately it is the mind that decides what is lumber—which part of our building will shadow the things of common experience, and which has no such counterpart." The result is the world of science, or symbolic knowledge, which gives a "shadow performance of the drama enacted in the world of experience."[17]

Eddington viewed the influence of the mind as the essential forming force in the world of science, since it is "the innate hunger for permanence in our minds which . . . directs the world-building."[18] Thus, the mind has by its power fitted the processes of nature into a system of laws "largely of its own choosing," and during the process the "mind may be regarded as gaining from Nature that which the mind has put into Nature."[19] Eddington maintained that "all the laws of nature that are classified as fundamental can be foreseen wholly from epistemological considerations. They correspond to *a priori* knowledge, and therefore are *wholly subjective.*"[20] Hyperbolizing Einstein's comments about science being a "free creation of the human mind," Eddington wrote that just as an untracked moor has no paths until a person creates them, the physical world has no configuration until a scientist gives it one. "Without the mind," Eddington observed, "there is but formless chaos."[21] And, indeed, he went so far as to observe that "the stuff of the world is mind-stuff."

We can now see that to Eddington physical science was derived from two essential ingredients: a small amount of empirical data coming from pointer readings, and the human mind. The significance of relativity theory was, to him, its clear illustration of how this reduction of science to these two ingredients occurred, and particularly its indication of how very little experiential data was involved. By emphasizing that science takes from all of external human experience only that which is found in pointer readings, Eddington was trying to show how inadequate science is for an explanation of the major questions of human existence. He was attempting to create a thirst in his readers for knowledge taken from nonscientific realms. Again and again he emphasized that science "omits so much that is obviously essential, there is no suggestion that it is the whole truth about experience."[22] He admonished that if the restriction of science to pointer readings "has taken away all reality," then "I am not sorry that you should have a foretaste of the difficulty in store for those who hold that exact science is all-sufficient for the description of the universe and that there is nothing in our experience which cannot be brought within its scope."[23]

One might have thought that Eddington, a scientist who obviously took pride in his profession, would have found unsatisfying a description of science that made it irrelevant to human concerns. To the contrary, it was precisely by creating a picture of science as a spare and noninformative system of mathematical schemes that Eddington prepared the way for his extrascientific message that was to follow. He was quite explicit in admitting that his description of science served a further goal:

> If you want to fill a vessel with anything you must make it hollow. . . . Any of the young theoretical physicists of today will tell you that what he is dragging to light on the basis of all the phenomena that come within his province is a scheme of symbols connected by mathematical equations. . . . Now a skeleton scheme of symbols is hollow enough to hold anything. It can be—nay it cries out to be—filled with something to transform it from skeleton into being, from shadow into actuality, from symbols into the interpretation of symbols.[24]

Eddington then pointed out that there was another whole realm beyond that of science: the world of the spirit, the world of intimate knowledge, whose existence is as real and probably even more important than the world of science. This spiritual realm lay beyond the world of physical knowledge. It was an "unseen world" that "will not submit to codification and analysis; or, rather, when we attempt to analyse it the intimacy is lost and it is replaced by symbolism."[25]

There were ways of learning about this spiritual world, Eddington believed, although these modes of cognition were entirely separate from the procedures of science. These other channels were the "eye of the body or the eye of the soul." When we use them, he said, we should dispel the feeling that "we are doing something irrational and disobeying the leading of truth which as scientists we are pledged to serve."[26]

In order to learn more about this intimate world, we should rely on our sense of "mystical contact with Nature," and we should recognize that these moods for modern man are "no longer blind alleys but open out into a spiritual world—a world partly of illusions, no doubt, but in which he lives no less than in the world, also of illusion, revealed by the senses."[27] Eddington called for the building of an intimate world taken from the human personality, a world that would be equal in standing to the scientific world taken from the symbols of the mathematician. "We all know," he wrote, "that there are regions of the human spirit untrammeled by the world of physics. In the mystic sense of the creation around us, in the expression of art, in a yearning toward God, the soul grows upward and finds the fulfillment of something implanted in its nature."[28] We should assert the validity of these yearnings, for without them the problem of experience ends in a veil of mathematical symbols. With their help, on the other hand, we lift "the veil in places, and what we discern through these openings is of mental and spiritual nature."[29]

Although Eddington was a deeply religious man, it is important to see that he opposed all efforts to justify religion by direct reference to scientific findings. He believed that in the past such efforts had led to a series of harmful collisions between science and religion, events that he hoped the future would avoid. Thus, he affirmed, "The religious reader may well be content that I have not offered him a God revealed by the quantum theory, and therefore liable to be swept away in the next scientific revolution."[30]

Throughout his writings on the topic of science and religion, Eddington wrestled with a major problem: on the one hand he wished to affirm that science had nothing to do with religion, while on the other he wanted to show that the new physics harmonized with religion. The difficulty of overcoming the problem is evident in his writings, although in his lectures he felt a bit freer to support his inner religious beliefs.

Several of Eddington's most popular books began as lectures. The Gifford Lectures of 1927 became *The Nature of the Physical World*; the Swarthmore Lecture of 1929 was published as *Science and the Unseen World*; the Messenger Lectures at Cornell in 1934 appeared later as *New Pathways in Science*; and the Tarner Lectures of 1938 were published as *The Philosophy of Physical Science*. Eddington usually revised his lectures (which were written out) thoroughly before publication, and he acknowledged the difficulty of changing the lectures into books. In the preface to *The Nature of the Physical World* he observed, "In the oral lectures it did not seem a grave indiscretion to speak freely of the various suggestions I had to offer. But whether they should be recorded permanently and given a more finished appearance has been difficult to decide."[31]

I have not been able to locate the texts of all of the original lectures, but I have found enough original and manuscript material to indicate that the relationship of science to religion was one of Eddington's most demanding problems in revising the lectures for publication. The handwritten manuscript of *The Nature of the Physical World* is in Trinity College Library, Cambridge, and in some instances one can compare portions that were later struck out with the published version. For example, at one point in the manuscript Eddington wrote:

> I do not claim spiritual power only for the religious mystic; but the closing of the eyes to the overwhelming supremacy of that power over material and even intellectual power is the gross error of the widespread materialism of this and other ages—a materialism far more deadly than the philosophic doctrine of that name which I have been combatting in earlier parts of these lectures.[32]

Eddington struck a line through this sentence, and it did not appear in the published version, although the terminology indicates that it may have been included in the Gifford Lectures themselves. These sentences are far more militant and assertive than the cautious Eddington normally permitted himself to commit to print. The sentence was undoubtedly, in his mind, more strongly supportive of his form of religious mysticism than he wished to be in published form.

In yet another section of the same manuscript the following lines were deleted before publication:

> The positive evidences for mystical religion, being in the part of our experience outside physical science, are unaffected by the recent changes in physical theory, which have formed the main theme of this book; they remain as adequate or inadequate as they used to be, and we have nothing essentially new to say about them.[33]

It is my opinion that the reason Eddington deleted this section is the opposite of the reason for the deletion of the previous one. Whereas the previous quotation asserted too much in favor of his religious viewpoint, the latter did not assert enough. It did not give Eddington enough room to say what he wished to say, namely, that the tenability of religious beliefs *was* affected by the recent changes in physical theory.

How was Eddington to point to the effects of the new physics on religious values without falling into the old trap of natural theology? He found a very intelligent answer: the new physics did not directly support religion, free will, and idealism (three of his commitments in the value realm), but, he said, it removed obstacles to them, it cleared the way for them. Eddington illustrated his answer by facing directly the vexed problem of the relationship of science and religion:

> This brings us to the view often pontifically asserted that there cannot be any quarrel between science and religion because they belong to altogether different realms of thought. The implication is that discussions such as we have been pursuing are altogether beside the mark. But the statement is one which requires a careful examination even if in the end we give a guarded assent to it. To avoid a quarrel they must confine themselves to their proper sides of the boundary and that involves a definite understanding of the boundary.[34]

Perhaps no scientist ever insisted as strongly as Eddington that the boundaries of science embrace such a restricted area, that science deals with so little of what is really important to humans. He restricted science to a small portion of human concerns, and he then beckoned everyone to fill the remaining space with value systems taken from nonscientific areas of thought. He realized that the space could be occupied by all sorts of value systems, including that of the atheist or the religious dogmatist, but he asked his readers to consider the virtues of his own. Far too sophisticated and subtle a person to engage in proselytizing for his own faith of Quakerism, he nonetheless pointed out that it had no creed and could never fall into conflict with science; it was therefore a faith particularly compatible with science. As he observed, "Quakerism in dispensing with creeds holds out a hand to the scientist."

Eddington from the Standpoints of Personality and Political Culture

To understand Eddington's interpretation of relativity theory, it is helpful to view it both as a personal creation and as a manifestation of political culture. On the personal side, Eddington was obviously heavily influenced by his own religious background. Quakers traditionally based their religion on experience, not on dogma. George Fox, one of the founders of Quakerism, emphasized that the Quaker way was empirical. He believed that he could come to know Christ on the basis of his inner consciousness, not through reasoning or analysis. Referring to his religious perception, Fox wrote, "This I knew experimentally."[35] The word "experimentally" here would probably best be described today as "experientially." Eddington echoed Fox's sentiments when he wrote that those who believe in the unseen world of intimate knowledge "are just as much facing the hard facts of experience" as those who build their knowledge on the indications of spectroscopes. Eddington's loyalty to the Quaker way of knowing, the belief in an "Inner Light," was indicated by his observation that "in the mystical feeling the truth is apprehended from within and is, as it should be, a part of ourselves."[36] We know from Eddington's handwritten journal, held in Trinity College Library, that the silent worship periods of Quaker summer schools were important events in his life.[37]

Eddington's roots in Quakerism could hardly have been more deep or authentic. His mother, Sarah Ann Shout, was a descendant of John Camm and John Audland, missionaries of Quakerism at the very beginning of the "Quaker explosion" in the seventeenth century who had worked directly with George Fox. Eddington's father, Arthur Henry Eddington, was also a member of a strong Quaker family. At the time of the birth of his son, A. H. Eddington was headmaster of Stramongate school in Kendal, an institution controlled by Quakers since the late seventeenth century.

Arthur Stanley Eddington's emergence as a popular figure and as a defender of traditional values contained its ironies. Quakerism has never been a view widespread in the social establishment of England and America but instead is a little understood eccentricity of a tiny minority. Furthermore, Eddington himself—conscientious objector in the First World War, bachelor, and social isolate, described by his contemporaries as often almost incoherent in conversation—was hardly a prime candidate for public adulation. He was a man who preferred to be alone or on a long hike with his intimate friend of forty years, C.J.A. Trimble. Similar adulation and distortion had come to Einstein himelf, who could not let the words "God" or "the Old One" tumble from his lips without reporters seizing upon them as evidence for his religiousness, ignoring his opposition to all existing creeds and much of what is usually considered to be religion. Religious faith was much more important to Eddington than to Einstein, and Eddington's beliefs fitted within an existing organized group, the

Society of Friends. Furthermore, Eddington actively promoted his interpretation of religion, considering it an important part of his activities. Nonetheless, his religion sharply contrasted with the faiths of well over 90 percent of those who found comfort in his writings. He must have been unprepared for his own popularity.

But if we wish to understand Eddington's prominence as an interpreter of relativity theory, we must go beyond his personal background to a consideration of political culture. When Eddington died in 1944, *Time* magazine called him "one of mankind's most reassuring cosmic thinkers."[38] Let us consider for a moment what about Eddington's view was reassuring to the culture in which he lived.

The theory of relativity was seen by many laypeople as the denial of absolute standards and values. Coming at a time when the global political and economic environment was exceedingly unstable, and at a moment when the successful Russian Revolution had thrown into question the entire range of assumptions and values upon which European order was based, the new physics seemed to be one more threat to reliable standards. That Einstein was a political radical and Jew did not help allay the misgivings of those who felt that the old order—political and scientific—was under attack from similar quarters. Einstein repeatedly emphasized his belief that physics did not directly relate to his social views, but this reassurance often only increased the bewilderment of laypeople.

Against this background, the message that Eddington brought was an exceedingly comforting one. An article in *Current Biography* observed of Eddington, "To the public he became important first for reconciling the average man to relativity and later for reconciling religion with science."[39] His insistence that science (particularly the *new* science) dealt with only a small part of reality increased the confidence of those people who understood nothing of that area of thought but could see, with Eddington's help, that there was still a very large realm where they were as well prepared as anyone to handle the challenges that constantly appeared. True, there was a momentous change in physics, but life could go on elsewhere as before. Indeed, the new physics seemed to show more clearly than the old physics that traditional values still reigned; the threats of determinism and materialism that had often been identified with the mechanistic Laplaceian world-view were now in retreat. When Eddington in 1932 gave a lecture titled "The Decline of Determinism," the Smithsonian

Institution asked and gained permission to reprint it and gave the lecture wide publicity on both sides of the Atlantic.[40] Thus, established institutions threw their support behind Eddington's defense of idealism.

Eddington's interpretation was sophisticated enough to avoid the danger of directly linking science to religion, but simple enough to be understood as defending traditional values, including religion. An English-speaking public that had learned from the debates over evolution in the nineteenth century that bringing the content of science into direct contact with religion was a risky endeavor that might damage both was ready to appreciate an interpretation of science that cleared the way for religion without making any direct assertions about it. Eddington's effort to reconcile science and religion was entirely within the tradition of British thought. As Charles Gillispie wrote in another context, "Ever since Newton, natural theology, if not quite a distinctively British approach to God, had at least been elaborated in far greater detail and with much more enthusiasm in Britain than in other countries."[41]

In the 1920s and 1930s there was a particularly strong conjunction between the conception of science and values held by Arthur Stanley Eddington and the needs of his English-language audience. That conjunction of elements has now passed, but without seeing how complementary and appropriate it was at the time, one cannot understand the role played by his writings.

VLADIMIR ALEKSANDROVICH FOCK

Any discussion of V. A. Fock's interpretation of Einstein's theory of relativity must begin with the unambiguous recognition that Fock accepted the mathematical formulation of both the special theory of relativity and the general theory. Furthermore, in a series of publications spread over a forty-year period he defended and praised Einstein's achievements. In 1939 he wrote a popular article on Einstein's birthday, calling him "the great physicist of modern times," and in the last years of Stalin's life, when ideological pressures in science were particularly strong in the Soviet Union, he castigated what he called "ignorant criticism" ("nevezhestvennaia kritika") of relativity theory and quantum mechanics.[42] To question the validity of relativity theory, wrote

Fock in response to Stalinist philosophers, was on an intellectual level with questioning the roundness of the earth.[43]

Nonetheless, Fock was an authentic "internal" Marxist, and he believed that it was his obligation to produce an interpretation of relativity physics that would counter those of Eddington, Jeans, and other physicists in the West who tried to depict relativity theory as the cancellation of materialist philosophy. "Many foreign physicists in their roles as authors and especially popularizers," he wrote, "love to provide philosophical commentaries of an idealistic spirit along with the physical theories which they expound."[44] His goal, he maintained, was to create "a correct materialistic understanding of the theory of relativity."[45] He believed that the way to do this was to show more directly than others had done that relativity theory was not merely a mathematical innovation but also a description of physical and material reality. By emphasizing that Einstein had indissolubly linked matter and energy and had created an objective description of the universe uniquely dependent on the distribution of this matter-energy, Fock hoped to rid relativity theory of the idealistic aura that some of his fellow Soviet Marxists lamented and Western writers like Eddington celebrated.

Fock pursued this goal within the framework of an explicit Marxism. In 1938 he wrote, "I am convinced that materialist philosophers will be able, sooner or later, to bring the newly discovered properties of matter into agreement with the basic formulations of materialism. But this has so far not been achieved. . . ."[46] It was necessary, Fock continued, "first of all, to strike from the hands of the idealists their basic weapon," namely, the belief that modern physical theories such as quantum mechanics and relativity "contradict materialism." The path to this achievement, he asserted, was a study of Lenin's writings, particularly his view that the valuable achievements of "bourgeois physicists" can be distinguished from their philosophical viewpoints.[47] Fock expressed the hope that Soviet philosophers would defend modern physics from the standpoint of Marxism, but he knew that by 1938 most of the Marxist philosophers who were capable of such a defense had been replaced by Stalinists who did not understand science.[48] The defenders of relativity among the philosophers were being eliminated in factional political struggles. Therefore, said Fock, "the physicists must take the initiative themselves."[49] And Fock accepted his own challenge.

In order to achieve his goals, Fock was willing to go so far as to suggest changes in Einstein's terminology and to supplement Einstein's theories with his own elaborations, while preserving the mathematical core of relativity. The aspect of Einstein's system to which Fock objected most strenuously was the term "general relativity," since it seemed to Fock to imply the absence of any objective reference standards in nature, a conclusion that he insisted could not be derived from the essential part of Einstein's theories. Indeed, Fock asserted that the relationships of material bodies described so brilliantly by Einstein were "just as objective as all other properties of bodies."[50]

Although Fock rejected the term "general relativity," he accepted the term "special relativity" as having concrete meaning with reference to Newtonian physics. Fock agreed completely that Einstein showed that the old absolutes of "space" and "time" were relative to frames of reference and therefore could no longer serve as reliable and unchanging standards. However, the new theory that Einstein presented was also, thought Fock, based on absolute standards, just as Newton's had been. Echoing the views of Max Planck,[51] Fock remarked, "The theory of relativity, after showing that a whole series of concepts earlier considered absolute were actually relative, at the same time introduced new absolute concepts. The majority of critics of the theory of relativity forget this."[52] Einstein found new absolutes in the constant velocity of light and in the concept of space-time—the conjoining of Newton's terms into a four-dimensional manifold. The theory of general relativity could thus be just as easily called the theory of absolute space-time or the theory of gravitation (Fock's preferred term). Fock thought a strong argument could be made that the new title would be more accurate in a philosophical sense than was Einstein's title. The one thing that the theory of general relativity was not, maintained Fock, was a purely relativistic theory.[53]

Fock believed, furthermore, that relativity theory should be seen as an explanation of nature that reflects the properties of physical reality.[54] Referring to the same "railroad track" explanations of special relativity theory that were important to Eddington when he maintained that all that counts in physics is measurement, Fock explained that:

Einstein, the founder of relativity theory, and, subsequently, other authors, based their expositions of its consequences on descriptions of the perceptions of imaginary observers, or, in the best cases, on the results of measurements carried out by these observers. At the same time, Einstein and the others accepted the sensations or measurements of these observers as being primary; what was actually primary here was objective relationships of material bodies and objective properties of space and time.[55]

Fock continued that the positivistic emphasis on the results of measurement gave many "Western physical idealists" the opportunity to subordinate the concept of scientific theory as a reflection of reality to the concept of theory as an "auxiliary construction" for the coordination of sensations.[56] Thus, Eddington could say that the physical world has no configuration until a scientist gives it one, just as an untracked moor has no path until a person creates one. To this philosophical viewpoint, Fock responded with the view that the word "relativity" in the "theory of relativity" has nothing to do with the form of philosophical relativism that says "our knowledge is so conditional and relative that it has a purely subjective character."[57] On the contrary, Fock continued, "the theory of relativity is a correct reflection of the essential properties of space and time."[58] And, finally, "the theory of relativity is a brilliant confirmation of dialectical materialism, particularly its principle that matter, movement and space are organically connected."[59]

If Fock had not gone beyond this purely philosophical rephrasing of the theory of relativity, he would not be remembered by Soviet physicists as a critic of Einstein (although he would still be remembered by Stalinist philosophers as a defender of Einstein). But Fock did go beyond a materialistic verbal framework for relativity theory. He thought that an insistence on paying attention to the physical content of Einstein's interpretation of inertial and gravitational mass would yield an insight different from that usually seen by scientists and philosophers. Fock believed that the equivalence of inertial and gravitational mass as defended by Einstein was mathematically valid, but physically true only in a local sense. Fock based his view here on the famous thought-experiment cited by Einstein in developing his interpretation. Einstein noted that changes in the extension of a spring by which a weight is suspended in an elevator can be equally well explained by an acceleration of the elevator in one direction or a gravitational field in the opposite direction. Thus, in the elevator pictured below it is impossible for an observer within the compartment to decide whether a stretching of the spring is caused by the acceleration of the elevator in direction *b* or a downward gravitational field in direction *g*.[60] The two forces are entirely indistinguishable, and thus gravitational and inertial mass are equivalent.

Einstein extended the analogy to gravitation on a planet's surface, a task that might at first seem impossible, since any acceleration involving the planet as a whole would have very different effects at different spots on the planet's surface. By considering only infinitesimal regions, however, as permitted by the use of differential equations, the analogy can be maintained. An individual infinitesimal region is to be considered in the same way as our earlier elevator. If we think of a particular cell *a* in a grid

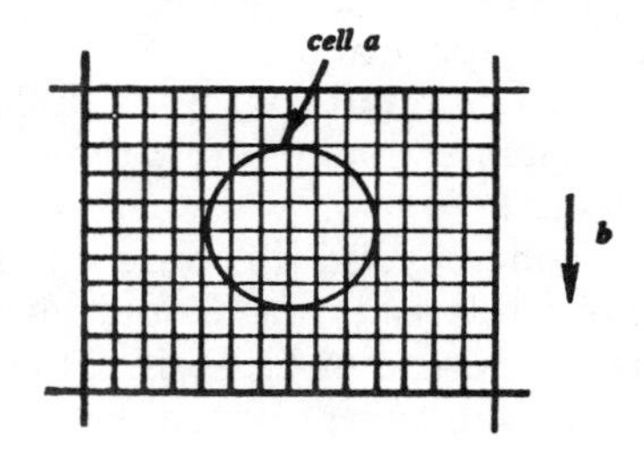

of infinitesimal cells as being like our elevator, it becomes clear once again that the gravitational field in any one spot can be transformed away by imagining an appropriate acceleration (in the case of the earth's surface, 32 ft/sec^2) in direction *b*. In Einstein's theory of general relativity, any gravitational field can be similarly replaced by an appropriate acceleration. Another way of stating the equivalence is to say that, given any two bodies, one cannot be accelerated more easily than the other and yet cause a stronger gravitational field.

Fock agreed with the local equivalence of acceleration and gravitation, but added that when one wishes to consider the universe, one must set initial or boundary conditions that reflect the properties of space as a whole. As Fock put it:

In order to construct a theory of gravitation or to apply it to physical problems it is . . . insufficient to study space and time only locally, i.e., in infinitely small regions of space and periods of time. One way or another one must characterize the properties of space as a whole. If one does not do this, it is quite impossible to state any problem uniquely. This is particularly clear in view of the fact that the equations of the gravitational, or any other field, are partial differential equations, the solution of which are unique only when initial, boundary or other equivalent conditions are given. The field equations and the boundary conditions are inextricably connected and the latter can in no way be considered less important than the former.[61]

Fock characterized the boundary conditions in two different ways. In the first case, he assumed space to be homogeneous at infinity in the sense of being characterized by the Lorentz transformations. Masses and their associated gravitational fields were then envisioned as being implanted in homogeneous Galilean space. The second case assumed a space-time that is only partially uniform, with the spatial part of it conforming to Lobachevskiian geometry. Usually termed the space of Friedmann-Lobachevsky, it contains well-defined gravitational fields when the mean density of matter contained within it is not equal to zero.

The important conclusions from these considerations, and the ones that reveal most graphically Fock's unorthodox position, concern the question of preferred or privileged systems of coordinates. His position here was connected with his desire to show that relativity theory was not entirely relativistic, that even it contained reference standards determined, ultimately, by matter. In each of the types of space envisioned by Fock—that is, Galilean space, space uniform at infinity, and Friedmann-Lobachevsky space—he said that there "probably" exists a preferred system of coordinates.[62] The word "probably" indicates Fock's continued hesitation in the case of Friedmann-Lobachevsky space; in the case of Galilean space and space uniform at infinity, he was confident of the existence of preferred systems of coordinates. The existence of such preferred systems of coordinates in each case would be, of course, contrary to Einstein's concept of the complete relativization of motion. Just as the special theory of relativity is associated with the relativization of inertial motion (and therefore the equivalence of inertial reference frames), so the general theory of relativity is associated with the relativization of accelerated motion (and therefore the equivalence of accelerated reference frames). But now Fock questioned whether the general theory actually was a generalization of the special theory in this sense.

Fock devoted much of his research to the task of proving that in space uniform at infinity there is also a preferred system of coordinates that is well defined apart from a Lorentz transformation. Fock's preferred system was harmonic coordinates, which, true to his commitment to the existence of physical reality, he believed reflected "certain intrinsic properties of space-time."[63] Yet it should be noted that Fock's reliance on harmonic coordinates was one of the most controversial aspects of his approach. Many physicists who found his criticism of the term "general relativity" interesting remained dubious of the preferential status of harmonic coordinates.[64]

Fock's stout defense of the preferred status of harmonic coordinates is responsible more than anything else for his becoming known among Soviet physicists as a critic of Einstein. His insistence on harmonic coordinates is now widely regarded as an unsuccessful appendage to relativity theory emanating from his eccentric criticism of Einstein's general relativity. An insight into Fock's position among other Soviet physicists in the mid-fifties is given in Leopold Infeld's memoirs, where he describes a 1955 discussion of general relativity in Moscow's Lebedev Institute:

> Most interesting for me was the discussion between Fock and the Soviet physicists, in which I also took part. . . . First Fock spoke and formulated his objections to Einstein's theory by defining his own conception of harmonic coordinate systems. Then I spoke in favor of Einstein's theory and was followed by Landau, Tamm and Ginzburg—all well-known physicists who preferred Einstein's theory to Fock's. They said that his additional equations for defining the coordinate system are unnecessary and provide nothing of value. Fock still stuck to his interpretation, as he does today, but I learned during my visit to Moscow that he was fairly isolated at the time.[65]

Isolated as Fock may have been among Soviet physicists in the fifties because of his attachment to harmonic coordinates, he was celebrated by the official press as the best example of a Marxist interpreter of relativity theory. Perhaps the high point in his influence with the general educated public came in this same year of 1955, when his interpretation of relativity was featured in the *Large Soviet Encyclopedia.* Two years before he had won his second Order of Lenin.

After the relative relaxation of the political atmosphere in the Soviet Union in the late fifties and sixties, Fock's influence rapidly declined. Perhaps the most graphic illustration of that shift can be seen by comparing the articles on relativity theory in the second and third editions of the *Large Soviet Encyclopedia*, published, respectively, in 1955 and 1974.[66] The first article, written by Fock, starts and ends with quotations from Lenin and presents the theory within the materialistic framework already described here. The article does not have a section on the general theory of relativity at all, but has instead one on the theory of gravitation. Nonetheless, it contains an accurate description of relativity and treats the theory as fully established.

The 1974 article, written by I. Iu. Kobzarev shortly before Fock's death, contains no mention of Marx or Lenin. It drops Fock's insistence on the term "theory of gravitation" and accepts the international terminology of "special" and "general" to describe the theory. All imprint of Fock's interpretation seems to have disappeared, except for two quite small details: the article still places an unusual emphasis on viewing relativity theory as an expression of the real properties of space and time (no subjectivism here), and Fock is one of the authors cited at the end.

Fock from the Standpoints of Personality and Political Culture

As we did with Eddington's interpretation of relativity, we shall also view Fock's interpretation both as a personal creation and as a manifestation of political culture. On the personal side, we have fewer firsthand descriptions of Fock than we do of Eddington, but Fock was well known to physicists in many countries. Infeld described him in the following way:

> He is an outstanding Soviet theoretical physicist, on a very high level, who came out for a change in Einstein's theory while retaining its mathematical skeleton, and he believed that he had much improved the theory. . . .
>
> Who is this Fock? Black hair, fairly fat, around sixty, pleasant, somewhat like Einstein personally. He had an attractive way about him, laughed gaily, was deaf, and wore a hearing aid which—so the spiteful said—he turned off when the argument of his opponent did not please him. He was dogmatic and very much believed in his own convictions. I may say that, during this time, I became friendly with Fock. Our relations were better than correct. I argued with him, tried to convince him that he was wrong, but of course without result.[67]

From all reports, Fock was brilliant, dogmatic, brave, witty, and ambitious. He did not shrink from polemics, and he got involved in a priority dispute with Infeld. He confidently maintained that he had persuaded Niels Bohr to change his opinion on the philosophic question of realism.[68] He was proud of the West European origins of his family and liked to tell people that he was descended from Dutch shipbuilders imported to Russia by Peter the Great. Like his ancestors, he also brought knowledge from the West to Russia. They had brought nautical crafts, and Vladimir Fock brought the new physics from Göttingen and Paris, where he studied for a year in 1927-1928.

Although there is no reason to doubt Fock's basic commitment to Marxism, it is clear that he used Marxism as a vehicle to promote his personal scientific viewpoints. We find in this complex physicist a shrewd awareness of both physical and political reality. Before we accuse him of opportunism, however, we should recognize that Fock was a brave leader in the effort to prevent Soviet physics from being subjected to the kind of perversion that occurred in Soviet genetics with the victory of Lysenkoism.

Fock first began discussing Marxism and physics in the 1930s, after it had become clear that those philosophers who wished to defend relativity physics from the standpoint of Marxism (for example, S. Iu. Semkovskii, A. A. Bogdanov, A. Goltsman, and B. G. Gessen) were disappearing. Some of the best disappeared during the purges. A new breed of Stalinist philosopher was coming to power, people who were often hostile to modern science.[69] A shift in Soviet political culture was occurring. Fock was one of several Soviet physicists who tried to develop a form of self-defense to make up for the loss of these "middle men," the sophisticated Marxists who appreciated modern physics. Some Soviet physicists emphasized the practical value of modern physics as a means of defense, an approach that became very powerful after the development of atomic weapons. Fock's method was different: he became a philosopher of science himself, developing and promoting his own materialistic interpretation of the theory.

Fock resolutely rejected the attacks on relativity theory by ignorant Stalinist philosophers and third-rate physicists like A. A. Maksimov and R. Ia. Shteinman.[70] Maksimov wrote in the Soviet Union's leading philosophical journal that Fock's defense of "reactionary philosophical ideas" in the "so-called theory of relativity" was holding back Soviet science by chaining it to idealistic views inherited from physicists of the capitalist countries. Fock replied in print that Maksimov's call for a rejection of relativity theory was an example of confusion and error and was based on an antiscientific spirit. And he asked Maksimov why he was so ready to take the word of Western physicists like Eddington and Pascual Jordan that accepting relativity meant rejecting materialism. The theory is brilliantly confirmed by experiments, he continued, and points to a new understanding of materialism.

Even some of the Soviet physicists who were critical of Fock's harmonic coordinates recognized that Fock's defense of relativity theory was a defense of Soviet physics at a moment when most other physicists refused to participate in the public debate. In later years some of them recognized this debt. After Fock's death in 1974 the prominent Soviet physicists M. G. Veselov, Peter Kapitsa, and M. A. Leontovich wrote of the situation in Soviet science in the late forties and early fifties: "More than once there flared up obviously false scientific tendencies connected with opportunistic maneuvers. At this moment Fock displayed great decisiveness and undeviating devotion to principle; he bravely and openly stood up for the purity of scientific ideas."[71]

By not only defending relativity physics but also developing an interpretation of it in terms of Marxist materialism, Fock merged his personal preferences with those of his political culture.[72] In the late forties and early fifties Fock won the approval of the leaders of the regime, and he was rewarded with the right to travel widely in Europe and America. Soviet intellectuals were told in university classrooms and philosophical journals that V. A. Fock had cleansed relativity theory of the idealistic trappings placed upon it by bourgeois scientists and had developed a version of the theory that accorded thoroughly with Marxism. He had done this, they were told, because of his careful reading of Marx, Engels, Lenin, and Stalin. And Fock agreed by praising Marxism even in his scientific works.[73] Fock benefited from the government, and he repaid it with political praise and cooperation.

CONCLUSION

Einstein's theory of relativity was such a radical break with common-sense notions possessed by laypeople of all societies—whatever the prevalent philosophies, religions, or ideologies—that it caused great public concern. In England and America in the twenties, underneath the fascination with Einstein and the titillation with popular lectures on relativity, there lurked an anxiety about how this new theory could be fitted with conventional beliefs, particularly religious ones. Did the theory of relativity mean that all secure intellectual and religious foundations were destroyed, all absolute standards invalidated? Citizens in those countries witnessed the comic adventures of newspapers offering prizes for the best explanation of relativity theory in a few hundred words and of telegrams being sent to Einstein asking "Do you believe in God?" and demanding an immediate reply. Against the background of such concerns, Arthur Stanley Eddington's writings on relativity theory accomplished a dual purpose, for they not only made the theory sound vaguely comprehensible but they also allayed these anxieties.

The theory of relativity was impartial in its ability to raise the fears of orthodox thinkers. Many Marxists, too, worried about relativity. Its linking of matter and energy seemed to some people to imply the disappearance of matter and materialism, its relativism seemed to undermine the concept of steady movement toward truth in which Marxists believed, and its use by popular writers like Eddington in order to support religious mysticism alarmed atheists and agnostics. All these features of the discussions of relativity theory disturbed Marxists. Against this background Vladimir Fock's interpretation of relativity theory helped to make the theory intellectually acceptable to a Marxist audience.

The comparison of the views of Eddington and Fock is obviously interesting to me, or I would not have brought them together in one frame. I do not, however, wish to be understood as maintaining that their interpretations of relativity theory are true opposites—mirror images, as it were—or that "just as Fock reacted to his public, so did Eddington." There are analogical qualities between Eddington and Fock, but there are also some very large differences. First of all, Fock was living in an oppressive society where the relationship between science and ideology was dramatically different from that in England. No one would have been able to publish Eddington's defenses of religion in the Soviet Union, whereas some of Fock's writings on Marxism *were* published in the West. Furthermore, Eddington was reacting to relativity theory; Fock was reacting *both* to relativity theory and "idealistic interpretations" of it like that of Eddington. Thus, Fock's interpretation was, in a sense, a reflection of both cultures, whereas Eddington's was much more closely rooted in his own. It is true that Eddington criticized Marxist materialism in several of his lectures, but his concern with an "adversary" was much less direct than Fock's. Fock's interpretation did not make sense without the supposition of opposing idealistic interpretations. Eddington's views could stand independently on the foundations of traditional idealism, modern operationalism, and the Quaker faith.

Nonetheless, both men were concerned with the same intellectual issues, and these issues can be summarized as two dualisms: relativism versus absolutism, and idealism versus materialism. The first dualism concerns knowledge and certainty, that is, epistemology. The second concerns the actual content of reality, that is, ontology. It is obvious that the two men were in radical opposition on the question of idealism versus materialism. Fock proclaimed himself a materialist; Eddington defended idealism. With regard to the other dualism, however, their positions were much closer. Both men displayed a yearning for the absolute. Again, Fock was the more confident. He affirmed that relativity theory supported the proposition that all that exists in nature is matter-energy. His preference for harmonic coordinates was based on a desire to rescue an absolute standard. Eddington also reached for an absolute, what he called the "yearning toward God." His belief that all the laws of nature have an a priori mental character was another assertion of absolute idealism. Fock and Eddington, then, had one characteristic in common: an unwillingness to live in a totally relativistic world.

Both Fock and Eddington ignored Einstein's warning not to try to use physics as a vehicle for social, philosophical, religious, or political messages. If they had paid more attention to Einstein's admonition, the quality of their writing would undoubtedly have been improved.[74] Yet one should admit that deep underneath the distortions of relativity theory that both Eddington and Fock committed, each for his own goals, there remains an unanswered question. Is it really possible to divorce knowledge of the physical universe entirely from human concerns? Although Einstein maintained that his scientific and his social views were entirely separate matters, he never answered the question in a definitive way. One can maintain that Einstein's own life was a witness to a relationship between his scientific views and his social ones. His hopes for order, justice, and rational explanation in the social order were too similar to his striving for order, simplicity, and causality in the world of physics to be a mere coincidence. But examining the relationship between Einstein's vision of the physical world and his vision of the social world is another paper.

The examples of Eddington and Fock give us interesting illustrations of ways in which relativity theory and political culture interacted in the hands of distinguished physicists. I have tried to view the work of these men both from the standpoint of their own personalities and as manifestations of their political cultures. Nonetheless, each of these two men demands independent examination in addition to the comparison I have drawn here. They would, of course, have been surprised at the thought that they would some day be compared. Any such comparison emphasizes a few features and ignores others. My hope is that the features emphasized here give at least a partial insight into the popular reception of relativity theory in different political cultures.

NOTES

1. A. Einstein, "Dialog über Einwande gegen die Relativitätstheorie," *Naturwissenschaften* **6** (1918), p. 697.

2. "Epilogue: A Socratic Dialogue," in Max Planck, *Where is Science Going?*, New York, Norton, 1932.

3. N. Bohr, "Natural Philosophy and Human Cultures," *Nature* **143** (1939), pp. 268-272.

4. Quoted in R. S. Shankland, "Conversations with Albert Einstein," *American Journal of Physics* **31** (1963), p. 50.

5. Quoted in A. Vibert Douglas, *The Life of Arthur Stanley Eddington*, London, Thomas Nelson and Sons, 1956, p. 104.

6. A. A. Friedmann, "Über die Krümmung des Raumes," *Zeitschrift für Physik* **6** (1922), pp. 377-387; A. Einstein, "Bemerkung zu der Arbeit von A. Friedmann, 'Über die Krümmung des Raumes,' " *Zeitschrift für Physik* **11** (1922), p. 326; A. Einstein, "Notiz zu der Arbeit von H. Friedmann 'Über die Krümmung des Raumes,' "*Zeitschrift für Physik* **16** (1923), p. 228. Friedmann also wrote a popularization of relativity, *Mir kak prostranstvo i vremiia* (1923), rpt. Moscow, 1965.

7. *O. Iu. Shmidt, Zhizn' i deiatel'nost'*, Moscow, 1959; O. Iu. Shmidt, *Izbrannye trudy*, 4 vols., Moscow, 1959-1960.

8. S. I. Vavilov, *Sobranie sochineniia*, 4 vols., Moscow, 1954-1956.

9. M. G. Veselov, P. L. Kapitsa, and M. A. Leontovich, "Pamiati Vladimira Aleksandrovicha Foka," *Uspekhi fizicheskikh nauk* **117**, no. 2 (1975), p. 375.

10. A bibliography of Fock's works from 1923 to 1956 is in *Vladimir Aleksandrovich Fok*, Materialy k biobibliografii uchenykh SSSR, seriia fiziki, vyp. 7, AN SSSR, Moscow, 1956.

11. Quoted in Douglas, *Life of Eddington*, p. 185.

12. A. S. Eddington, *The Nature of the Physical World*, New York, Macmillan, Cambridge, Cambridge University Press, 1928, p. 275.

13. Einstein, *Relativity, The Special and the General Theory: A Pop-*

ular Exposition, trans. Robert W. Lawson, London, Methuen, 1920, esp. chaps. VI–X.

14. M. Born, *Die Relativitätstheorie Einsteins und ihre physikalischen Grundlagen, gemeinverständlich dargestellt*, Berlin, Julius Springer Verlag, 1920, esp. pp. 166-167.

15. A. S. Eddington, *New Pathways in Science*, New York, Macmillan, Cambridge, Cambridge University Press, 1935, p. 11.

16. Ibid., p. 22. Also see Eddington, *The Nature of the Physical World*, pp. 230-231.

17. Eddington, *The Nature of the Physical World*, pp. 230-231.

18. Ibid., p. 242.

19. Ibid., pp. 243-244.

20. A. S. Eddington, *The Philosophy of Physical Science*, Cambridge, Cambridge University Press, 1939, pp. 56-57.

21. Eddington, *The Nature of the Physical World*, p. 335.

22. A. S. Eddington, *Science and Religion*, London, Friends Home Service Committee, 1931, pp. 9-10.

23. Eddington, *The Nature of the Physical World*, p. 254.

24. Eddington, *Science and Religion*, pp. 7-8.

25. Eddington, *The Nature of the Physical World*, p. 321.

26. A. S. Eddington, *Science and the Unseen World*, London, Allen and Unwin, 1929, p. 49.

27. Eddington, *The Nature of the Physical World*, p. 324.

28. Ibid., p. 327.

29. Eddington, *New Pathways in Science*, p. 322.

30. Eddington, *The Nature of the Physical World*, p. 353.

31. Ibid., p. vi.

32. Manuscript copy, "The Nature of the Physical World," Trinity College Library, Cambridge, pp. 309-311.

33. Ibid., p. 317.

34. Ibid., p. 339.

35. See the discussion of Fox in D. Elton Trueblood, *The People Called Quakers*, New York, Harper and Row, 1966.

36. Eddington, *The Nature of the Physical World*, p. 321.

37. "Eddington Journal," Trinity College Library, Cambridge, p. 56.

38. *Time* **44** (4 December 1944), p. 50.

39. *Current Biography*, 1941, pp. 254-256.

40. A. S. Eddington, "The Decline of Determinism," *Annual Report of the Board of Regents of the Smithsonian Institution*, Washington, D.C., 1932, pp. 141-157, reprinted from the *Mathematical Gazette* **16** (May 1932).

41. Charles Coulston Gillispie, *Genesis and Geology*, New York, Harper and Brothers, 1959, p. 222.

42. V. A. Fock, "Velikii fizik sovremennosti" (K 60-letiiu A. Einshteina), *Leningradskii universitet* III/23, no. 13 (1939); V. A. Fock, "Pro-

tiv nevezhestvennoi kritiki sovremennykh fizicheskikh teorii," *Voprosy filosofii* no. 1 (1953), pp. 168-174. All of the translations from Fock's works that appear in the text are my own unless otherwise noted.

43. Fock, "Protiv nevezhestvennoi kritiki sovremennykh fizicheskikh teorii," p. 172.

44. V. A. Fock, "K diskussii po voprosam fiziki," *Pod znamenem marksizma* no. 1 (1938), p. 151.

45. V. A. Fock, "Sovremennaia teoriia prostranstva i vremeni," *Priroda* no. 12 (1953), p. 13.

46. Fock, "K diskussii po voprosam fiziki," p. 153.

47. Ibid., p. 155.

48. See Jehoshua Yakhot, "The Theory of Relativity and Soviet Philosophy," *Crossroads* (Israel Research Institute of Contemporary Society), Autumn 1978, pp. 92-118.

49. Fock, "K diskussii po voprosam fiziki," p. 155.

50. "Es ist nicht überflüssig, zu unterstreichen, dass das Verhältnis von Körpern oder Prozessen zum Bezugssystem ebenso objektiv ist (d.h. unabhängig von unserem Bewusstein) wie überhaupt alle physikalischen und anderen Eigenschaften der Körpern." V. A. Fock, "Über philosophische Fragen der modernen Physik," *Deutsche Zeitschrift für Philosophie* no. 6 (1955), p. 742.

51. Planck commented, "The concept of relativity is based on a more fundamental absolute than the erroneously assumed absolute which it has supplanted." M. Planck, *The New Science*, Greenwich, Conn., Meridian Books, 1959, p. 146. Planck expressed the same idea in earlier publications, for example, *Das Weltbild der Neuen Physik*, Leipzig, 1929, p. 18.

52. Fock, "Protiv nevezhestvennoi kritiki sovremennykh fizicheskikh teorii," p. 172.

53. Fock liked to put his viewpoint in French: "(1) La relativité physique n'est pas général; (2) la relativité générale n'est pas physique." V. A. Fock, "Les principes mécaniques de Galilée et la theorie d'Einstein," in *Atti del convegno sulla relatività generale: Problemi dell'energia e onde gravitazionali*, Florence, 1952, p. 12.

54. See V. A. Fock, "Sistema Kopernika i sistema Ptolemeia v svete obshchei teorii otnositel'nosti," in *Nikolai Kopernik*, Moscow, 1947, pp. 180-186; also his "Nekotorye primeneniia idei Lobachevskogo v mekhanike i fizike," in A. P. Kotel'nikov and V. A. Fock, *Nekotorye primeneniia idei Lobachevskogo v mekhanike i fizike*, Moscow and Leningrad, 1950, pp. 48-86.

55. Fock, "Sovremennaia teoriia prostranstva i vremeni," p. 18.

56. V. A. Fock, "Osnovnye zakony fiziki v svete dialekticheskogo materializma," *Vestnik Leningradskogo Universiteta* no. 4 (1949), esp. pp. 44ff.

57. Fock, "Protiv nevezhestvennoi kritiki sovremennykh fizicheskikh teorii," p. 171.

58. Fock, "Osnovnye zakony fiziki v svete dialekticheskogo materializma," p. 36.

59. Fock, "Protiv nevezhestvennoi kritiki sovremennykh fizicheskikh teorii," p. 171.

60. See Hans Reichenbach, *The Philosophy of Space and Time*, trans. M. Reichenbach and J. Freund, New York, Dover, 1958, p. 223.

61. V. A. Fock, *The Theory of Space, Time and Gravitation*, trans. N. Kemmer, New York, Pergamon Press, 1959, p. xv.

62. Ibid., p. xvi.

63. Ibid., p. 351. Also see V. A. Fock, "Poniatiia odnorodnosti, kovariantnosti i otnositel'nosti," *Voprosy filosofii* no. 4 (1955), p. 133. The reasons for using harmonic coordinates are not, of course, all philosophical. As John Wheeler pointed out in discussion at the Jerusalem Einstein Symposium, some Western physicists have recently shown interest in harmonic coordinates for physical reasons. Fock himself recognized the physical significance of his approach. He wrote in his *The Theory of Space, Time and Gravitation*, "All concrete problems in gravitational theory discussed in this book are solved in harmonic coordinates. This ensures that the solutions are unique." And, further, he remarked, "Our discussion shows that the conditions ensuring the uniqueness of the coordinate system arise directly from the formulation of the physical problem." For these statements and examples of his use of harmonic coordinates on the equations of motion problem, see V. A. Fock, *The Theory of Space, Time and Gravitation*, trans. N. Kemmer, 2d ed., New York, Macmillan, 1964, pp. 4, 263-267, 365-375.

64. See my brief discussion in *Science and Philosophy in the Soviet Union*, New York, Alfred A. Knopf, 1972, p. 135.

65. Leopold Infeld, *Why I Left Canada: Reflections on Science and Politics*, trans. Helen Infeld, ed. Lewis Pyenson, Montreal and London, McGill-Queen's University Press, 1978, p. 88.

66. V. A. Fock, "Otnositel'nosti teoriia," *Bol'shaia Sovetskaia Entsiklopediia*, 2d ed., vol. 31, 1955, pp. 405-411; and Iu. Kobzarev, "Otnositel'nosti teoriia," *Bol'shaia Sovetskaia Entsiklopediia*, 3d ed., vol. 18, 1974, pp. 623-628.

67. Infeld, *Why I Left Canada*, pp. 75-77.

68. V. A. Fock, "Ob interpretatsii kvantovoi mekhaniki," in P. N. Fedoseev et al., eds., *Filosofskie problemy sovremennogo estestvoznaniia*, Moscow, 1959, p. 235.

69. Yakhot, "The Theory of Relativity and Soviet Philosophy."

70. See the discussion in my *Science and Philosophy in the Soviet Union*, esp. pp. 116-121.

71. Veselov, Kapitsa, and Leontovich, "Pamiati Vladimira Aleksandrovicha Foka," pp. 375-376.

72. For a strong defense by Fock of dialectical materialism as a way of viewing physics, see his "comments," *Slavic Review* **25** (September 1966); see also Loren R. Graham, "Quantum Mechanics and Dialectical Materialism" and "Reply" and Paul K. Feyerabend, "Dialectical Materialism and Quantum Theory," ibid., pp. 381-417.

73. Fock, *The Theory of Space, Time and Gravitation*, p. xviii.

74. Ernst G. Straus reported that Einstein expressed the following opinion of Eddington: "Er ist ein Windhund, aber ein Windhund grossen Formats." Personal communication, 17 March 1979.

Einstein's Impact on Scholarship and Twentieth-Century Culture

Roman Jakobson

EINSTEIN AND THE SCIENCE OF LANGUAGE

In an address broadcast in September 1941 to a meeting of the British Association for the Advancement of Sciences and entitled "The Common Language of Science,"[1] Einstein reminded his listeners that at a most advanced stage of development language, despite all its deficiencies, "becomes an instrument of reasoning in the true sense of the word." One may add that for Einstein himself language, from its rudiments to the various stages of increased development, became, especially during the American, the most retrospective, period of his life, a favorite theme of intense metalinguistic reasoning. The scientist's heightened attention to these questions and his startling gift for thoughtful, eloquent testimonials on the different topics of this area should be confronted with the data conveyed about Einstein's childhood by his biographers.

Thus, for instance, the lines devoted to "little Albert" in Philipp Frank's most instructive volume assert: "Indeed, it was a very long time before he learned to speak, and his parents began to be afraid that he was abnormal. Finally the child did begin to speak, but he was always taciturn. . . ." Even when nine years old and in the last grade of elementary school, "he still lacked fluency of speech, and everything he said was expressed only after thorough consideration and reflection."[2]

A number of biographers have commented on Einstein's incapacity or reluctance to talk until the age of three and on his lifelong difficulties in learning and mastering foreign languages. In addition, Gerald Holton first published the written assertion of Einstein's sister, Maja, that in his childhood the acquisition of speech "proceeded slowly, and spoken language came with such difficulty that those around him were afraid he would never learn to talk."[3]

The prominent mathematician Jacques Hadamard—at the time, dean of sciences in the Ecole Libre des Hautes Etudes, created in New York by French refugees, and a visiting professor at several American universities—pursued his inquiry into the process of mathematical discovery, a research he had begun in Paris and developed during 1943-1944 in connection with his extensive course of lectures at the Ecole Libre. His systematic work on this subject resulted in a book published in 1945. On various occasions he approached me to discuss problems linking this attractive project with the science of verbal and other signs. In accordance with Hadamard's proposal, I sketched, and he inserted into his study, my brief linguistic outlook of those days on the puzzle of wordless deliberations:

> Signs are a necessary support of thought. For socialized thought (stage of communication) and for the thought which is being socialized (stage of formulation), the most usual system of signs is language properly called; but internal thought, especially when creative, willingly uses other systems of signs which are more flexible, less standardized than language and leave more liberty, more dynamism to creative thought. . . . Amongst all these signs or symbols, one must distinguish between conventional signs, borrowed from social convention and, on the other hand, personal signs which, in their turn, can be subdivided into constant signs, belonging to general habits, to the individual pattern of the person considered and into episodical signs, which are established ad hoc and only participate in a single creative act.[4]

At the very moment of sending his book to the printer, Hadamard received, as he states in a footnote, "a letter from Professor Einstein . . . containing information of capital interest." This late "Testimonial" was adjoined to the volume as its second appendix. Both of us subjected the "circumstantial and thorough" answers of Einstein's message to a close examination and confronted his introspection with the aforementioned linguistic summary. The innermost and nearly wordless character of Einstein's creative process was described in his replies to the questions about the kinds of signs that emerge in his mind when absorbed in scientific discoveries: "The words or the language, as they are written or spoken, do not seem to play any role in my mechanism of thought."

The psychologist Max Wertheimer narrates how he used to sit for hours alone with Einstein while the latter disclosed to him "the story of the dramatic development which culminated in the theory of relativity." Einstein affirmed here (decades before his letter to Hadamard!) that his thoughts on this subject did not arise in any verbal formulation: "I very rarely think in words at all. A thought comes, and I may try to express it in words *afterward.*" The belief of certain people that "their thinking is always in words" made him laugh.[5] Evidently, the development of Einstein's thought forestalled the consolidation of his language.

As Einstein testified in the letter appended to Hadamard's book, "certain *signs* and more or less clear *images*" (italics added), the two kinds of "psychical entities which seem to serve as elements in thought," can be—already in this preverbal period—deliberately reiterated and reordered and thus become a personal repertory of significative devices. The question of joint reproduction and recombination indicates that the identification and rearrangement of components, or, in other terms, the complementary ideas of invariance and contextual variability, actually obsessed Einstein with regard to a prelinguistic, individually semiotic stage. For him, as he states in his "Testimonial," it was evident that the "desire to arrive finally at logically connected concepts is the emotional basis of this rather vague play with the above mentioned elements."

Three subjective factors—desire, emotion, and "pure intuition"—underlie Einstein's conception of creative thought as selective, assertive, and combinatory play. His repeated reference to "this rather vague play" is connected with his *profession de foi* launched at the conclusion of the same testimonial: "what you call full consciousness is a limit case which can never be fully accomplished."

It is quite symptomatic for Einstein's mentality and for his acute memory of a child's lingering struggle with the unassailable language that in the replies he offered to Hadamard's and Wertheimer's astute questions, both the laborious search for "conventional words" and their interference with the original "associative play" are relegated—whether by unwillingness or by inability—to an evidently later, "afterward" stage, a "secondary" phase aimed at a "sufficiently established" system of standardized words and regular constructions, namely, words and constructions capable of being reproduced at will and, above all, "communicated to others." Einstein's testimony that "in a stage when words intervene at all, they are," in his case, passive—that is, *"purely auditive"*—fully corresponds to the child's correct perception of environmental speech paired with the still defective production of his own utterances.

Similar evidence appears in "Conversations with Albert Einstein," recorded by the physicist R. S. Shankland: "When I read, I *hear* the words. Writing is difficult, and I communicate this way very badly."[6] It is notable that in Einstein's case, as elucidated by Hadamard, the primordial elements of usual thought, "before the words intervene," seem to be of the visual, as well as of the muscular, apparently gesticulatory, type.

In his "Autobiographical Notes," Einstein draws a distinct line of demarcation between personal thinking and interpersonal *communication.*[7] In the latter process, by means of verbalization and syntactic rules, the conceptual systems become "communicable," whereas the process of *thinking* itself creates what he calls "a free play with concepts," which may even develop for the most part without the use of sensorily cognizable and reproducible signs, and beyond that may develop "to a considerable degree unconsciously." As Einstein had posited a decade earlier, all that is necessary is to fix a set of rules, comparable with the arbitrary rules of a game, whose rigidity alone makes the game possible, whereas "the fixation will never be final."[8]

The relation between "the concepts which arise in our thought and in our linguistic expressions" acquires two disparate treatments in Einstein's writings. In his "Remarks on Bertrand Russell's Theory of Knowledge," he insists on the impossibility of either the conceptual or the verbal "free creations of thought" being inductively gained from sense experiences: "we do not become conscious of the gulf—logically unbridgeable—which separates the world of sensory experiences from the world of concepts and propositions"[9]—briefly, the raw empirics from the theory of science. On the other hand, Einstein repeatedly assailed language for compelling us to work with words importunately tied to inadequate, prescientific concepts and for turning our conventional instrument of reasoning "into a dangerous source of error and deception." For instance, the essential equivalence of two concepts is easily hidden when use is made of disconnecting misnomers.

With regard to Einstein's personal and primary inclination to attribute to the act of thinking complete independence from language, it is evident from his own testimony that emotional yearnings not only are at work in guiding inventive thought in his role as philosopher-scientist but also underlie—in such tragic experiences as the events of the Second World War—his "passionate striving for clear understanding" of mankind and for supernational "general truths." At such moments "the words of the language" suddenly emerge to the foreground. In his broadcast of 1941, quoted above, Einstein concluded: "the mental development of the individual and his way of forming concepts depend to a high degree upon language" and upon "the verbal guidance of his environment." The discoverer still insists, however, that scientific concepts "have been set up by the best brains of all countries and all times," and, *of course*, as he remembers to add, it is done in the *solitude* of the creative process. Yet this time he takes into account also the "*cooperative* effort as regards the final effect," a joint effort that "in the long run" may overcome the contemporary "confusion of goals."

∞

Aside from Einstein's intimate, one might even say innate, conversance with the fundamental questions of the place assigned to language in the human mind, deep spiritual bonds tied the physicist to a remarkable forerunner of modern linguistics, the Swiss scholar Jost Winteler (1846-1929). Winteler's dissertation, issued in 1876, displays a challenging methodological novelty and acuity in his approach to the sound system of languages, with his fundamental distinction between its "accidental features" (variations) and "essential properties" (invariants).[10] But the author's theoretical fundamentals were received among academic bureaucrats with biased distrust. Hence the courageous seeker was doomed to sacrifice his far-sighted scientific plans for the gloomy lot of a lifelong, first active but early retired schoolmaster.

In 1895 the adolescent Albert Einstein, having failed the entrance examination to the Federal Institute of Technology in Zurich, took a year's refuge in the cantonal school at Aarau and there became a student, boarder, and young friend of Jost Winteler, who later also was to be the father-in-law of Albert's sister, Maja. A good deal of evidence shows how auspicious this sojourn turned out to be. Thus, Miss Helen Dukas has kindly supplied a quotation from the short biographical memoir written by Maja Winteler in 1924:

In the family of a teacher of the [Aarau] school and scholar in the linguistic-historical field, [Einstein] found acceptance and sympathy for his manner, and he therefore felt at once very much at home. . . . Thus the time in Aarau became for him in many ways significant, and one of the best of his whole life.

The young student and the older man evidently saw political matters in a similar light. Professor Elmar Holenstein of Ruhr-Universität Bochum refers in his comprehensive paper, "Albert Einsteins Hausvater in Aarau: Der Linguist Jost Winteler,"[11] to an unpublished letter by Einstein to Winteler of 1901, condemning the German "worship of authorities" (*Authoritätendusel*) as "the greatest enemy of truth." A month before his death Einstein continued to praise his Aarau teachers, "who based themselves on no external authority."

The daily conversations with the lucid preceptor probably acquainted the responsive teenager with the essential principle and term of Winteler's dissertation—the "situational relativity" (*Relativität der Verhältnisse*)[12]—and with the indissoluble interconnection of the concepts *relativity* and *invariance*, which underlie Winteler's linguistic theory and which competed for a while as tentative names of Einstein's primary discovery. Particularly instructive among Holenstein's unpublished sources is the letter written 10 April 1942 to the director of the Swiss Landesbibliothek about Jost Winteler, the Aarau teacher, by one of the latter's sons, Dr. Jost Fridolin Winteler, seeking to demonstrate how steadfast remained the memory of the relations between the teacher and his students and the appreciation of the former's clear and perspicacious judgment:

Von ihm habe ich auch erstmalig Ausführungen über Relativität gehört, die dann Einstein, der die Kantonschule in Aarau Besuchte, da die Matura bestand und bei uns wohnte, mathematisch entwickelt hat (1895-96) [It is from him (Jost Fridolin's father) that I heard for the first time statements about relativity, which then were developed mathematically by Einstein (1895-96), who studied at and graduated from the Aarau canton school and stayed as a boarder in our house].

According to Einstein's own acknowledgment, "the germ of the special relativity theory" was already contained in those paradoxical reflections (*Gendankenexperiment*) that first inspired him throughout his Aarau school year and appeared to him "intuitively clear."[13]

Winteler's empathic style strikes the readers of the preface to his doctoral thesis:

My work in its essence is addressed solely to those who are able to grasp the verbal form as that revelation of the human mind which stands to the mind itself in much more inner and sweeping relations than even the best products of a most consumate literature. Thus the addressees of my work must conceive of the inquiry into the latent powers which determine the continual motion of the verbal form as a task which, in its interest and relevance, competes with any other field of knowledge.[14]

A thorough affinity seems to link this passage with the fervent lines of the address that Einstein delivered in 1918 in honor of Max Planck: "The supreme task of the physicist is to arrive at those universal elementary laws from which the cosmos can be built up by pure deduction. There is no logical path to these laws; only intuition, resting on sympathetic understanding of experience, can reach them."[15]

The Aarau disciple preserved forever an exalted memory of Winteler's "clairvoyant mind." Among the numerous precious letters exchanged between 1903 and 1955 by the two eternal friends, Albert Einstein and Michele Besso,[16] one encounters a staggering message sent 16 February 1936 from Princeton to Bern with feverish images, such as "a tormenting mathematical demon," "the desperate state of human affairs," and "die Narren in Deutschland." The dramatic message bears a momentous reference to "Professor Winteler's prophetic spirit, who recognized the imminent danger so early and so thoroughly." The whole letter abruptly concludes with words of hope that the transient autarchy of bare statistical physics will finally be overcome by the universal speculative mind.

∞

We examined Einstein's ideas on language in light of his diverging attitudes toward cognition, on the one hand, and toward communication, on the other. We also touched upon the question of Einstein's boyhood proximity to an outstanding linguistic pathfinder of that epoch. Let us turn to the question of Einsteinian impulses reflected in contemporary linguistic theories, or at least to analogies between modern physics and linguistic trends.

Despite the variety of relativistic ideologies in the diverse provinces of artistic and scholarly activities, the common denominator of their main slogans, devices, and attainments is beyond doubt. I quote a nearly twenty-year-old attempt to delineate the international strivings that have animated our generation:

Those of us who were concerned with language learned to apply the principle of relativity in linguistic operations; we were consistently drawn in this direction by the spectacular development of modern physics and by the pictorial theory and practice of cubism, where everything "is based on relationship" and interaction between parts and wholes, between color and shape, between the representation and the represented. "I do not believe in things," Braque declared, "I believe only in their relationship."[17]

Notwithstanding the somewhat different forms that the notion of "fundamental affinities" takes in the arts and in the sciences, the dominance of the search for relationship over that for the related items themselves knits together the topological nucleus of this century's art and Einsteinian science. Whatever private alienation the revolutionary scientist may feel toward some forms of artistic innovation, one cannot bypass eloquent documents of solidarity, such as the professions of faith of the great modern seekers in art—for example, Piet Mondrian's declaration of 1920, published in *Néo-plasticisme* (Paris): "Les plans colorés, tant par position et dimension que par la valorisation de la couleur n'expriment plastiquement que *des rapports* et non des formes."
When recollecting and rereading the various evidence of the close intertwining among the Moscovite artistic, literary, and scientific avant-garde of the 1910s and 1920s, I realize how great and productive the fascinated acquaintance with the writings of Einstein and his adherents was. Both the Moscow Linguistic Circle, a young experimental association struggling for a revised theory of language and poetry, as well as the later historic ramification of the same trend, the so-called Prague structural school, explicitly referred to Einsteinian methodological endeavors in attempting to link the focal problems of relativity and invariance. One of the examples illustrating the professed relationship is the Project of Standardized Terminology, prepared and published by the Prague Linguistic Circle for the Phonological Conference of 1930.[18] In the list of "fundamental notions," the first place belonged to the *phonological opposition*, and this entry was followed by a reference to the opposites themselves, termed *phonological units*. This architectonic hierarchy became ever closer to "the new physical view" insofar as the structural analysis of the verbal universe superseded the previous mechanical approach. In "Einstein's Theory of Relativity" as conceived by the philosopher Ernst Cassirer, "there exists only the unity of certain *functional relations*, which are differently designated according to the system of references in which we express them."[19]

The appreciation of the relativity of the form of thought attempted by two of the most original American linguists, Edward Sapir (1884-1939) and Benjamin Lee Whorf (1897-1941), and in particular the former's direct reference to "the physical relativity of Einstein,"[20] offer another significant example of a daring linguistic initiative that purposely bordered upon Einstein's conceptual framework and upon the direct, albeit restrictive, question posed in Einstein's broadcast of 1941: "to what extent the same language meant the same mentality." Any impact necessarily implies not only similarities but also instructive cleavages of opinion.

Perhaps the most telling concordances between innovation in physics and that in contemporary linguistics are those coincidences that seem to be due to purely convergent, independent development. Such latent correspondences reveal a substantially parallel course in these different sciences. Both Einstein's demand of the theoretical physicist to strive for the highest possible standard of rigorous precision in the description of pure *relations*, as in his address at Max Planck's sixtieth birthday (1918), quoted above, and the close counterpart of this demand—namely, the ever stricter inquiry into the physical world as a network of interrelated components—stand in eloquent correspondence with the tasks of advanced linguistics. Careful comparisons between the fundamental concepts of relativistic physics and the constituents of language as analyzed and defined by contemporary linguists disclose a salient isomorphism that could be easily exemplified on the different levels of verbal structure.

A few widespread phonological cases may suffice to reveal the generality of the problem. The *distinctive features* that fulfill the main task of speech sounds are, as Einstein would term them, rigorously relational ideas, intuited as binary oppositions. Thus, for instance, in those consonantal systems that make sense-discriminative use of the so-called "flatness" feature, flat consonants are phenomenologically equivalent; in our perception they are mainly distinguished by a particular lowering of their inferior formant. In diverse languages we observe certain differences in the sensimotor modalities of this process. For example, a fairly similar auditory effect is obtainable by labialization and by pharyngealization, or, in other terms, by the narrowing of the frontal or of the buccal ends of the mouth cavity. But since the difference between these two special cases is never used for sense-discriminative purposes, the common denominator outweighs the difference (as well as some other, likewise superficial modalities). The typology of languages asserts the structural invariance of the feature in question, and the universal laws of language prove to admit no more than one single opposition of present and absent flatness.[21] In linguistics the principle of equivalence (instead of mechanical sameness) puts limits on the significance one can expect from a search for separate, uncoordinated particulars of experience and yields instead the gradual discovery of law-governed paucity in the fundamental relationships that underlie the verbal (as well as the physical) universe.

"Nun fiel mir ein" (Now it came to me)—that is the item from Einstein's "Autobiographical Notes"[22] that sounds like a first glimpse of the general theory of relativity and like a joint slogan of the contemporary sciences, all tending to transform an abundance of raw stuff into a parsimony of general laws. The problem of *equivalence* proved to be as pertinent for the principle of relativity as for the discovery of linguistic universals. A substantial revision of the time-space model, notwithstanding the differences in the statement of these questions with respect to diverse sciences, leads us far from the previous mechanical routine. Among such new linguistic vistas calling for a vital interdisciplinary discussion, one may bring to the foreground the notion of dynamic synchrony, the reversible course of current events, and the conception of any change in progress as an intrinsic simultaneity of sensible *oscillations.*

Niels Bohr repeatedly insisted on the deep links that at present tie together physics and linguistics, to whose interrelation both of us devoted a joint MIT seminar at the end of the 1950s. The "exigencies of relativistic invariance," in Bohr's favored term,[23] were intently discussed with respect to the search for and structure of the ultimate constituents of both the physical and the linguistic universe, the "elementary quanta," as they were termed in physics and were picked up from physics by linguists. The endeavor of our linguistic generation to conceive of the verbal mass as a "discontinuous" matter, which is composed of elementary quanta and hence reveals a "granular" structure, partly continues an older set of efforts. At the same time, however, it exhibits an evident dependence on the development of the exact sciences, which was, I testify, a genuine source of inspiration for the linguistic avant-garde of the first third of our century, both in the Western and the Eastern scientific centers.[24]

Let us finally mention that two polar and inseparable problems—namely, symmetry (with its various transformations) and asymmetry, on the one hand, and the breaking of symmetry, on the other—permeate the diverse sciences. In his *Thematic Origins of Scientific Thought* Gerald Holton pointed out the initiative, vital role that symmetry arguments have acquired in Einsteinian physics. Substantially analogous concepts find an ever wider application in the analysis of any linguistic structure. Yet the entire symmetry-asymmetry complex in linguistic research, both in its ontological commitment and in the role of a pure formal device, must be seen to belong more to the victories of tomorrow rather than to the solutions of yesterday and today. However, we may perhaps console ourselves on that point with a thought that Einstein wrote down only four weeks before his death: "For us . . . the distinction between past, present, and future is only an illusion, albeit a stubborn one."[25]

An epochal scientific theory may regenerate in contemporaneous poetry into an elemental myth. Thus, for instance, Vladimir Majakovskij, the Russian avant-garde poet, from his first, anxious glimpses of 1920 into the theory of relativity and until the eve of his suicide in 1930,[26] praised "the futurist brain of Einstein" and devoted his ultimate drama of 1929, *Banja* (Bathhouse), to the crushing victory of such an unusual brain over the alleged absolute of time.

ACKNOWLEDGMENTS

For her critical remarks on several passages of this paper, the author's deep gratitude is due to Dr. Amelia Rechel-Cohn of Harvard University.

NOTES

1. Published in *Advancement of Science* **2**, no. 5 (1941), pp. 109-110.

2. P. Frank, *Einstein: His Life and Times*, trans. George Rosen, ed. and rev. Shuichi Kusaka, New York, Alfred A. Knopf, 1947, pp. 8, 10.

3. G. Holton, *Thematic Origins of Scientific Thought: Kepler to Einstein*, Cambridge, Mass., Harvard University Press, 1973, p. 367.

4. J. Hadamard, *An Essay on the Psychology of Invention in the Mathematical Field*, Princeton, N.J., Princeton University Press, 1945, pp. 96-97.

5. M. Wertheimer, *Productive Thinking*, New York, Harper, 1959, pp. 213-228.

6. In *American Journal of Physics* **31** (1963), p. 50.

7. A. Einstein, "Autobiographical Notes," trans. P. A. Schilpp, in Schilpp, ed., *Albert Einstein: Philosopher-Scientist*, Evanston, Ill., Library of Living Philosophers, 1949.

8. A. Einstein, "Physics and Reality," trans. J. Piccard, *Journal of the Franklin Institute* **221** (1936), pp. 349-382.

9. A. Einstein, "Remarks on Bertrand Russell's Theory of Knowledge," trans. P. A. Schilpp, in Schilpp, ed., *The Philosophy of Bertrand Russell*, Evanston, Ill., Library of Living Philosophers, 1946, pp. 277-291, esp. p. 287.

10. Jost Winteler, *Die Kerenzer Mundart des Kantons Glarus, in ihren Grundzügen dargestellt*, Leipzig and Heidelberg, 1876.

11. In *Schweizer Monatshefte* **59** (March 1979), pp. 221-233.

12. Winteler, *Die Kerenzer Mundart*, p. 27.

13. "Autobiographical Notes," p. 53.

14. Winteler, *Die Kerenzer Mundart*, p. viii.

15. A. Einstein, "Principles of Research," in Einstein, *Ideas and Opinions* (based on *Mein Weltbild*, ed. Carl Seelig), new trans. and rev. Sonja Bargmann, New York, Crown Publishers, Inc., 1954, pp. 224-227.

16. Albert Einstein and Michele Besso, *Correspondance 1903-1955*, ed. P. Speziali, Paris, 1972.

17. These lines were written on the threshold of the 1960s for the conclusion to my *Selected Writings*, The Hague, Mouton, 1962, vol. 1, p. 632.

18. In *Travaux du Cercle Linguistique de Prague* **4** (1931), pp. 309-323, esp. p. 311.

19. E. Cassirer, *Substance and Function & Einstein's Theory of Relativity*, New York, Dover, 1923, p. 398.

20. E. Sapir, *Selected Writings*, Berkeley, University of California Press, 1949, p. 159.

21. Cf. R. Jakobson and L. Waugh, *The Sound Shape of Language*, Bloomington, Ind., Indiana University Press, 1979, pp. 113ff.

22. "Autobiographical Notes," pp. 64-65.

23. N. Bohr, *Atomic Physics and Human Knowledge*, New York, Wiley, 1958, pp. 71-72.

24. A. Einstein and L. Infeld, *The Evolution of Physics*, New York, Simon and Schuster, 1942, pp. 263-313.

25. Einstein and Besso, *Correspondance*, p. 537.

26. See Victor Erlich, *Twentieth-Century Russian Literary Criticism*, New Haven, Yale University Press, 1975, p. 151.

Erik H. Erikson

PSYCHOANALYTIC REFLECTIONS ON EINSTEIN'S CENTENARY

PART I
THE VICTORIOUS CHILD

WHOEVER HAS READ Albert Einstein remembers one or the other statement of his that affirmed his belief in the revelatory order of the Universe. This he was apt to express in such assurances as that God does not play dice ("Er wuerfelt nicht"). But it seems to me that we mortals cannot help playing a kind of card game with the lives of our famous men. We all have before us a by now standard series of biographic data reported by Einstein or by others. We can only put these data on cards, shuffle them, and spread them out before us to see whether we can discern an order fit for the particular game *we* think we know how to play.

In joining in the celebration of what in Einstein's existence may help each of us to elucidate our respective work, I shall first reflect on the *victorious child* in Einstein. To begin with, there is the astonishing, if not sensational, fact that most summaries of Einstein's early childhood refer to some early fears on the part of his parents and others that there might be some defect in this, of all children. Relatives as well as maids and teachers seem to have repeatedly attached nicknames to him denoting retardation, isolation, or inertia. Such "data" are hard to assess, however, since freely expressed and more or less friendly worries can be part of folkways. And then, all these data were recalled and reported when that child had made a singular mark in the world. However, the dominant fact in all these stories was later reported by Einstein thus: "My parents were worried because I started to talk comparatively late, and they consulted the doctor because of it. I cannot tell how old I was at that time, but certainly not younger than three. . . . Also, I never exactly became an orator later."[1] Einstein added that his development was later completely normal, except for the "peculiarity" that up to the age of seven he used to repeat softly his own words (and others' words as well?).

I shall leave to specialists any attempt to diagnose little Albert's symptoms in the light of today's knowledge (on dyslexia, say, or on the two sides of the brain) and discuss them rather in the context of his overall development. The cards before me make me ask: Were the boy's symptoms due to an outright *defect* or to a systematic *difference* in development; or were they also reinforced by a mighty *diffidence*—or, eventually, even some *defiance*?

His younger sister, Maja, in the loving biographical sketch written in her adulthood (which on a visit to the Einstein Archives I was kindly permitted to see), ascribes the pecularity of repeating words as if aftertasting them to his lifelong *Gruendlichkeit*, that is, *thoroughness*.[2] She reports that at two and a half years of age little Albert, on being shown this very newborn sister, asked where her little wheels were—for he had been promised a sister to play with.[3] But here I would sort the cards a little differently. Let me say at this point that little Albert's life must always be seen within the context of the provincial Swabian-Jewish folkways in a semirural setting. This background gave his family some lasting characteristics, even though they moved to Munich when he was one year old. Now, as to the little wheels, the German word for it is *Raedchen*, or, in Swabian, *Raedle*, which rhymes with that for little girl (*Maedchen*, or *Maedele*). Is it not possible that a play with rhymes began early in this thoughtful child, even as it continued throughout his life as a humorous need at the oddest moments "ein Gedichtle zu mache"—to make a little poem or "ditty"? More, one might consider a special preoccupation with "the way things rhyme" to be an important trend throughout Albert's development.

Einstein's biographer Philipp Frank agrees with the explanation of thoroughness but adds that "when Albert was nine years old . . . he still lacked fluency of speech. . . . he was called *Biedermeier* (Honest John) by his classmates."[4] In many situations the child would rather not speak or communicate at all, although he seemed to be quite actively observing and, above all, playing. Einstein's own interpretation, however, offers an additional perspective:

As a pupil I was neither particularly good nor bad. My principle weakness was a *poor memory* [italics mine] and especially a poor memory for words and texts. . . . Only in mathematics and physics was I, *through self-study* [words we must underline], far beyond the school curriculum, and also with regard to philosophy so far as it had to do with the school curriculum.[5]

As to his family, Jewish families like the Einsteins (and they had been registered in this general area since the 1750s) could feel at home as a traditional Jewish minority among regional Germans and, above all, as an educated elite, devoted to German literature and music with an almost religious fervor. To be Swabian, furthermore, meant to take *Gemuetlichkeit*, a certain comfortable and comforting mood in daily living, very seriously indeed. Albert's family was sophisticated, with a general freethinking orientation that, apparently, could permit their little son to behave differently from other children. At any rate, though his early pictures show him to be serious, he also seems calm—and handsome.

His parents did not hesitate to send Albert to a Catholic public school where he was the only Jewish pupil and, of course, heard a lot about God—a personage that remained, as my opening quotation exemplified, a constant source of humorously philosophical recourse. ("The dear Lord may be tricky [*raffiniert*]," he once said, "but he is not mean [*boshaft*].") But the boy had private instruction in the Jewish religion, the dietary rules of which he insisted on obeying meticulously for a while. And, indeed, the atmosphere of his home was so freethinking as to permit him to take religion seriously. In later life he did not hesitate to use Jesus' name in the same breath, as it were, with those of the Jewish prophets,[6] while he, typically, objected to the *Buchstabenkultur* in which, so he claimed, "the pure doctrine is obscured by much worship of the letter."[7]

∞

Obviously, then, little Albert had it in him to object to having to learn anything in any but his own way. In his early childhood this could be expressed in a sudden rage (against a private teacher, for example) that was attributed to a disposition inherited from his maternal grandfather. Later, the resistance against enforced instruction, far from ever being "broken," became a deep and basic character trait that permitted the child and the youth to remain free in learning, no matter how slowly or by what sensory or cognitive steps he accomplished it. I see a connection here with what he later emphasized as *Begreiflichkeit* (comprehensibility), that is, an active and intuitive "beholding" as a necessary step in thinking. Could it be that the need to wait for such moments—that is, a delayed "I see!" phenomenon—had not permitted him to accept too early and too glibly the ways language has of prescribing meanings not really "grasped"? And remember, one of his later most childlike and yet wisest sayings is that the most incomprehensible aspect of the world is its comprehensibility.[8]

His childhood religiosity, so Einstein reports, came to an abrupt end at the age of twelve. It seems, however, to have been fused with a certain readiness for revelatory experience, which in its earliest form he called *Wunder* (awe, wonder) when he spoke of the experience when his father showed him a compass:

> A wonder of such nature I experienced as a child of 4 or 5 years, when my father showed me a compass. That this needle behaved in such a determined way did not at all fit into the nature of events, which could find a place in the unconscious world of concepts (effect connected with direct "touch"). I can still remember—or at least believe I can remember—that this experience made a deep and lasting impression upon me. Something deeply hidden had to be behind things.[9]

He then returns to a later wonder, namely, at that critical period of a new religious direction at the age of twelve, his acquaintance with the "heilige Geometrie-Buechlein"—the holy geometry booklet of Euclid.

Here, however, it is not possible *not* to ask in passing: What, in looking so early and with such veneration so far away, did he look away *from*? For later, in adulthood, he could make the statement, *"I sold myself body and soul to Science—the flight from the I and WE to the IT"*[10]—a statement to which we certainly must return. In the meantime, a medical student who from young Albert's tenth to his fifteenth year witnessed and "fed" his apparent "exceptional intelligence" also testifies that "in all those years I never saw him in the company of other boys of his age."[11] Maja, in turn, reports how he loved solitary *games* of patient persistence, such as jigsaw puzzles by which he trained his capacity to fit forms to one another, complicated block buildings, and card houses of unbelievable height. And here Maja thought to detect the influence of his mother's tireless needlework, even as later his mother inspired and cultivated his lifelong devotion to musical activity—here, too, letting him find his own way to play Bach and Mozart and to "think" musically on the piano.

With his fiddle, he certainly became one of the truest amateurs among the greatest minds.

The importance of playfulness, of course, later reached way into his most intimate thinking. Inspired by Gerald Holton, I have had a chance to quote Einstein extensively in a book called *Toys and Reasons.* I mention the title here not by way of advertisement but in order to point to its origin, William Blake's saying that "The Child's Toys and the Old Man's Reasons Are the Fruits of the Two Seasons." For, as Holton reports, toward the end of his life, Einstein mused "that he was brought to the formulation of relativity theory in good part because he kept asking himself questions concerning space and time that only children wonder about."[12] And in 1944 Einstein declared that from a psychological viewpoint, *combinatory play* of a visual type, which he also referred to as "associative play" (a point to remember), "seems to be the essential feature in productive thought—before there is any connection with logical construction in words or other kinds of signs which can be communicated to others. The above-mentioned elements are, in my case, of visual and some of muscular type."[13]

Thus, the cards tell me that Einstein succeeded in saving the child in himself even when, increasingly, he had to accept, with a kind of nonviolent resistance, isolation and even punishment rather than submit to standardized formulations. At any rate, in the Gymnasium, the German type of classical high school, compulsory learning seemed to him to be more and more of one style with the then increasingly compulsory Prussian kind of military training, which was resented in the south German states that had become part of the German Reich. And over the years his resistance to the "system" stiffened even as teachers' tolerance of such a pointedly "reserved" student began to wane. And so, when in his fifteenth year his family moved to Milan, leaving him behind in Munich so he might properly graduate (which also meant eventually to do his one-year stint in soldierdom), he became deeply depressed and even took refuge in a doctor's certificate that he should be permitted to leave—but not before his homeroom teacher had asked him to leave the school because his "mere presence spoiled the respect of the class." And, Einstein added, "Because of my poor memory for words, this ["the dull, mechanized method of teaching"] presented me with great difficulties *that it seemed senseless for me to overcome*" (italics mine).[14] Was it senseless to try because it was impossible, or because, to his by now decided diffidence, it was simply not worth it? At any rate, in order to complete his liberation he asked his father to have him disengaged from formal Jewish affiliations as well as from German citizenship.

What followed was as liberating a *Wanderschaft* as any of us erstwhile *Gymnasiasten* have experienced—a *Wanderschaft* in which the southern landscape ("Kennst Du das Land . . ."), the Italian people, and their language seemed to offer the liberated northerner exactly that sensual humanity that, according to our principal literary prophet, Goethe, fate owed to us. And if this move permitted young Albert to choose the "neutral" as well as the ethnically multiple Swiss identity, this identity seemed truly to combine promisingly the elements of the *gemuetliche* Swabian, the freethinking Jew, and the budding cosmic scientist.

This much about Einstein's childhood. I think one can celebrate today all at once a childhood leading to unique creativity, a man's singular understanding of what the link of childhood and maturity can mean, *and* his time's public and almost mystical appreciation of the child in him, no matter how much or how little people understood of his work. This phenomenon seems to happen once in a while in history. Newton said, near the end of his life, "I do not know what I may appear to the world; but to myself I seem to have been only like a boy playing on the sea-shore, and diverting myself in now and then finding a smoother pebble or a prettier shell than ordinary, whilst the great ocean of truth lay all undiscovered before me."[15]

How young Einstein, perhaps not without his passive collusion (for he did not want to go into engineering as his father wished him to do), failed to pass the entrance examination into the Polytechnic Institute in Zurich and found one most significant year of final schooling in a provincial Swiss institution with a Pestalozzi tradition—all this is well known. And there it suddenly occurred to him to ask what light waves might look like to an observer who was keeping pace with them. On his return to the Polytechnic Institute, young Einstein was ready, as it were, to choose his own identity—as scientist, universal citizen, and Jew. How fate, as he put it, "to punish me for my contempt of authority . . . made me an authority myself"[16]—to reflect on that, I will leave to others.

But I must return once more to that small boy who could not—or would not?—speak when most children do. We have a right and the duty to draw certain conclusions from the data of this extraordinary life. We know only too well that the child described here today would be subjected to specialized examination and, perhaps, to treatment. My emphasis here is on "specialized," since it seems that Albert's speech development, when noted down only on its own terms, could indeed be suspected to be symptomatic of a developmental defect. And I certainly believe that similar cases demand and merit cautious attention. But the goal of such observation, as I have tried to indicate in my manner of reporting, should be the overall assessment of a child's range of traits and interests and of his capacity to integrate that range in a way that permits some playfulness and peacefulness at whatever pace of progress. All too standardized demands on all children for the sake of stereotyped curricula—curricula that often fail to be convincing in their relation to one another—may be apt (so this life history warns us) to endanger such talents as will be all the more needed in an increasingly mechanized world.

But it behooves a psychoanalyst also to return, if ever so briefly, to what Einstein himself describes and a variety of observers confirm as the price of his creativity, namely, a certain sense of isolation. Frank unhesitatingly describes Einstein as a "lonely person among students, colleagues, friends, family," and he comes to the sweeping conclusion that "neither professional activity nor a family had a great significance" for him.[17] If Frank intended to include marital intimacy, then Einstein in some of his most touching letters (such as his final word to the son of his friend Besso after the latter's death) seems to have sadly agreed[18]—even as he went all out in suggesting that his biographer should not overlook "the irrational, the inconsistent, the droll, even the insane, which nature, inexhaustively operative, implants in an individual, seemingly for her own amusement."[19]

Yet, I must say that when one reads some of his letters and sees his reflection in some of the surviving members of his inner circle, one cannot doubt that this man knew, and knew how to give, some intense sense of intimacy. A certain alternation of isolation and outgoingness (especially toward children) seems to have retained here, too, the character of a dynamic polarization. And one must not forget, in the life of such a one who so mysteriously (and to himself mystifyingly) came to mean so much to so many, to account for the role of his public—and of mankind itself—in the totality of his human relatedness. (Here one recalls the way in which this man learned to look into cameras as if he were meeting the eyes of the future beholders of his image.)

When Einstein, then, made the statement of his turn from "I" and "We" to "It," he testified only to the vividly existential sense with which he experienced a certain polarity between work and intimacy. This, as a certain imbalance, has been described in a number of studies as typical to scientists. But consider what he made of the IT: no less than a rendezvous in the Here and Now between the universal Logos and man's scientific insight.

∞

If I should now indicate the way in which I think I could fit these scanty data cards into our scheme of the early stages of psychosocial development, I would begin with little Albert's open-eyed wonder as necessarily rooted in an early *basic trust* such as cannot emerge without the experience of that mutual "recognition" of mother and child, or what Joan Erikson has described as the eye-to-eye relationship. This, it seems, is later transferred to the world at large, so that Einstein could eventually say: "The mental grasp of this extra-personal world within the frame of the given possibilities swam as highest aim half consciously and half unconsciously before my mind's eye.[20] This statement, in turn, is reminiscent of Meister Eckhart's more mystical saying: "The eye with which we see God is the eye with which God sees us."

But we also postulate for early childhood a dystonic *basic mistrust* without which basic trust could not be syntonic, because it would remain unguided and unselective. And we may, in fact, recognize such a diffident and even defiant selectivity in Albert's early resistance to a verbal communication that he may indeed have found unreliable because of some weakness of verbal memory. Be that as it may, my admittedly scanty references to Einstein's parents can here be summarized in the impression that they did maintain throughout a trustful attitude, fostering in the boy an extraordinary self-trust that found renewal at every stage.

Our second childhood stage calls for the development of the rudiments of *will power* in the emergence of a sense of *autonomy*, which must be balanced by, and yet must maintain itself against too much impairment by, *shame* and *doubt*. We saw how nicknamed Albert turned his doubt against those who, he felt, behaved unduly authoritatively and at the end (so I would surmise) could shame all-too-sure experts by his laughter. Here Frank's description is telling:

Einstein's conversation was often a combination of inoffensive jokes and penetrating ridicule, so that some people could not decide whether to laugh or to feel hurt. Often the joke was that he presented complicated relationships as they might appear to an intelligent child. Such an attitude often appeared to be an incisive criticism and sometimes even created an impression of cynicism. Thus the impression Einstein made on his environment vacillated between the two poles of childish cheerfulness and cynicism. Between these two poles lay the impression of a very entertaining and vital person whose company left one feeling richer for the experience.[21]

Or again:

To those about him his laughter was a source of joy and added to their vitality. Yet sometimes one felt that it contained an element of criticism, which was unpleasant for some. Persons who occupied an important social position frequently had no desire to belong to a world whose ridiculousness in comparison to the greater problems of nature was reflected in this laughter. But people of lesser rank were always pleased by Einstein's personality.[22]

As to an early sense of *initiative*, Einstein always claimed that he owed his insights to no other special drive or talent than *curiosity*—an outgoing attitude that must certainly combine trust, autonomy, and initiative. But we also ascribe to the stage of initiative the rudiments of a sense of guilt, which, no doubt, can in many a curious child be aggravated by the awareness of some sinful curiosity. I shall not, and I could not, specify the fate of such early conflicts in Einstein's childhood. I can only conclude that when he made the statement of his turn from "I" and "We" to "It," he was aware of some of the interpersonal conflicts that he thus learned to avoid and yet also to sublimate in his concentration on the phenomenal.

The early conflict of infantile initiative versus a sense of guilt is settled in the play age; and as we noted, the playfulness of Albert's gifted senses continued to provide him with a grasp of the lawfulness of things, from melodies to the rhyming of language and eventually to the mathematical nature of factuality. This he could carry over, as he told us, into the school age, saving a sense of *competency* against all the evidence of some *inferiority* in matters of classroom learning. And so, to summarize what already is much too summary, young Albert became ready in adolescence to actualize his gifts of superior play in the setting chosen by him—a setting where he had learned to find confirmation in the actuality of mutual enlightenment as well as in the mathematical proof of his way of thinking.

As to the mysterious universal reaction to the child in Einstein, we may mention in passing that the polarity of being an eternal child and yet also a wise old man at the same time is in fact a traditional configuration, if of a rare and august type: there are gods, saints (St. Francis), and true savants like that. There are Jesus' astounding references to the eternal promise of becoming like children. Laotse's very name seems to refer to the fact that he was born with a little white beard. And in Einstein's home culture, Goethe once declared himself a *Weltkind* (a world-child) sitting among the prophets. Such pointers are not lost on us: we live in the century of the child as well as in that of Einstein—and, we may now add, of Freud, who taught us the meaning of all the fateful pathology traceable to man's prolonged childhood, and yet also spoke mournfully, for it is lost so often, of the "strahlende Intelligenz des Kindes"—the radiant intelligence of the child.

PART II
ON THE "I"

Mankind, then, seems to greet decisive changes in the contemporary world image with a strange mixture of deep shudder and yet universal veneration of the childlike courage of the savant who convinces his time that man can afford to abandon childish illusions and accept even shocking insight as a revelatory gift from a trusting (and trustworthy) universe. This suggests that we turn from one unique child's and future world-image builder's early curiosity to the investment that *all* human beings (and thus all children) may have in the traditional or newly integrated world image of their times. In my second reflection, then, I shall attempt to relate mankind's esthetic and moral "gut" reactions in response to scientific formulas to some basic needs for orientation as they arise step for step in every human being's development.

World Perspectives

In 1917, in an essay called "A Difficulty in the Path of Psychoanalysis,"[1] Freud proposed "to describe how the universal narcissism of men, their self-love, has up to the present suffered three severe blows from the researches of science."[2] The first blow, he contended, had come from the Copernican world view, for in the early stages of his researches, man believed at first that his dwelling-place, the earth, was the stationary centre of the universe, with the sun, moon and planets circling round it. In this he was naively following the dictates of his sense-perceptions, for he felt no movement of the earth, and wherever he had an unimpeded view he found himself in the centre of a circle that enclosed the external world.[3]

This, Freud claimed, fit well into man's wish to feel like the lord of the world; and thus, the general acceptance of Copernicus's work was the first, the *cosmological* blow to his all-human narcissism.

Darwin caused the second, the *biological* blow, for man had begun to deny any reason to his fellow creatures, the animals, while "to himself he attributed an immortal soul, and made claims to a divine descent which permitted him to break the bond of community between him and the animal kingdom."[4] And finally, Freud claims for himself the initiation of the third, the *psychological* blow, which "is probably the most wounding," for "these two discoveries—that the life of our sexual instincts cannot be wholly tamed, and that mental processes are in themselves unconscious . . . amount to a statement that *the ego is not master in its own house.*"[5]

In this "up-to-the-present" statement, Freud does not mention Einstein or relativity. In order to look at this whole matter developmentally, that is, with reference to the slow, complex, and conflictuous process of every human being's "growing up," I must briefly discuss some ways in which Einstein and Freud use the German word *Ich*—that is, I—and also, how differently their respective uses of the word are translated into English. Freud's occasional use of *Ich* as the seat of consciousness, which "gives mental processes an order in time and submits them to 'reality testing,' "[6] corresponds to Einstein's use of *it,* as when the latter says: "The experiences of an individual appear to us arranged in a series of events; in this series the single events which we remember appear to be ordered according to the criterion of 'earlier' and 'later.' There exists, therefore, for the individual, an I-time, or subjective time. This in itself is not measurable."[7] In translation, however, Freud's references to the conscious *Ich* is apt to get lost in its habitual translation as "ego," which is the major psychoanalytic term for a central and powerful, but by no means always conscious, synthesizing "inner agency" in the person.

My reflection, however, must begin with an emphasis on this conscious "I," the most obvious and yet most incomprehensible fact of our existence—for the moment the "I" tries to comprehend itself, it becomes a self. As it emerges in the course of a long childhood, the "I" acquires what (doing Einstein one better) we may call an I-Space-Time, which grows slowly and, at the end, must confirm many prerational forms of certainty as it also secures against distinct forms of early dread. Every *Homo erectus* (look at the riddle Oedipus was confronted with) has emerged from a variety of postural space-times and, at the end, needs the double reassurance of some absolute standpoints, even as *Homo ludens,* in gradually renouncing much of infantile fantasy, must secure for himself a sense of continued leeway, of free play. *Homo faber* has learned to learn the most intricate skills in the construction of a technological universe but forever needs ideologies to tell him for whom and for what he produces. And *Homo sapiens* must, at all cost, have unified theories. All this, then, can be seen to contribute to the prerational investment that man has in existing world images, and therefore to the shudder that accompanies revolutionary changes such as those effected in our time by Einstein as well as by Freud.

What we have learned to discern in psychoanalysis are the *developmental risks* built into the ontogeny of the human person just because—as we saw exemplified a moment ago in the great life celebrated today—a firm sense of "I" is acquired only gradually during the stages of a long childhood. The postural perspectives mentioned lead from a kind of prone embeddedness to the stepwise conquest of an upright position and a goal-directed locomotion—each stage associated with a particular state of sensory, cognitive, and verbal development as well as an expanding radius of human relationships. Only together do all these perspectives determine, at a given stage, the center as well as the horizon of the total experience.

A psychoanalyst who is usually preoccupied with the unconscious aspects of the drives and defenses that endanger and secure our mental coherence is almost embarrassed to name dimensions of the human "I" that are as conscious and as obvious as anything can be, dimensions that in fact escape us only by their very omnipresence. Central, indeed, is the criterion of *centrality*: for each of us is not only in the center of the universe but also, in fact, central to the universe as he or she experiences it. But as we are together on a special occasion such as this, we are also aware of the importance of the "We" as a communality of "I's" who share a belief in some perspectives that can be confirmed in daily ritualizations and jointly confessed to in moments of ritual celebration—even if such communality should reveal to us the relativity of the space-time centeredness of each of us and yet, once we comprehend it together, some shared existence. For could one not say that the "I," in comprehending the relativity of its selves, confirms the invariance of its continuity, even as the "We," in granting the relativities of its "I's," realizes the invariance of its communality?

But all such confirmations also underscore the specific dread that lurks behind each quality of "I"ness: our very need for a sense of being central to our own experience is always at risk and can be lost in a sense of being shunted to the periphery, forgotten, abandoned, forsaken. I use the word "dread" here rather than "anxiety" because the clinical approach has induced us to interpret all adult upsets as untoward "regressions" to infantile anxieties. It is one of the very functions of human childhood that certain forms of dread must be experienced and counteracted by qualities of the "I" that fit, as we grow into it and as we recreate it, the world image of our culture and technology. So we understand ontogenetically, too, why mankind underwent a certain crisis when the cosmic elaboration of centrality had to be abandoned.

The qualities of "I"ness are all interrelated. Let me mention a few more. We need to maintain a sense of selective *activity* against a sense of paralyzing inactivation, of *wholeness* as against fragmentation, and of *awareness* as against confoundedness. You will recognize the probability that all these must be taken care of in a given world image.

But now let me go back, way back, to the very beginnings of ontogeny, when the "I" emerges from the early symbiosis with the mother and realizes its differentiation from what is then duly recognized as the maternal person—in my term, the *primal other.* I admit that many aspects of these earliest infantile experiences can only be postulated on the basis of what we clinicians have learned from the primary defects and deficits resulting when something goes wrong with them. On the other hand, the human being never "overcomes" some of its earliest critical experiences: no, they demand to be reexperienced and "confirmed" on all later and higher levels. Thus, if in later life a God image or a universal atman becomes the *ultimate other,* this can give a certain eternal unity to the beginning and the end of existence. And there seem to be at least some configurational affinities between man's original experience of feeling embedded in the maternal presence and such all-surrounding matrices as mass or matter—not to speak of Oliver Lodge's "full-bodied ether," the belief in the ubiquity of which is said to have been abandoned only with much "dismay."

In pointing this out I hope that I will not be suspected of suggesting that any such persistent and pervasive image as a maternal matrix is the "cause" of any viable scientific concept that, at its historical moment, is suggested by facts and serves science as a whole in replacing some earlier and less "probable" concepts. Rather, what I am suggesting here are certain contextual meanings adhering to various discoveries at a given time. That early ontogenetic stages are in the air, as it were, at all times we may see, for example, from the myth created by Copernicus after he had dethroned the earth and man from the center of the universe. In his *De revolutionibus orbium coelestium* he developed the vision of an elemental interplay among the now dethroned earth, the now enthroned sun, and the moon by which the earth once a year was impregnated with new life. Thus, he suggested a new unity in the form of a kind of cosmic family life.

Such mutuality is reflected in the Platonic theory of vision, which, so Holton reminds us, postulated an inner light emitted by the eye. To effect perception, however, "there must be outside the eye a 'related other light,' . . . that allows rays to come from the objects. Once more, a coupling between the outer and inner world is clearly attempted."[8] After Kepler's *Dioptrics*, however, gone is "the 'recognition' of soul by soul in Neoplatonist discussion of optics."[9] And Holton concludes:

> The decoupling between *lux* and *lumen*, between subject and object, observer and the observed, and with it the destruction of the earlier, holistic physics, was a painful and lengthy process. The reason why it was ultimately victorious is the reason why the same process in all other parts of science worked: once the separation was made, there ensued a dazzling enrichment of our intellectual and material world.[10]

This seems to be accompanied also by a new and larger unity, a reconciliation in unification. As Einstein put it:

> The nonmathematician is seized by a mysterious shuddering when he hears of "four dimensional" things, by a feeling not unlike that awakened by thoughts of the occult. And yet there is no more commonplace statement than that the world in which we live is a four-dimensional space-time continuum.[11]

Returning to Freud's remarks on Darwin, we can now see that he did more than offend our self-love by postulating our descent from the animal world. Darwin obviously put in question also the "I" 's "natural" belief in existing within a process permitting some self-determination, even if borrowed from a divine intention of making man a creature of fateful choices.

Now, with regard to all the surprises and shudders attending changing world views, Freud was the first to work with the matter-of-fact conviction that his studies, to prove their own worth, *must* lead to major resistances, the "working through" of which was in fact part of his historical job. Freud suggested that we subject ourselves and one another to a process of "analysis" (which certainly implies some temporary sacrifice of *wholeness*) and predicted that such analysis would expose the fact that we by no means possess all the strategic sense of *awareness* by which we are trying to remain masters in our choice of means and ends. On the other hand, he also offered coherent insights that would restore a sense of *ego-synthesis*, permitting some transcendence of the most delusional of our passions and some reconciliation of our intentions and of our capacities. About this, more later.

The greatest innovators of world images eventually are venerated. It is almost as if the discoverer of new laws confirming some verifiable intentions within creation thereby come closer to the creator himself, as Einstein thoughtfully and humorously mused on many occasions. However, the resulting enrichment of the material world has by now led to the invention of such gigantic powers of destruction that one may well wonder where the Devil had his hand in scientific progress. And as to intellectual enrichment, we know how new concepts can also be misused as pseudo-ethical positions—for example, the "survival of the fittest," the ubiquity of the libido, or, for that matter, relativity. But there we clinicians, preoccupied as we usually are with human conflict, crisis, and failure, can put in a good word for the human race, for we can observe in our daily work that enlightenment does heal. And we can witness to a readiness also deeply rooted in the nature of human childhood to sacrifice childish reassurances for the sake of more mature and lasting insights, if these indeed are conveyed convincingly at the right moment in the right way, that is, somehow all at once logically, esthetically, and ethically.

Relativity in the Psychoanalytic Method

In my final reflection I want to claim that even though Freud's so-called metapsychological theories were steeped in the quantitative and mechanistic terms of the science of his century, his *daily method of observation*, and thus his central deed as a great healer, is really "unthinkable" without some recourse to—to put it cautiously—a concept of relativity. Let me assure you, however, that I have not invented this trend of thought for this festive occasion: the word "relativity" occurs in the title of the first chapter of my first book, when I knew even less what I was talking about.[12]

But first a word on the two masters. What I know of Freud's and Einstein's correspondence gives me the impression that it is dominated by a certain ambivalent discomfort over the fact that together they dominated so much of the Zeitgeist, the spirit of the times, and yet really did not know what to do with each other. But there is also a trend in their correspondence that would indicate that to each the other's theory may have contained a specific risk to his own theory. Freud's theory, of course, represents a gigantic effort at a true and up-to-date science of the mind. Yet in his collected works relativity is mentioned only twice, and both times with a warning against *relativism*, because, as he says, putting the blame on Einstein's "believers": "Upon many of those who admire without comprehending" Einstein's theory may have the effect of "diminishing their belief in the objective trustworthiness of science."[13] And then there is a deeply ironic reference to those who seem to relativize all-important contradictions. Here, Freud mockingly makes his own opponents say: "After all, who can possibly know who is right? In spite of our antagonism, pray allow us to represent your point of view in our publications. We hope that you will be kind enough in exchange to find a place for our views which you deny."[14] And he muses, "In the future, when the misuse of Einstein's relativity has been entirely achieved, this will obviously become the regular custom in scientific affairs."[15]

Einstein, in turn, emphasized, as we saw, the role of "associative play" in scientific thinking—if, alas, as only one step in the conceptual process. Yet (or is it therefore?) what Freud called "free association" seemed dangerous to him. In writing to Freud he, too, refers to the other man's followers, who, he says, "use his terms without understanding them." But he goes further. He writes that even Freud's unbelievers (*Unglaeubige*) cannot resist Freud's ideas, so that they become "used to thinking and talking in your concepts when they let themselves go [*sich gehen lassen*]."[16] I was kindly permitted to see some of these letters in their handwritten original, and it so happens that this one is marked by what is popularly called a Freudian slip—a *Fehlleistung*. Einstein intended to refer to those nonbelievers as *sie*, that is, "they"; but he inadvertently capitalized *Sie*, so that it comes to mean "you." The text thus accuses Freud himself of "letting himself go" in the use of his concepts. I think that Einstein would have been amused if shown this bit of irrepressible honesty, for he seems indeed to have seen as Freud's main fault what in a letter (to A. Bacharach) he called the old man's "exaggerated trust in his own associations [*Einfaelle*]."[17]

Freud, of course, was Einstein's senior by a quarter of a century. And Einstein was always frank in acknowledging his inability to follow Freud's theories, while he sincerely admired his grand style, which to Freud was not more than a compensatory compliment. When they met for the first time in 1927 in the house of Freud's son Ernst in Berlin, Freud reported to Ferenczi, "He [Einstein] is cheerful, sure of himself and agreeable. He understands as much about psychology as I do about physics, so we had a very pleasant talk."[18] It was on his eightieth birthday that Freud received from Einstein the most affirmative letter yet, referring to a few instances ("not very important in themselves") that "in my judgment exclude any other interpretation than that provided by the theory of repression." And, he added, "it is always delightful when a great and beautiful conception proves to be consonant with reality."[19] To this, Freud responded, "You are so much younger than I am that I may hope to count you among my 'followers' by the time you reach my age. Since I shall not know of it then I am anticipating now the gratification of it."[20]

With regard to their dialogue *Why War?*, it is interesting to note that it was Einstein's suggestion that Freud "bring the light of [his] far-reaching knowledge of man's instinctual life to bear upon the problem . . . for such a presentation might well blaze the trail for new and fruitful modes of action."[21] Ernest Jones reports that Freud wrote to his friend Max Eitingon "that he was conducting a discussion with Einstein for a League of Nations publication. He added drily that he did not expect to get a Nobel Peace Prize for it. Three weeks later he reported that he had finished writing 'the tedious and sterile so-called discussion with Einstein.' "[22]

Yet Freud obediently, if unenthusiastically, enlarged on his theory of love and death instincts, and then apologized for what he called his mythology, only to challenge Einstein concerning *his* scientific mythology: "All this may give you the impression that our theories amount to a species of mythology and a gloomy one at that! But does not every natural science lead ultimately to this—a sort of mythology? Is it otherwise to-day with your physical science?"[23] But then Freud ended with a most serious formula for an *indirect* method of eliminating war and nominated *love* ("all that produces ties of sentiment between man and man") and *identification* as the mainstays of it: "All that brings out the significant resemblances between men calls into play this feeling of community, identification, whereon is founded, in large measure, the whole edifice of human society.[24]

Now, I know as well as anybody the intellectual games and modish practices that psychoanalysis can be made to serve. Yet it seems to me that in its essence this method works with a kind of *caritas* and depends for its insights on the power of *identification*. In fact, from the point of view of an ethos of healing, I would consider Freud's most revolutionary contribution to be the insight that the healer cannot understand in his patient what he has not learned to understand as potential in himself. Whatever its association with established professions, this commitment offers an enlightening experience that Freud treasured far beyond its therapeutic implications, for it is based on a fellowship in existence, deepened by the awareness that we all share in mankind's conflictuous unconscious.

My point is that the "classical" psychoanalytic procedure, which is at the same time a cure, a method of observation, and an instrument of training, provides a prime instruction in human relativity. Let me sketch here some of its main trends,[25] well known as they are to most of you. To begin concretely, in prescribing that the *psychoanalyst* sit in a comfortable chair at the head end of a couch on which he has instructed the *analysand* to recline and to say what comes to mind, Freud seems to have created one of the most static situational contrivances ever invented, for it specifically excludes any eye-to-eye contact and all conversational give-and-take. And yet, this situation facilitates in the analysand a "free floating" of "associations," which move about with varying speed through the distant past, the immediate present, and the feared or wished-for future and concurrently through the spheres of concrete experience, of thought and fantasy, and of dream life. At the same time, the *analyst*, who during his training has himself undergone a full psychoanalysis, listens with a perpetual but (at its best) disciplined and unobtrusive awareness of his own associative responses to his analysand's verbalizations.

If I now compare the movements of the analyst's and the analysand's thoughts with the relative motions of two coordinate systems, I know only too well that this is, at best, a suggestive parallel, for we are speaking not of two model railroad cars but of aware, self-observing as well as observing, organisms capable of verbal communication and engaged in a therapeutic encounter. How surprisingly this method works could, I admit, be illustrated only by a detailed example from a case history, which, however, would immediately call for comparison with other examples. Instead, I must attempt to abstract a few essential aspects of the process at work.

The analysand suffers more or less explicitly from an arrest in his present development. Before long, however, the very interplay of the struggling present with the unlived past and its repressed experiences will make his associations revive, if often in symbolically disguised form, conflicts intrinsic to a range of previous stages and states of development. Often, this does not become clear before the associations reveal that the analysand has begun to project some important themes and affects of the past on the analyst's person and life: for example, by reacting and responding to the psychoanalyst alternately as if he were, say, an oversolicitous or punishing parent of the same or the other sex, or some other once strategic person loved ambivalently. This we call *transference*, a form of love attachment indispensable for therapy and yet (most instructively) distorted by the very conflicts that caused the disturbances under observation.

In the meantime, the analyst's reflections on the analysand's associations are searchingly moving in a wide range of professional experiences and interpretive possibilities as suggested by the theoretical state of the field. Also, he becomes aware that here or there the patient's present state and past conflicts evoke feelings and images from the corresponding stages of his own past that he, in turn, is apt to project on the analysand. Freud called this the analyst's *counter-transference*, and the analyst must become aware of it in order to avoid any possible collusion of his own irrational affects and resistances with those of the patient. I would like to think that within the limits of our analogy transference and counter-transference somehow parallel what in the theory of relativity may be called *transformation* from one system to another system at the same stage or to itself at another stage. At any rate, such transferences elucidate what we may call the *developmental and historical relativity* of human experience.

I hear you asking, how can anybody be fully cognizant of all these details? And I agree: he cannot and, by God, he should not. But I submit that only on the basis of a trained awareness and acceptance of the developmental and historical relativity governing the processes he deals with can the psychoanalytic observer's insight be relaxed enough to let the most acutely important items lift themselves out of the mass of "material"—until they join in what Einstein described as comprehensibility. Only this, in turn, permits the psychoanalyst to offer from time to time an interpretation of the analysand's associations that fits the therapeutic moment. That is, at its best the interpretation will be unique in tone and wording for this moment in the encounter of this pair of analysand and psychoanalyst; and yet (or for that very reason) it will convincingly explain some invariant lawfulness of psychic functioning. Thus, gradually—very gradually—systematic insight into unconscious processes will effect a comprehension of the relativity of developmental stages and historical states and thereby confirm the invariance of the "I" 's continuity. Thus restored in its *developmental momentum*, it is also liberated in the *choice of commitments* to the interplay with other lives. This, at any rate, is "the idea." . . .

My reflections, so I was instructed, were to illustrate some relations of my field with Einstein's work. In my third reflection I have, in fact, used our most concrete and daily method in an attempt to demonstrate how Einstein's search for invariants in different coordinate systems may sanction an analogous form of thinking in another discipline. No doubt, both my colleagues and yours will have severe reservations in regard to this. I present it because I am convinced that it has a potential validity for which we must try to find the right words. For what I suggested extends to the applications of psychoanalysis and especially also of that new phenomenon that, until it grows up, will be called psychohistory. A sense of historical relativity certainly demands an aware involvement of the observer in the historical process and thus, in fact, calls for a new kind of Hippocratic responsibility. Whatever general influence these newer methods will have on human enlightenment, I hope to have made it appear probable that to have the courage of one's developmental and historical relativities, as well as of the invariants of human existence, must become part of that wider human ethos that fosters what Freud in his response to Einstein called love and identification—that is, a sense of "We" based on the awareness of a common humanity in the most insightful form possible. Such an ethos must help to counteract the more frightful implications of science and technology. It is, at any rate, the endeavor in which Freud and Einstein were closest to each other.

ACKNOWLEDGMENTS

Work on this essay was, in part, supported by the Maurice Falk Medical Fund.

NOTES

Part I

1. Quoted in Banesh Hoffmann, with the collaboration of Helen Dukas, *Albert Einstein: Creator and Rebel*, New York, Viking Press, 1972, p. 14.
2. Maja Einstein, "Albert Einstein, Beitrag für sein Lebensbild" (1924), unpublished, Einstein Archives, Institute for Advanced Study, Princeton, p. 10.
3. Ibid., pp. 9-10.
4. Philipp Frank, *Einstein: His Life and Times*, trans. George Rosen, ed. and rev. Shuichi Kusaka, New York, Alfred A. Knopf, 1947, p. 10.
5. Hoffmann and Dukas, *Einstein*, pp. 19-20.
6. Albert Einstein, *Ideas and Opinions* (based on *Mein Weltbild*, ed. Carl Seelig), new trans. and rev. Sonja Bargmann, New York, Crown Publishers, 1954, p. 184.
7. Ibid., p. 187.
8. Ibid., p. 292.
9. Albert Einstein, "Autobiographical Notes," trans. Paul Arthur Schilpp, in Schilpp, ed., *Albert Einstein: Philosopher-Scientist*, Evanston, Ill., Library of Living Philosophers, 1949, p. 9.
10. Einstein to Hermann Broch, 1945, quoted in Hoffmann and Dukas, *Einstein*, p. 254 (italics added).
11. Max Talmey, *The Relativity Theory Simplified and The Formative Period of its Inventors*, New York, Falcon Press, 1932, pp. 164-165.
12. Gerald Holton, *Thematic Origins of Scientific Thought: Kepler to Einstein*, Cambridge, Mass., Harvard University Press, 1973, p. 356.
13. *Ideas and Opinions*, pp. 25-26.
14. Hoffmann and Dukas, *Einstein*, p. 25.
15. Ibid., p. 257.
16. Ibid., p. 24.
17. Frank, *Einstein*, p. 49.

18. "But what I admired most in him as a human being was that he succeeded to live many years not only in peace but also in continuing consonance [*dauernder Konsonanz*] with a woman—an undertaking in which I have failed twice rather shamefully." Albert Einstein and Martin Besso, *Correspondance, 1903-1955*, ed. Pierre Speziali, Paris, Hermann, 1972.

19. Albert Einstein, Foreword, in Anton Reiser, *Albert Einstein: A Biographical Portrait*, New York, Albert and Charles Boni, 1930.

20. "Autobiographical Notes," p. 5.

21. Frank, *Einstein*, p. 77. One would assume that governesses and maids played a significant role in attempting to draw little Albert into ordinary and shared childhood activities—and failing, denounced him as *Pater Langweil* (Father Bore). There is one bit of mean writing in which the anger stored up against criticizing and name-calling women may well have come through, namely, Einstein's response to American women who, for patriotic reasons, wished to censor and, in fact, exclude him from the U.S. In one single paragraph Einstein refers to "the beautiful sex," the "good tasting Greek virgins consumed by Minotaurus," "clever and patriotic little women," and the "quacking of the faithful capitolinian geese," while he characterizes himself as a man who is against war in all forms except the "unavoidable one with his own wife." "Reply to the Women of America," in Albert Einstein, *Mein Weltbild*, ed. Carl Seelig, Amsterdam, Querido Verlag, 1934, p. 61.

22. Frank, *Einstein*, p. 77.

Part II

1. Sigmund Freud, "A Difficulty in the Path of Psychoanalysis," *The Standard Edition of the Complete Psychological Works of Sigmund Freud*, ed. James Strachey, 24 vols., London, Hogarth Press, 1953-1974, vol. 17, p. 137.

2. Ibid., p. 139.

3. Ibid., pp. 139-140.

4. Ibid., p. 140.

5. Ibid., p. 143.

6. Sigmund Freud, "The Ego and the Id" (1923), *Standard Works*, vol. 19, p. 55.

7. Lincoln Barnett, *The Universe and Dr. Einstein*, New York, Bantam Books, 1957, pp. 45-47.

8. Holton, *Thematic Origins*, p. 123.

9. Ibid., p. 124.

10. Ibid., p. 125.

11. Barnett, *The Universe and Dr. Einstein*, p. 67.

12. "Relevance and Relativity in the Case History," in *Childhood and Society* (1950), rev. ed. New York, W. W. Norton, 1963.

13. Sigmund Freud, "Psychoanalysis and Telepathy" (1921), *Standard Works*, vol. 18, p. 178.
14. Sigmund Freud, "New Introductory Lectures" (1933), ibid., vol. 22, p. 144.
15. Ibid., p. 144.
16. Einstein to Freud, 30 July 1932, Einstein Archives.
17. Einstein to Bacharach, 25 July 1949, ibid.
18. In Ernest Jones, M.D., *The Life and Work of Sigmund Freud*, 3 vols. New York, Basic Books, 1953-1957, vol. 3, p. 131.
19. Einstein to Freud, 21 April 1936, Einstein Archives.
20. Freud to Einstein, 5 March 1936, in Jones, *Life and Work of Freud*, vol. 3, p. 204.
21. Albert Einstein and Sigmund Freud, *Why War?*, Paris, International Institute of Intellectual Co-operation, League of Nations, 1933, pp. 13, 20.
22. Letters to Max Eitingon, 18 August and 8 September 1932, in Jones, *Life and Work of Freud*, vol. 3, p. 175.
23. Einstein and Freud, *Why War?*, pp. 46-47.
24. Ibid., p. 49.
25. For a previous attempt to spell out this connection, see my "Elements of a Psychoanalytic Theory of Psychosocial Development," in Stanley Greenspan and George H. Pollock, eds., *Psychoanalytic Perspectives on Human Personality Development Throughout the Life Cycle*, National Institute for Mental Health, 1980, vol. 1, pp. 11-61.

Nathan Rotenstreich

RELATIVITY AND RELATIVISM

THE SAYING *habent sua fata* applies not only to the fate of books but also to that of words and formulas by the very fact that they are liable to become stereotypes and turn into slogans, are exposed to misconception, or aid in the reinforcement of certain trends. The terms "relativity" and "theory of relativity" as Einstein employed them to characterize his theories, are a case in point. Before commenting upon these terms, let me make some general observations on the framework of the present analysis, though I am aware of being ill-qualified to appreciate all aspects of Einstein's theory. I shall start with a few comments on the concept taken to be antithetic in meaning to "relative" or "relativistic," namely, that of "the absolute."

∞

The concept of the absolute connotes in the first place—and this indeed is its literal meaning—something separate or separated, something existing independently of any other entity or context. In this sense the absolute implies a primary merger of three constituent aspects: the aspect of existence, the aspect of being the cause of that existence, and the aspect of the essence inhering in the self-enclosed character of that existence or reality. To put it negatively, in terms of the considered entity there is no cause extraneous to it, and its existence corresponds fully with its features or essence. The three aspects, in their self-enclosed interaction, articulate the absolute character of the entity or the existence and hence its separation, separation being a negative or, if we prefer, a comparative or relative articulation of the position of the entity or existence. To be absolute qua separate is not only to be detached from other entities but also to be self-enclosed and thus self-sufficient and totally independent.

We find that this meaning of the term "absolute" was first applied in the classical period to the concept of the absolute good. We might add that the various aspects of the self-sufficiency of the absolute made the concept applicable in the terminology of ethics, since to be absolutely good, that is, perfect, is to be independent, that is, totally self-sufficient. I should add that the ethical connotation of goodness is related to completeness, since the reality or the entity that is designated as good emanates goodness out of itself and is, in the ethical sense, forthcoming, relieving, delivering, and so on. In this sense an entity that is complete because it is totally self-sufficient is good by definition. Thus the aspect of being perfect designates, on the one hand, the independent position of the entity and, on the other, its overflowing essence, namely, its ethical quality. In view of those considerations, the term "absolute" was applied to the position and essence of God, as found in the terminology introduced by Nicholas of Cusa.

These comments concerning the absolute should suffice as a frame of reference in our attempt to elucidate some aspects of what is called relative.

∞

In the first place, we can say that to be relative indicates a position of being dependent, in contrast to the independence inherent in the position of the absolute. Yet the aspect of dependence may in turn have different meanings, which can be hierarchically organized. A dependent entity can be connected with another entity, as, for instance, right is connected with left. The position of being at the right is not caused by the position of the left, or the other way around. The two positions are correlated, and their mutual dependence amounts essentially to correlation.

But dependence may have a stronger meaning, as of something conditioned, not in the sense of being correlated but in the sense of being brought about. We can indeed describe *B* only by considering it as related to *A*, and we are stepping beyond the time sequence of *A* and *B* by positing the independence in terms of the reality, or the emergence, of *B* as an effect of a causative *A*. Its causality or causativity is taken to be a paradigmatic case of dependence qua conditioning. Obviously, neither in terms of correlation nor in terms of conditioning can one assume separate entities. This is so in terms of correlation even less than in terms of conditioning, since the position of left has a meaning only in relation to right, whereas the position of being a cause may have a meaning prior to and outside of the context of *A* and *B*, though it is possible that by separating *A* from the causal context, we attribute to it the position of being a cause. The relatedness amounts here, among other features, to the impossibility of abstracting *B* from the context as against the possibility of abstracting *A* from it.

The asymmetry inherent in the various aspects of relations or relatedness is emphasized when we introduce an additional term in the description and say that the dependent entity is a derived one. Being derived underscores the position of lacking independence, of being involved to such an extent that not only is the derived entity a separate one, but also that its very locus and essence come into prominence in its derivative status. To be sure, the notion of being derived again has different meanings. In a sense an effect derives from its cause, but as an effect it may in turn become a cause, and thus the position of *B* vis-à-vis *A* changes vis-à-vis *C*. To be derived can refer to being a copy. We say that an art student makes a copy of a Leonardo da Vinci painting; but the product of his activity is a derivation with a meaning opposed to that of making a carbon copy, whereby a text corresponding to another text is reproduced simultaneously, in the usual meaning of the word, by means of a mechanical process.

A more strongly marked aspect of the position of being derived comes to mind when we speak of an entity's being contingent. Such entities are dependent, conditioned, derived, and so on. In addition, however, they lack an inherent cause for their existence, and therefore their existence is merely contingent. They might not have existed at all; their factual presence is accidental and as such may perhaps be explainable or perhaps merely encountered and observed. The paradoxical character of a contingent entity lies in its being dependent but not totally explainable and in the possibility that the explanation for its presence may lie within itself, in which case it would therefore resemble, dimly at least, one aspect of an absolute entity. However, this explanation is impossible, because it has been assumed that to be absolute implies a total conformity between a self-cause and existence, whereas that conformity cannot be assumed in the case of a contingent entity. A contingent entity lies within a context, but there is no line of derivation from the context to the particular entity, or, at least, we do not know of such a line.

An additional example of a special kind of relationship on the borderline of correlation is the position of a name, as described by John Stuart Mill. A name, Mill says, is relative in that it relates to one thing. Its signification can be explained only by mentioning something else. A case in point would be the name father, which can be explained only by referring to a son or a daughter. This is not exactly a correlation like that of left and right; rather, it is a contextual structure, due to what Mill calls the signification of relative names.[1] Probably there are additional meanings to the notion of being related qua dependent, but for our analysis of relativity the meanings outlined above should suffice.

∞

When we move one step further from the description of the various components of the very broad and vague expression "to be related" in its antithetic position to that of "being absolute or separated," our first comment must concern the level of the analysis. We are exploring statements or propositions pointing to situations or states of affairs that are viewed as basically related, or, to put it negatively, as not permitting the separation or elimination of an element from a context. But in attempting such a description, we also recognize that this context is only temporary and tentative. Thus it is bound either to become obsolete or to pave the way for something else, something that will indeed be related as well but will occupy a new position compared with that of the previous context. Tentativeness, of course, can be understood strictly in terms of time sequence, when one context is pushed aside and another is placed, or suggested, to occupy its position. Yet when we look at tentativeness as a basic quality, we have to add that it can be viewed as related to a certain end, which would perhaps be the perfection of the statements, or, to put it more broadly, the perfection of theories or systems. On the way to perfection, the theories and systems are indeed tentative. But perfection implies a finality beyond tentativeness and thus, at least retrospectively, enables us to see the related contexts, viewed in themselves, in their position of bringing about, even against their intention, the final end. In this case the perfect would not imply an absolute qua separate position, because eventually the tentative positions are immersed or absorbed in the perfect one. Here the absolute would mean a final product related to the steps leading to it. It goes without saying that this model of interaction between relatedness and finality served both Aristotle and Hegel in their surveys and analyses of the history of philosophy. Hence we must distinguish between a fragmentary or sporadic relatedness and a relatedness that goes beyond the boundaries of certain relata and may eventually bring about a two-way flow of relatedness from the partial context to the end and from the end to the context. Paradoxically, within this structure the partial contexts lead to the final end but are not dependent upon it, whereas the final end, though final, is by the same token dependent upon the partial contexts or structures.

As we are already concerned with statements about relational contexts, we may move one step further. Many philosophical explorations are interested in what might be described as spheric themes or, parallel to these, spheric attitudes and hence are concerned with spheres of related elements. The most obvious instances of this thematic interest concern the spheres of knowledge and science, ethics and normative behavior, and aesthetics and evaluations of products or objects according to a criterion of beauty. Concerning the parallel attitudes, we may say of them that the attitude of knowledge is exhibited in conceiving statements about objects, factors, or elements, structuring them within contexts, and asserting about them propositions held to be adequate to the state of affairs comprised in them. The attitude of conducting or determining one's behavior according to norms is understood as intervening in a given situation and ordering it according to a certain preference, which is guided in turn by a normative principle—friendliness, mercy, frankness, justice, and the like. The attitude of appreciation of a certain object, something produced or encountered that elicits a particular sensitivity or sensation, such as liking or disliking, is taken to be an aesthetic attitude, a response according to a certain norm, which is usually described as one of beauty or, as the standard antithesis of beauty, one of ugliness.

I mention the end of attitudes not only in order to bring into prominence the correlation between a sphere, which has a thematic core, and a stand, which has a psychic core, but also in order to emphasize that in regard to both themes and attitudes we encounter a certain basic correlation or dichotomy. The dichotomy of truth and falsehood is apparent in terms of the sphere of knowledge and its corresponding attitudes, the dichotomy of right and wrong and that of beauty and ugliness, in terms of taste or sensitivity. This is to be emphasized, because such a dichotomy or correlation dualistically structured does not apply in all the cases that must be taken into account. Niels Bohr is quoted by Werner Heisenberg as having said that the opposite to a correct assertion is a false one but that the opposite to a profound truth can again be a profound truth. Hence we have to be cautious when applying dualistic correlations in regard to the thematic schemes, since, for instance, we can say that justice in terms of *fiat justitia* is a norm, as are mercy, forbearance, and the like. *Mutatis mutandis* we say that a renaissance painting is beautiful, as is an impressionist one. Therefore we face here the question of structures of relatedness within one context, such as to be profound or good, and the structure of relatedness, which is, by the same token, an attempt to draw lines of separation, as for instance to be true or false, though in certain cases one thematic description oversteps its original boundaries. We therefore speak of being truthful as a quality of moral behavior and not only as an attribute or a statement about a fact. We speak about being right as encompassing both *gerecht* and adequate in the sense of *richtig*. Sometimes we may express satisfaction with the results of our reasoning or syllogism by referring to it as "beautiful."

∞

Within the different contexts we have described, we may discern one common feature, namely, the intentionality to what is called validity. Validity would be a principle that is related to the different contexts of relatedness. It extends beyond the boundaries of one context only. This internal and external quality of validity expresses itself in the fact that within the sphere of knowledge validity takes the shape of truth, within the sphere of behavior it takes the shape of goodness, and within the sphere of taste it takes the shape of beauty. Validity is a comprehensive principle and thus not primarily implicated in any of the thematic contexts. By becoming involved in those contexts, it is shaped by them thematically.

The question arises here whether validity is transcontextual only in the nominal sense. In the thematic sense it is not only integrated in contexts, and not even just dependent upon them, but it is also eventually dependent because it is derived from elements outside the contexts. At this point let us analyze some of the possible interpretations of validity as a principle in what might be described as an ascending order, that is to say, from a minimal to a maximal interpretation of relatedness. For this part of the analysis I am greatly indebted to a presentation by Heinrich Scholz, which I shall summarize with some changes in his order of presentation as well as in the substance of his argument.[2]

The first mode of relatedness can be described as that pertaining to the concept of validity (*Geltungsbegriff*), which is indeed a characteristic concept of relation, since validity presupposes the existence of subjects that are capable of conceiving and acknowledging. By the nature of the situation, to be valid is always to be valid for somebody, that is to say, for a subject. To put it differently, validity implies or calls for worthiness of being acknowledged (*Anerkennungswürdigkeit*). In this sense, to be true is always to be true for a subject who has already conceived the demand for truth or who is, or can be, in possession of a criterion of truth. In other words, the truth of a content of a proposition implies that it is something we cannot create arbitrarily. We entertain that content independently from any arbitrariness; this situation is, to say the least, one aspect of the position of being true or worthy of being acknowledged. We distinguish here explicitly between the relative position of truth and validity and their relational position, and indeed, that distinction is the guiding thread in the various distinctions that follow. Scholz observes that even Bolzano's "truths in themselves" necessarily carry the character of being related, since the divine intellect is their essential or necessary correlate.

We may elaborate on this analysis of the relational character of truth or validity by suggesting the distinction between the principle of truth, or even the criterion of justifying the truth of a statement, and the truth of a particular statement that cannot be separated from its specific content. On both levels we may employ the term "correlation" rather than "relation" or "relatedness," since "correlation" denotes in the first place the presence of two relata and the essential tie obtaining between them, whereas the notion of relation or relatedness applied to the situation may lead to an interpretation that lacks the basic symmetry inherent in the notion of correlation. The second observation is more than terminological; it broadens the scope of the structure of correlation by including in it, in addition to the principle of truth, a principle of another sort, namely, the principle of good. We could say that the quality of being worthy of acknowledgment can be attributed both to the principle of truth and to other principles, goodness being a major case in point. We deliberately leave open the question whether other principles, for example, beauty, should be included in that spectrum.

The emphasis placed upon the aspect of correlation is more than an issue of terminology. By stressing correlation we bring into prominence a two-directional dependence. In it the response of the subject—the empirical subject, and not only the divine intellect—is determined by the principle and by the configuration, that is, by the statement, maxim, or action having a moral content. Through that determination, acknowledgment proper as an act or as an intentionality becomes an affirmation or a response. It brings about convictions, consummation of the search after truth or goodness, and awareness—for instance, of the essential difference between vindicating a true statement and performing a good deed. For the statement continues to be considered as true unless it is proved false, and it can thus be taken for granted, whereas a good deed, although it is an accomplished fact, cannot be assumed to be necessarily a step in the direction of additional good deeds. True statements, even if they are interrelated, are capable of being isolated to a certain extent, whereas a good deed, once accomplished, belongs for better or for worse to the past. The problematic situation obtaining between the course of events and their determination by a moral maxim is faced anew with each deed. This last point brings into prominence not only the dependence of the subject upon the principle but also *pari passu* a kind of dependence of the principle upon the subject. Through the subject the principle gains material content and by the same token is incorporated into the structure of thinking and knowing and of shaping deeds by maxims or imperatives. The whole process of vindication and the tension between the principle and its exposure to verification, that is, to being checked in certain situations, is made possible by the presence of the subject. Thus in the area of truth and goodness the subject, even if the principle has no meaning for him, is nevertheless part and parcel of what might be described as the noematic situation, and, if we may use that term, of the deontic one also. Here again we go beyond correlation in the sense of pointing to a context of a correlation in terms of the fundamental phenomenological structure of the noematic and the deontic spheres.

∞

At this juncture, again following the distinction suggested by Scholz, we encounter a different aspect of the concept of being related, which, as a matter of fact, already implies the concept of relativity called the perspectivistic concept. This aspect carries with it not only the optical association but also what Friedrich Nietzsche called perspectivism. Nietzsche described perspectivism as a form of what he called specificity (*Form der Spezifität*). Every meaning depends necessarily on a certain perspective, and all interpretations are essentially perspectivistic evaluations. From the position of all our basic urges (*unsere Grundtriebe*) we arrive at a perspectivistic evaluation of all that occurs. Hence, interpretation of relatedness qua relativity depends on the specific stand taken by a human being. Only an awareness of the sum total of the different positions taken by human beings allows us to draw conclusions extending to the full spectrum of human potentialities.

We may enlarge on this aspect of stand or perspective by pointing to several other aspects, which indeed became current topics in popular, scientific, and philosophical discussions of the notion of relativity. We refer not only to perspectives of individuals but also to perspectives of groups of individuals, brought together or united by a certain prevailing perspective, which, in prevailing, ceases to be a perspective in the sense of a personal point of view and turns into a determining, causative factor. Thus, for instance, perspective in the optical sense would lead us to concepts like *Zeitgeist*, style of a generation, and presuppositions, as well as to certain modes of expression or approach, for instance, naive and sentimental poetry in Schiller's sense or what is called the Promethean spirit. Yet a stand as a unifying perspective may take a more deterministic form, as when applied to an interest, in the sense that the interest of a group brings about solidarity within the group or that solidarity is a unifying factor creating the group. It can lead to a concept like that present in Marx's theory of class interests, and even more so to the popular reception of the predominance of the economic factor in Marx's theory. Here the stand becomes what is called *Sitz im Leben*. The main difference between that interpretation of relativity and the previous interpretation of correlation is in the position of the exclusive or preponderant relevance of the empirical human subject or subjects in regard to certain attitudes or stands.

To elaborate this difference descriptively, we must go beyond Scholz's analysis. There is a structural difference between the correlation characteristic of the noematic situation or that pertaining to imperatives, and the dependence of models upon world views or theories, or codes of norms upon individual stands, historical situations, anthropological or cultural milieux, and the like. When we refer to correlations, we refer to the inner logic of the field. There is no validity detached from validation, and all validation, by virtue of its very intentionality, refers to the principle or content to be affirmed. When we refer to the determination of world views by situations, we assign to the situations an impact, or even a supremacy, upon which is predicated one's orientation in the world, the abstract kernel of which can be extracted as being dependent upon the behavioral context of a certain culture or historical period. In this case the structure of correlation is replaced by the structure of causal dependence, although the causal impact of the situation, taken either chronologically or culturally, on the views held or the norms adhered to may not be obvious. Even when accepting the vagueness of a particular causality, we are aware of the shift from the understanding of the internal structure of certain contexts, such as the cognitive or normative one, to the dependence upon an extraneous cause that is factually discerned but whose meaning is not expressed in the context as it is empirically discerned.

An attempt to draw an analogy between the human subject, in its correlated position in regard to principles or norms, and situations qua *Standort* cannot hold good. To be sure, a subject, a human empirical subject, is not identical even with the sum total of his acts of intentionality. In the first place, the subject is a comprehensive being, a psycho-physical entity. His acts of intentionality toward validity constitute only segments of his feelings, emotions, and cognitive acts. Yet there is no gap between the total profile of a subject and his intentionalities toward validity precisely because the process of validation as related to validity is part and parcel of the structure of certain acts of intentionality, such as affirmations, assertions, and evaluations, or even misgivings, hindsight, and the like.

This relation does not apply to the alleged causative position of situations. Their relatedness to principles or to norms is at most a de facto relationship that we discern when looking at certain models of the construction of theories or certain modes of behavior in various historical periods. Looking at these situations as observers, we may find a kind of relationship between a total culture and a particular mathematical theory, or between a broad general behavior and a religious outlook, which may express an attitude toward, for example, iconic presentations. Moreover, and this is certainly one of the lessons of contemporary exploration of the structures of cultural milieux, it is not always certain precisely where we find the causal preponderance. Does the cultural milieu determine the scientific theory, or does the scientific theory, with its assertions and limitations, determine, at least partially, the cultural milieu? To mention only one obvious example: Are tools determined by the cultural milieu, and do they as such determine the level of experimentation available or practiced in a certain culture? Or is a certain theory presupposed when tools are constructed, and do they in turn mold the culture milieu? Even if we apply the model of causality, the identification of what is a cause and what is an effect or of what is primary and what is secondary is by no means preestablished. Relativities, understood in the sense that versions of relativity are different from identifiable structures of relatedness, are by no means quite as unequivocal as is somehow "atmospherically" taken for granted. We shall confine ourselves to these remarks, since we have to revert to a major issue in this context, namely, the problem of situations being determined by principles and norms rather than principles and norms being applied from within or superimposed upon situations. At this point we may move to our further consideration connected with the theory of relativity as understood by Einstein, though that theory only serves as a *causa occasionalis* for these reflections, despite the obvious fact that a theory of such magnitude cannot and should not be merely a *causa occasionalis.*[3]

∞

Let us observe in the first place that a certain ambiguity pervades the language that Einstein applied to his theory. Thus, for instance, Einstein says that from the relativistic point of view the hypothesis that the universe is infinite and Euclidean at infinity is a complicated hypothesis.[4] Yet his next sentence refers to the general theory of relativity and not to the relativistic point of view (2). And indeed, we must distinguish between the theory of relativity and a relativistic approach. We shall follow Einstein himself and refer only to the relevant broader philosophical aspects. "The object of all science," says Einstein, "whether natural science or psychology, is to co-ordinate our experiences and to bring them into a logical system" (1). Einstein places his theory, or at least our customary ideas of space and time, in relation to the character of our experiences. It is obvious that a formulation of the program of all science in terms of coordination or of constructing a logical system either presupposes or brings into prominence the contextual character of our knowledge and understanding, which we previously called relatedness. To be sure, there are different structures of relatedness, and here the distinction between prerelativity physics and relativity physics becomes significant. Einstein says that in prerelativity physics an interval has a physical meaning independent of the choice of coordinates (5). There are preferred coordinates, which Einstein identifies with Cartesian systems (7). Moreover, in prerelativity physics, in order to specify relations in space, a body of reference or a space of reference is required and is used in addition to the Cartesian system of coordinates (15).

Venturing to interpret these statements, we may say that the basic program of coordination or systematization is realized in prerelativity physics, according to Einstein, by constructing systems that refer to independent coordinates. The system relates to the coordinates, but the position of the latter is not necessarily or essentially implicated in the coordinates. The move characterized by the theory of special relativity is in the direction of changing the position of the coordinates by integrating them into the structure of relatedness.

Einstein observes further that in prerelativity physics space and time were separate entities, as manifested in the independence of specifications of time from the choice of the space of reference. When the hypothesis of the absolute character of time was abandoned, the four-dimensionality of the time-space concept was immediately recognized (30). One can interpret this argument by assuming that the conception of space and time as separate entities and the hypothesis of the absolute character of time are two identical descriptions of the same basic or philosophical presupposition. The position of being absolute coincides with the position of being separate. Hence the elimination of the separate position, of time, space, or both, amounts to the conclusion that there is no absolute relation in space independent of the space of reference, as there is no absolute relation in time between two events (30-31). Consequently, according to the principle of relativity in its broadest sense, we give the laws a form in which they are valid in every four-dimensional system of coordinates (61).

It is not necessary to dwell on the further elaborations of the theory of relativity. What emerges from Einstein's statements is that the theory of relativity presents not a relativistic interpretation of phenomena but a structure of phenomena imbued with a relatedness among them, which is guided by laws referring to that structure. One can assume that perhaps even the title given to the theory—relativity—refers to the negation of absoluteness in the Newtonian interpretation of time and space, according to which absolute time and absolute space contain the meaning of being true and mathematical. As such, they connote a position that, by its own nature, flows equably without relation to anything external and remains immutable and immovable as applied to space. The negation of that position and the affirmation of an all-embracing relatedness of both time and space may perhaps be the underlying reason for the use of the term *relativity.*

This is the gist of the observation by Heinrich Scholz and of many other interpretations, among which it would be only proper to mention, for instance, that of Professor Banesh Hoffmann. According to him, there is no way of determining absolute rest or uniform motion, and this impossibility is inherent in the title *The Principle of Relativity.* To emphasize again, simultaneity is relative because it is dependent upon a frame of reference, that is to say, it has to be conceived within a structure of relatedness. There is no justification for introducing the notion of universal absolute time, which would mean that there are positions to be determined or identified outside the structure of relatedness. Hence we should not let ourselves be misled by the term "relativity," let alone by the use of "relativistic theory," which may be the result of some difficulty in rendering German expressions in English.

We must observe at this point that philosophical interpretations of the theory of relativity have emphasized this aspect, and the same must be said precisely when we point to the interpretations stemming from different philosophical schools. Let us first mention the interpretation of Ernst Cassirer, who emphasized that in the theory of relativity the conception of the constancy and absoluteness of the elements is abandoned in order to give permanence and necessity to the laws instead. We can disregard Cassirer's attempt to read a kind of compensation into the structure, that the elimination of one conception is neutralized by the emergence of another. But his actual interpretation seems to hold true, namely, that the structure is guided by laws that by definition shape contexts. What is abandoned in the new approach is the independent position of elements, let alone particular elements, including space and time in the physical sense. Cassirer adds that if one attempts to seek in the theory of relativity a kind of modern confirmation of the Protagorean principle that man is the measure of all things, then one has missed the decisive achievement of the theory. The theory teaches us that not everything as it appears is true. On the contrary, the theory warns us that phenomena that are valid only for one particular system are not to be taken as truth in the scientific sense, that is to say, as an expression of the all-embracing and ultimately valid lawfulness of experience.[5]

We should also mention in this context the interpretation presented by Hans Reichenbach, who, as is well known, represents a philosophical approach differing from Cassirer's. Reichenbach says that the word "relativity" should be interpreted as meaning "relative to a certain definitional system."[6] The plurality implied here is not a plurality of different views or of systems of contradictory contents. It is merely, as Reichenbach puts it, a plurality of equivalent languages and thus of forms of expression that do not contradict one another and have the same content. It is obvious that Reichenbach, according to his philosophical principles, places the emphasis on expression or language. He sums up his interpretation by saying: "Relativity does not mean an abandonment of truth; it only means that truth can be formulated in various ways."[7]

The difference in interpretation between Cassirer and Reichenbach is clear: Cassirer, faithful to his philosophical position, brings into prominence the necessity of lawfulness, whereas Reichenbach sees the paramount issue in the difference between content and expression, or, to put it in stronger terms, between truth and formulation. But from the critical point of view it is obvious that the object of both interpretations is the relativistic interpretation, and it is worth noting here that Reichenbach, too, mentions Protagoras in the course of his analysis and observes that to give a Protagorean interpretation of Einstein's relativity is utterly mistaken. Such an interpretation leads to the erroneous notion that the relativity of space-time measurement is connected with the subjectivity of the observer. That erroneous interpretation can be rendered differently by stating that the privacy of the world of sense perception is the origin of the relativity maintained by Einstein. "The definitional character of simultaneity . . . has nothing to do with the perspective variations resulting for observers located in different frames of reference."[8] All this can be said without pointing to broader concepts of Einstein's theory referring to the velocity of light as a constant magnitude or, as it is sometimes put, "an absolute signal velocity," because it is not our aim to question whether a kind of absoluteness comes back into the context of Einstein's theory but to show the basic difference between a relativistic attitude and the theory of relativity. That difference can be reformulated by stating that the relativistic theory attributes a special position to the *Standort,* or stand determining the context, to the extent that the context is dependent upon it. This applies, as we have seen, both to cognition and to normative attitudes. In the theory of relativity, on the other hand, the *Standort* is incorporated into the context and structure and constitutes precisely the axis of the theory. To be sure, in popular language the very term "relativity" has been relativistically interpreted or been used as a kind of ultimate justification of the relativistic mood, and moods do not discriminate and are not meant to be based upon precise interpretations. That is why they are moods and not systematic positions.

∞

One may ask whether, beyond and above the coordinated relatedness, a comprehensive form of time and space can be or should be assumed. Einstein says, "Just as it was consistent from the Newtonian standpoint to make both the statements, *tempus est absolutum, spatium est absolutum,* so from the standpoint of the special theory of relativity we must say, *continuum spatii et temporis est absolutum.* In this latter statement *absolutum* means not only 'physically real,' but also 'independent in its physical properties, having a physical effect, but not itself influenced by physical conditions' " (55). One could assume from that statement, and from the very fact that Einstein employs the term *absolutum,* that he refers to a kind of an a priori level. We can see that this is not so when we look into another statement of his: "The only justification for our concepts and system of concepts is that they serve to represent the complex of our experiences; beyond this they have no legitimacy. I am convinced that the philosophers have had a harmful effect upon the progress of scientific thinking in removing certain fundamental concepts from the domain of empiricism, where they are under our control, to the intangible heights of the *a priori.* . . . this universe of ideas is just as little independent of the nature of our experiences as clothes are of the form of the human body. This is particularly true of our concepts of time and space, which physicists have been obliged by the facts to bring down from the Olympus of the *a priori* in order to adjust them and put them in a serviceable condition" (2).

One can question whether the dichotomy implied between making experience serviceable and being conditioned or determined by experience is a legitimate dichotomy. Probably Cassirer adopted this reasoning when in his interpretation of the theory of relativity he distinguishes between the approach of the physicist and that of the philosopher. For the physicist, says Cassirer, what he calls "space" and "time" is a concrete, measurable manifold, which he attains as a *result* of the lawful organization of the particular points. For the philosopher, space and time signify nothing but forms and modi, and therefore the presuppositions of precisely that organization. For him they are not the results of this organization but are exactly this organization and its basic directions.[9] It is not my aim here to present a sort of *apologia* for the philosophical consideration, though it may be assumed, following the analysis of Gerald Holton, that this is not the only position Einstein held.[10] Be that as it may, even when there is no a priori level, the idea of relatedness is a central idea. As such, it can have different manifestations, but difference does not lead to a relativistic interpretation in terms of determination by a position or factor that lies beyond the structure of relatedness.

It is appropriate to mention at this juncture that an attempt to present a particular time structure as having a paradigmatic status cannot be germane to the theory of relativity. This is the line Henri Bergson took when he attempted to show that the plurality of times, which is inherent in the special theory of relativity, does not rule out the oneness of real time but, on the contrary, presupposes it. Bergson took as his primary point of departure the order of quality and its experiential correlate of meaning qua aging and duration. According to him, no work of analysis can resolve qualities or their experiential correlates into pure quantity. Not only does that experiential time occupy the primary position, but there is also a basic or ontological difference between time and space, since time cannot be exhausted by measurement, whereas space can.[11] Again, it is not essential to dwell here on the question of a possible separation between time and space, whether they are conceived of ontologically or understood as a priori forms. The essential point is that structures of relatedness basically imply pluralities of coherent structure and as such do not allow for a separation of one structure by attributing to it a primordial position, as Bergson's durée would imply. One could say that

Bergson's attempt confuses that which is first from our point of view (*pros hemas*) with that which is first in itself (*pros auton*). Thus Bergson, in his attempt to deny the constructive character of knowledge, falls back into what can perhaps be bluntly called the "anthropormorphic catch."[12]

∞

After this brief discussion of Einstein's theory, it is not necessary to repeat the basic issue, namely, that knowledge, represented here by a paramount scientific theory, is a structure of relations. We can sum up by saying that it is so, without reference to a particular theory or even to a particular discipline, like physics or history. We may say broadly that relatedness is just the other side of the coin in a cognitive attempt to present integrations of data and that integration is the essential direction of knowledge. To be sure, knowledge can be understood, as mentioned at the start of this analysis, from the point of view amounting to what Immanuel Kant described as *Fürwahrhalten.* But from the thematic point of view, knowledge as explanation and knowledge as employer of laws for the purposes of explanation eventuates through the establishment of contexts. These can be reinforced by laws or by what are called covering laws in the contemporary vocabulary. The aspect of contexts is essential for the momentum of knowledge, whether we ask "how" or "why." We can even go a step further and say that when we refer to a transcontextual position, as, for instance, in the cosmogonic argument for the existence of God, that is to say, when we refer to an entity that is absolute in its position, the fact that the transmundane cause is brought into the picture in order to explain processes is an indication of the basic fact that a transcontextual element is employed for the explanation of the context. To be sure, in different disciplines the contexts are of a different character. For instance, in the sphere of natural events we can refer to a statement by Moritz Schlick that laws of nature are abbreviated expressions for the order in which events succeed one another. In the sphere of historical events or phenomena we not only look for the order of discussion but can also see that the aspect of the "why" becomes more prominent. Now exchangeability, to refer at this juncture to one of the concepts put forward by structuralism, is again a manifestation of the contextual character, though that character may in a particular social structure take the form of a two-way flow, as indicated by the notion of exchangeability. We may therefore say that the cognitive attitude and structure is essentially imbued with relatedness and that the object of knowledge is to discover, find out, construct, or interpret structures of relations.

Coming now to what we have previously called deontic structures, we realize that here the situation is different, because integration cannot be taken as germane to the structure of the deontic realm and its objective. To be sure, integration is not totally missing from the deontic structure, but it cannot be the pivotal point. The reason for that difference lies first in the fact that in the deontic structure we do not encounter "mute" or "brute" data or facts. On the contrary, the facts call for a certain attitude or response, which may be broadly described as one of acknowledgment. Speaking in terms of direction, we may say that integration, as pertaining to knowledge, amounts to extracting data from their primary givenness and placing them in a context, whereas acknowledgment amounts to "descending" to the data concerned and recognizing them in their givenness. For example, the demands we make of children are different from those we make of adults. This is not just a matter of the factual potential of the partner or the addressee; it is a normative stand in itself. That is to say, we recognize the addressee as a partner; therefore we cannot simply, as it were, integrate him into an anonymous, or let us call it abstract, structure of norms. We recognize or acknowledge the norms. Yet, since the attitude of recognition or acknowledgment is here at least one of the guiding attitudes, it is confined not only to the thematic aspect but also to the positional aspect or to the given situation, be it in terms of age or in terms of social status, of the person approached. Hence, the very acknowledgment implies an understanding of the situation and the proper choice of the norm to be applied. This structure not only applies to demands, when we demand different things from different persons according to their situation, but it also applies to rights. The broad notion of rights is interpreted variously according to the particular case. To be sure, such an acknowledgment in terms of a situation, from whatever perspective we view it, has inherent limitations from the normative point of view. There is bound to be a thematic component in the acknowledgment of the situation. I take the attitude of demanding or of granting rights, not because the person encountered is John or Paul, but because the person can be identified according to a substantive criterion, not only that he is a human being but also that he is one of a particular age, ability, social situation, condition, and so on. To sum up, we may say that since an element of recognition of the person in his different aspects, or of factual situations, is inherent in the deontic structure, that structure exhibits a relatedness between thematic norms and given situations, recognized or acknowledged both in themselves and with a view to the norm to be applied. This makes the deontic structure different in its inner logic from the cognitive structure, though in both the aspect of relatedness is prominent.

We may now move to a second element of relatedness in the deontic sphere, one that may be analogous to the cognitive structure, though differing from it. If we refer again to the context guided by laws in the cognitive sphere, we may inquire into the relationship between one law and another, such as between the law of gravitation and the law of thermodynamics, or we may ask about laws controlling evolution and the physical rhythm of the cosmos. Unification, or the unified theory, would represent in this case an objective epitomizing the very direction of knowledge toward contexts. The all-embracing context would be relatedness *par excellence.* The logic of the deontic structures is different, and this difference lies in the aspect discussed previously. If a primary factor of acknowledging a situation must be taken into account, that is to say, if the situation as such has to be seen in its particularity, one may wonder whether there can be one single context of norms that must be applied to all situations and persons involved. In the deontic sphere there is a tendency to point to one norm, for instance, *fiat justitia pereat mundus.* But this tendency is often mitigated by an opposite tendency to apply against *justitia* the norms of forbearance, mercy, tolerance, understanding, etc. Hence we face the question whether there is a structure of relatedness in terms of the norms themselves in their plurality and variety or whether there is a variety in which each of the partial norms is self-enclosed and the application of the norms to the situations or persons involved is left to the judgment (*Urteilskraft*), of which Kant said there is no method to guide that human or cognitive faculty.

However, the structure of relatedness is even more complicated. In the first place we have to mention that we face the demand to be consistent, and that demand may have two focuses. First, if we apply the norm of demanding, not giving in, in one case, should we apply that norm in another case? Or perhaps the correlated norm of acknowledging the situation or person may lead us to the conclusion that, without giving up consistency, we may approach one case differently from another. Although we are consistent, we take into account not only the thematic norm but also the situational imperative. The second focus is that when we adhere to a norm of demanding, we may demand from ourselves an adherence to additional norms that seem to be germane to the norm of demanding. For instance, we may give precedence to a principle over a situation, in the sense that situations are only factual points of departure that have to be governed by thematic or defined norms. Within this scope we may take a step further and demand the utmost. For instance, we may even demand a sacrifice for the sake of a principle, though, as we know, for instance, from religious traditions, some exceptions are permitted or even prescribed. The parallel aspect would be that if we take the position of forbearance and remain within consistency, we should also take the position of being tolerant in matters of religious beliefs and behavior, day-to-day style of life, and the like.

However, it is also possible to take the view that precisely when on certain issues one adheres to "rigid" norms, on other issues one can be forthcoming and forgiving. Whether this differential attitude conforms to the notion of consistency is perhaps not all that important, for one can take the position that, because of one's acknowledgment of specific situations and persons, a variety of norms and their application is justified. If this is so, then, given a plurality of axes (situations, persons, norms), a mutual mitigation and adjustment of norms is essential or even unavoidable.

All these examples, and certainly more, can be brought forward to explain that in the ethical sphere we are entitled to refer to structures and thus to relatedness, though in this sphere the relatedness is more complicated than that pertaining to cognitive structures. This complication of relatedness will serve as at least one of the explanations for the fact that the ethical sphere is more exposed to the tendency toward relativity, as being different from relatedness. We shall now turn to this subject.

∞

As we have seen, the theory of relativity does not lead to a relativistic interpretation of structures, and, in the words of Adolf Grünbaum, it is patent that this theory is utterly noncommittal concerning the thesis of ethical relativism, even though, with all reservations on the analytical level, it has been understood as a reinforcement of the relativistic tendency. We have to observe again that relativism amounts to the view that structures are not self-enclosed; that is, they do not contain their own guiding principles but are determined by and dependent upon basic factors outside themselves. It is perhaps no accident that Grünbaum mentions ethical relativism and not cognitive relativism.[13]

To be sure, the sphere of knowledge has certain built-in devices that make it more immune to a relativistic interpretation in the sense indicated above. In the first place, it is accepted, for different reasons, that the sphere of knowledge does not present "ultimate results." There are several reasons for that reservation. Science—and let us confine ourselves to this manifestation of the cognitive attitude—can be and is viewed, at least in certain interpretations, cumulatively; that is, results are formulated against the background of previous results. Thus there is an interdependence or even a progressive process of results. We have to distinguish between the objective of knowledge as an ideal *telos* and the results. Moreover, it is accepted that scientific knowledge is transmissible, and thus does not have the character of an "insight," which could be attributed to philosophical statements or to artistic formulations in literature. Transmission implies not only communication among different human beings but also the checking of findings. The procedure of checking implies that a finding has to be taken *cum grano salis*, as being only tentatively valid, and is thus exposed to the procedure of refutation. We distinguish between claims to validity and modes of refutation and thus introduce into the sphere of knowledge or science certain reservations about findings as such. In addition, we distinguish between the formal or logical aspect of the scientific method and its material or concrete aspect. Thus we can acknowledge the validity of the formal aspect and still take a more reserved view of its material aspect. We distinguish between science as an ongoing process and science as a product, and the product can be taken not only as tentative but also as determined by the process. Thus the refutation of the product may conform to the same basic principle that guided the process. Moreover, since the context of discovery and that of justification are not and cannot be identical, we recognize that there is some leeway within the cognitive or scientific structure for discerning ingredients or even strata. That discernment makes it possible to distinguish between the tentative and the definite within the same structure and thus, even if we grant a certain degree of relativity qua dependence upon extraneous factors, we still do not undermine the structure as such.

This observation is reinforced by the basic distinction between knowledge and truth. Even when conceding, for the sake of argument, the relativity of knowledge in the sense advanced in the present analysis, we still maintain that truth is a principle guiding all knowledge and all its results, since knowledge aims at validity, and validity in this sense amounts to truth. Whatever our definition of truth, we can distinguish here, as is rather common, between the principle of truth and the method by which our propositions are established as true. Even if we say that to prove them true amounts to verification, we still assume that there is a formal or a nominal definition of truth that is not determined by procedures, let alone by contexts outside the particular context in question. When we claim that the truth of a proposition amounts to a consensus approving that proposition, we may wonder whether a consensus, which obviously amounts to an understanding and agreement among human beings, is to be regarded as an extraneous factor. This reservation is implicit because the persons involved in providing the consensus cannot be detached from the procedures, methods, experiments, and discussions related to it, that is, from the methodical steps related to the sphere of knowledge. When wishing to verify a certain hypothesis or scientific theory, we do not seek a consensus among the lay population; rather, we expect the so-called scientific community to be involved in the procedure of forming a consensus. That community in turn cannot be detached from the course of the scientific endeavor and from adherence to the principle of truth in its core meaning, that is to say, as a principle demanding a report on the actual state of affairs. We can always take refuge in the semantic notion of truth, applying it not in the elaborate sense presented by Alfred Tarski but in its everyday sense, which in a way adumbrates different philosophical renderings of the concept of truth.

As against the cognitive sphere, the normative and ethical spheres are more exposed to a relativistic interpretation, and, to be sure, the term "interpretation" in this context is an understatement. The difference between the cognitive and normative spheres is due to several factors. A few of them have to be identified, because they in turn throw light on the hidden presuppositions of the relativistic syndrome. When we referred to truth, we did indeed refer to what is taken to be the guiding principle of a certain structure, that is, the cognitive structure. We did not ask what are the motivations for adherence to that principle, that is, what motivations differ from the relationship or correlation between the proposition and the actual state of affairs whereby that principle determines the direction of that correlation. But when we look for motivations, we are already moving from the area of truth to the area described as that of truthfulness, from the cognitive to the normative structure. The discernment of motivation opens the door to the relativistic trend, because motivations may point to factors, salient or latent, that are subcutaneous to the structure we are analyzing from what might be called an immanent point of view. We should mention here Nietzsche's statement that philosophers believe in absolute knowledge, in knowledge for the sake of knowledge, in an association between virtue and happiness, and in the comprehensibility of human actions.[14] Nietzsche wanted to change the perspective and to show that philosophers are led by instinctive moral definitions in which former cultural conditions are reflected. To be sure, when we adhere to knowledge for the sake of knowledge, we detach ourselves from those alleged motivations and can present the principle of truth as pertaining to knowledge; or, to put it differently, the pursuit of truth is coterminous with knowledge for the sake of knowledge. But when we engage in the activity of so-called unmasking and take into account instinctive inclinations or cultural conditions, we give up *ab initio* the internal structure and expose the hidden background, of which the philosopher is intentionally oblivious or from which he intentionally tries to escape. Thus the affinity between a relativistic interpretation and the unmasking operation becomes visible. We shall have to look at this aspect and see its significance for the relativistic approach to norms and ethical values without pretending to present a full-scale criticism of ethical relativism. Discernment of the quasi logic of relativism and identification of some of the failures inherent in it may suffice for the objective of our analysis.

∞

As said before, the moral sphere, more so than the cognitive attitude, is exposed to a relativistic interpretation, for reasons that seem inherent in the structure of that sphere, and it is our duty to admit those reasons precisely because the conclusion we are driving at is meant, at the least, to show that the relativistic interpretation is too rash. Let us first mention some components of the moral sphere lending themselves to a relativistic interpretation or even perhaps eliciting it. The moral sphere refers to human behavior. Even if we distinguish between codes of behavior, that is, the normative aspect, and modes of behavior, that is, the aspect of performance, we still have to be aware that extraneous occurrences or situations are involved in both. These are by definition fragmentary and piecemeal. We distinguish between mores and norms, although it is often difficult to draw the line of demarcation between the two. Mores can be understood as codes of behavior, pertaining not to one particular individual but to a number of individuals or to groups. It is possible to point to a norm that is embodied, as it were, in the code or in the mores, but we may always wonder whether in adhering to the mores, the acting human being also adheres to the code said to underlie the mores.

An additional aspect related to the mores may be understood as having a broader connotation. There can be no question that codes of human behavior depend upon situations, historical epochs, or climates of opinion. This dependence holds good even in relation to the basic occurrences in human life and their interpretation. It appears, for example, in the relationship and evaluation of adulthood and old age, or in the attitude toward death, whether these are understood as occurrences located within the human species or interpreted in the individual sphere of self or of one's fellow man. Let us take another example: the controversy over experimentation with human life. It goes without saying that if there did not exist a contemporary background of opinion in which experimentation is taken for granted and understood to be a legitimate mode of intervention by human beings in the surrounding world, the present shift toward experimentation with human life would be inconceivable. We may add that the same attitude of experimentation regards life, even human life, as one of the data for a scientific approach. Thus no intrinsic limitation on experimental intervention in the course of investigation of data may be acknowledged. The above few examples should serve to illustrate the visible and acknowledged dependence of codes of behavior on changes in basic approach generated by historical circumstances.

We may take a step further. Human behavior, even when guided by norms, takes place within a setting of reality. Reality is presupposed. Again, behavior, when guided by codes of norms, is a mode of intervention in reality and thus a superstructure upon reality. Even if we take an extreme example, for instance, Kant's categorical imperative to treat human beings as ends and not merely as means, we presuppose the existence of human beings and also acknowledge them as such. Moral behavior does not create reality; at its optimum, it conditions or molds reality. In this sense the moral sphere fundamentally depends upon an ulterior factor or stratum, which renders the pretense of a self-enclosed normative structure ontologically impossible. Indeed, relativism, explicitly or not, somehow derives its argument, or even its power of persuasion, from this scale of dependence.

Looking now at the problem from the perspective of the norm, we have to observe that actually there is more than a single norm. That is to say, phenomenologically speaking, the deontic structure is basically infused with a variety or plurality of norms, even if we disregard the dependence of the normative prescriptions upon the strata of reality. We are implicated in dilemmas of selection between life and the good life, between mercy and justice, in the imperative sense of that term, between what we may regard as short-range benefits or concerns and long-term objectives. In the deontic sphere our situation is not similar to that in the cognitive one, where there is one norm, that of truth, and where the variety of hypotheses and theories is gauged according to that single norm. Facing a variety of norms involves us in a situation of choosing or giving preference. Here again relativism, or the relativistic interpretation of the ethical sphere, appears to focus on acts of deciding, which can be presented not as being guided by norms but as being motivated by factors extraneous to the normative structure in the precise sense of the term. The exposure to realities, predilections, mores, traditions, and so on seems to be a plausible factor in determining the choice or in tipping the scale in one direction rather than another.

Let us mention in this context the most significant factor, which can be seen as the axis of the relativistic approach. In the moral sphere we are concerned not only with behavior and its codes but also with motivations. Even if we do not subscribe to *Gesinnungsethik* in the theoretical or theological sense of that term, we are nevertheless concerned with its implications for the doer or agent. One's intentions, one's character, appear as a constitutive element of the moral sphere. We distinguish between the motivation and the deed and so may adopt a lenient attitude when the motivation accords with a certain standard of honesty or a concern for another's welfare, even while the performance of the act appears to defeat the intention. In abhorring or despising hypocrisy, we are taking into account a motivation and not the performance of an act itself. This is so even if we keep in mind

Nietzsche's dictum that when we forbid certain acts, we do not forbid the intention (*Gesinnung*) but only a certain utilitarian application (*Nutzanwendung*) of this intention.

To be sure, we may ask why we refer to intentions in the moral sphere even though we do not subscribe to the *Gesinnungsethik*. A possible answer might be that intentions refer to the doer's total profile, or at least to a broader aspect than his particular acts of behavior. Since behavior is by definition segmental and takes place within situations, being, at least to a certain extent, a response to the situations, intention by its very locus indicates the agent who takes a leap into the situation and within it. Thus intention constitutes an element or a stratum more constant in nature than the fragmentary, sporadic character of the particular deeds. Since there exists a plurality of norms in the ethical sphere, our consideration of intentions may amount to a kind of compensation for what we find missing in the thematic aspect of the normative sphere. When we add to this that in the moral sphere, from a systematic point of view, we are concerned not only with the norm and with the acts but also with the agent, the dimension of intentions seems to be closer in relation to the agent and also representative of him. Deeds are many, but the honest man is one. This significance of the aspect of intentions is paramount in the moral sphere and may lead us to the paradoxical conclusion that the intentions differ from the deeds: the deeds of an agent may be determined by circumstances, not by his intentions. In this sense the line of continuity does not lead from intentions to behavior but from extraneous motivations to acts. Motivations can be disregarded, or they can be suppressed. Eventually we can even adhere to truth without being truthful, and this adherence may be determined either by the momentum of our activity and not by our intentions or by the benefit that accrues to us because of our behavior. Strangely enough, even some of the examples cited by Kant concerning the merchant of honest behavior do not lack the element of the benefit that will eventually accrue to the honest merchant, though obviously Kant wanted to emphasize the *Gesinnung* and not the result.

So far, we have tried to present in an "honest" way some of the structural arguments, implicit or explicit, that give support to the relativistic interpretation of the normative sphere and make relativism more plausible within that sphere than within the sphere of knowledge. We shall now try not to demolish relativism totally but to show that some of the previous arguments must be limited. Thus relativism cannot possibly appear as a persuasive interpretation of the normative sphere.

∞

We cannot deny the dependence of the behavior of human beings upon situations. Situations are part and parcel of man's existence. Regardless of the organic substratum, it is banal to observe that persons grow in situations and that, at least to some extent, they internalize them. Acts of behavior such as frowning or smiling are customary reactions to the conditions eliciting them, and man's attitudes are often imbued with an interpersonal content, touching a basic dimension of ethical or normative behavior.

Moreover, normative behavior, which does not depend on custom but stems from deliberate decisions, is basically or initially concerned with situations. Normative behavior as such does not create the situations that may call for it or activate it. Yet the behavior stemming from a decision refers to situations, for otherwise it would be totally meaningless. A decision to provide help is elicited by a situation that we have discerned as calling for help; we would otherwise be totally unwarranted in extending a helping hand if, for instance, prevention rather than help is needed. To be sure, one can always interpret the preventive act as help, but to prevent someone from committing a major mistake is something different from giving him aid in the form of medical assistance. Hence, situations are inherent in ethical behavior and cannot be seen as extraneous factors determining or conditioning the code of behavior. Relativism proper presents a view that emphasizes the dependence of the code of behavior on factors such as instincts or even on cultural situations, which do not belong within the structure of the deontic sphere. At this point we have to distinguish between the application to situations of certain norms and the emergence of norms in the thematic sense from situations. Applicability, as an inherent component of the normative structure, is a mode of facing the situation and differs from the allegedly derivative character of the norm. To put it differently, relatedness qua relatedness to situations is indeed an essential quality of the ethical structure. But here again relatedness differs from relativity in the sense we are exploring in this context.

The essential issue at stake is whether certain norms or notions of behavior, such as to be honest, to be fair, to behave properly, or to be guided by respect, by a sense of justice, and so on, have an inherent capacity to direct situations, or, on the contrary, whether they are names only, which, for the sake of their application, may be said to rely not on situations and on the causative factors in them but rather may be said to produce situations. The suggestion presented here is that these notions and their cognate meanings have a generative capacity of their own. To face the application of these notions is in a sense to analyze them, to juxtapose them, to perceive some of their affinities as well as some of the potential conflicts among them; but at the same time, it is necessary also to be aware of the quality of obligation or demanding, which is basically related to these notions. To be sure, they reveal a whole spectrum of meanings, some of which we only "read" and some of which we "read" while being aware of their imperative aspect. We may be conceived of as open to certain demands and imperatives, but here the correlation between our accessibility and the intrinsic meaning of those notions must be recognized. There are certain notions toward which our attitude in knowing them does not exhaust itself but is *pari passu* acknowledging them, cognizing them, and recognizing them. Relativism, with all the difficulties in which it might be involved because it is bound, at least indirectly, to preserve the notion of obligation or responding, is essentially a sort of reductionism. For relativism reduces notions to conditions or situations without ultimately being able to give a proper account of the emergence of their imperative quality from their situations. Here again we have to distinguish between applicability and validation. We are bound to be aware of the significance of a situation in order to apply the norms "properly," let us say. But the existing distance, or at least some extent of the distance, between the situation and the norm remains.[15]

Having noted the reductionist direction of relativism, we may discern an additional feature in it. Suppose that our norms depend on cultural situations. To refer to the example cited by Kant, we recognize that the prohibition against eating the blood of animals is an outcome, that is to say, an effect, of the culture of a society based on hunting. But once a prohibition—and the same applies *mutatis mutandis* to injunctions of a positive character—emerges in a society, implicitly or explicitly it may become detached from the primary surroundings. It may become an imperative, irrespective of its socio-historical roots. Even when we apply the notion of invention to the sphere under consideration, the invented norm may have an imperative quality, and the same applies even if causally or genealogically it can be traced to a certain empirical situation. Relativism in this sense is much akin to what is known as the intentional fallacy or the genetic fallacy. For once we are aware of certain sources or causes, we immediately question the substantive position of the notions, which can be traced to the circumstances that engendered them. On this issue we again have to distinguish between the situation as an occasion for the formulation of a norm and the situation as a causative factor for the formulation of the norm itself. In addition, we have to distinguish between the position of the norm in the process of its emergence or formulation and its position once its meaning has become explicit, when it may elicit certain responses or call for the formulation of certain demands. In perceiving this distinction, we notice the difference between relativism as a conception of causation referring to norms, and the meaning of norms in both their thematic and their imperative parameters.

∞

In summing up, we may say that relativism, in trying to make us aware of the empirical circumstances of the emergence of the norms, or even of their invention, falls prey to its own trend toward exclusiveness. Thus, and this is a banality in the criticism of relativism, it becomes a kind of absoluteness in terms of the circumstances within which the norms are formulated, and these circumstances, in turn, are interpreted as extraneous or are the sum total of what goes by the name of ulterior motives.

In our criticism of relativism we adopted the position of confronting the subject and perceiving its weaknesses from within, as it were, rather than of criticizing it from without. Thus, to some extent we are interpreting here the Hebrew dictum "Know how to respond to the heretic." But at the same time, we are applying the reversal of another principle and saying, "To understand everything is not to accept everything."

NOTES

1. J. S. Mill, *A System of Logic . . .*, London, Longmans Green and Co., 1941, p. 27.

2. Heinrich Scholz, "Zur Analysis des Relativitätsbegriffs—eine Skizze," *Kant-Studien* **27** (1922), pp. 369-398.

3. Max Born observed of Lord Haldane's *The Reign of Relativity* (1921) that it "merely enlarged upon the trivial proposition that 'everything is relative.' " See *The Born-Einstein Letters: Correspondence between Al-*

bert Einstein and Max and Hedwig Born from 1916 to 1955 with Commentaries by Max Born, trans. Irene Born, New York, Walker and Co., 1971, p. 78. The observation appears in Born's comments on his letter to Einstein of 7 April 1923. I am obliged to my friend Yehuda Elkana for calling my attention to that letter. Professor Elkana was kind enough to read the draft of the present paper and make several suggestions for its improvement.

4. Albert Einstein, *The Meaning of Relativity* (Stafford Little Lectures of Princeton University, May 1921), 5th ed., Princeton, Princeton University Press, 1955. Page references in the text will be to this edition.

5. Ernst Cassirer, *Zur Einstein'schen Relativitätstheorie—Erkenntnistheoretische Betrachtungen*, Berlin, Bruno Cassirer Verlag, pp. 51, 56.

6. Hans Reichenbach, "The Philosophical Significance of the Theory of Relativity," in Paul Arthur Schilpp, ed., *Albert Einstein: Philosopher-Scientist*, Evanston, Ill., Library of Living Philosophers, 1949, p. 295.

7. Ibid., p. 295.

8. Ibid., pp. 294-295.

9. Cassirer, *Zur Einstein'schen Relativitätstheorie*, pp. 84-85.

10. Gerald J. Holton, "Mach, Einstein, and the Search for Reality," in Robert S. Cohen and Raymond J. Seeger, eds., *Ernst Mach, Physicist and Philosopher*, Boston Studies in the Philosophy of Science, vol. 6, Dordrecht, D. Reidel Publishing Co., 1970, pp. 165ff.

11. Henri Bergson, *Duration and Simultaneity, With Reference to Einstein's Theory*, trans. Leon Jacobson, with an intro. by Herbert Dingle, Indianapolis, Library of Liberal Arts, 1965, pp. 157ff.

12. This point has been well taken by Samuel Hugo Bergman, "On the Philosophical Significance of Einstein's Theory," included in his *Hogei Hador* [Contemporary thinkers], Jerusalem, Magnes Press, The Hebrew University, 1970, pp. 15ff. Bergman stresses the emancipation from anthropomorphic predilections as central in Einstein's theory.

13. Adolf Grünbaum, "The Bearing of Philosophy on the History of Science," *Science* **143** (1964), pp. 1406-1412.

14. Friedrich Nietzche, *The Will to Power*, trans. Walter Kaufmann and R. J. Hollingdale, ed. Walter Kaufmann, New York, Random House, 1967, p. 407.

15. Compare my earlier analysis, "On Ethical Relativism," *Journal of Value Inquiry* **11**, no. 2 (1977), pp. 81ff.

Yehuda Elkana

THE MYTH OF SIMPLICITY

INTRODUCTION

THERE ARE TWO alternative dramaturgic approaches to history: the perspective of Greek drama, and the perspective of the epic theater. Since theater, good theater, is indeed a mirror of all that there is, an analysis of these two world-views will lend us the two general perspectives for viewing history.

Greek drama is a development of the inevitable. Fate is immutable, and man can influence only in minor details the when and where of his own destiny. The very tension that we experience in drama is caused by our knowledge of the inevitable. It is only our sense of pending doom that makes us fear what is coming, thus creating dramatic tension. The Greek tragic view of history is that the future will unfold as preordained, and after an event has occurred, it becomes clear that this was the only way it could possibly have occurred.

There is a tradition in Western culture to identify fate in Greek tragedy with the order of nature and thus to view natural occurrences and events as inevitable, just as the gods' reaction to human action must run its preestablished course. Already for Epicurus, the physicist's destiny was a symbol for utter servitude to blind forces: "It would be better to follow the myths about the gods, than to become a slave to the physicist's destiny. Myths tell us that we can hope to soften the gods' hearts by worshipping them, whereas destiny involves implacable necessity."[1]

It was a later development to apply inevitability and preestablished harmony not only to god-supervised nature but also to the unfolding of human knowledge of the world. The conviction emerged and grew, leading up to its positivistic absoluteness in the Victorian frame of mind, that not only is there one reality, with its immutable laws, but also that we humans are on a sure course to find out all, or at least cumulatively more and more about that reality: one nature, one truth about nature. Science, the chief glory of Western culture since the scientific revolution, is an inevitable unfolding of knowledge; what we know, we had to know—if not here, then there, if not now, then at another time, if not discovered by one man, then by another. Moreover, the twentieth-century scientist-philosopher, still living among typical nineteenth-century images of knowledge, does not even distinguish between the inevitability of natural occurrences and the inevitability of our knowledge of nature.

Alfred North Whitehead actually derives the spirit of modern science from Greek tragedy. According to him,"The pilgrim fathers of the scientific imagination as it exists today are the great tragedians of ancient Athens: Aeschylus, Sophocles, Euripides. Their vision of fate, remorseless and indifferent, urging a tragic incident to its inevitable issues is the vision possessed by science. Fate in Greek Tragedy becomes the order of nature in modern thought."[2] And, he would have added had he distinguished between the order of nature and our knowledge of it, fate in Greek tragedy becomes scientific knowledge in modern epistemology. As Whitehead continues the analogy, it becomes clear that for him knowledge of the order of the world is conflated with order of the world itself: "The absorbing interest in the particular heroic incidents as an example and a verification of the workings of fate, reappear in our epoch as concentration of interest on the crucial experiments."[3] To illustrate his point, Whitehead describes the joint meeting of the Royal Society of London and the Royal Astronomical Society when the Astronomer Royal announced the verification of Einstein's predictions in the famous eclipse expedition:

> The whole atmosphere of tense interest was exactly that of the Greek drama: we were the chorus commenting on the decree of destiny as disclosed in the development of a supreme incident. There was dramatic quality in the very staging:—the traditional ceremonial, and in the background the picture of Newton to remind us that the greatest of scientific generations was now, after more than two centuries, to receive its first modification.[4]

How different was Einstein's attitude to experiments! Unlike the Greek chorus at the Royal Society, for whom the outcome of the experiment would prove or disprove Einstein's theory—crucial experiments verifying the workings of fate-knowledge—Einstein had a Spinozistic certainty in his theory: it carried reality; it already involved the proper selection of facts and their interpretation. This reality had as much inevitability about it as any theory could have; the order of the world and its experimental expression counted much less. There was no artificial separating out of facts. The "facts" of those who separated facts from theories did not seem relevant to him. Ilse Rosenthal-Schneider reports the story:

> Once when I was with Einstein in order to read with him a work that contained many objections against his theory . . . he suddenly interrupted the discussion of the book, reached for a telegram that was lying on the windowsill, and handed it to me with the words, "Here, this will perhaps interest you." It was Eddington's cable with the results of measurements of the eclipse expedition [1919]. When I was giving expression to my joy that the results coincided with his calculations, he said quite unmoved, "But I knew that the theory is correct"; and when I asked, what if there had been no confirmation of his prediction, he countered: "Then I would have been sorry for the dear Lord—the theory is correct."[5]

Lest the analogy between the Greek dramatic conception of human destiny and our modern, nineteenth-century-influenced epistemology be misunderstood, let me make it clear that I do not mean the emotional aspect of tragedy here. And not accidentally so: the essence of dramatic tragedy, as Whitehead and others clearly saw, is not in unhappiness; it is in the "solemnity of the remorseless working of things," the inevitability of destiny.

Such a view of science leads us to believe that there is only our science to be discovered—that the great truths of nature, had they not been discovered by a Newton or an Einstein, would have been discovered by someone else sooner or later; that unlike religion, art, music, or political ideology, there is no such thing as "comparative science" among different cultures, and that any attempt at creating it is meaningless; that the temporal unfolding of our knowledge of the world, though it may be an infinite process, is nevertheless fixed and inevitable. This is the style of Greek tragedy. As mentioned, this attitude became part of our epistemology during the scientific revolution, reached its glorious climax with the success of the Newtonian world-view in the nineteenth century, and characterizes the positivistic view of science. Einstein was among the first and perhaps greatest twentieth-century thinkers to reject science as a Greek tragedy.

Epic theater is quite different. The idea of epic theater has been developed by Bertolt Brecht and Walter Benjamin, and its main historical thesis, as formulated by Benjamin, is very simple, in glaring contrast to Greek drama: "It can happen this way, but it can also happen quite a different way."[6] What does this "it" refer to? The "it" is the event. Epic theater, in order to make its point, purposefully avoids historical facts that the audience is aware of, lest they lapse into the tragic mood of knowing what is inevitably coming. Life is unpredictable, and events can go in any direction; therefore life is unsensational. What is true of historical inevitability also holds for psychological inevitability, and this, too, is avoided. In short, epic theater is a relaxed, nonsensational, reflective attitude to unpredictable events. To put it in another formulation: the historical question is not what were the sufficient and necessary conditions for an event that took place, but rather, what were the necessary conditions for the way things happened, although they could have happened otherwise.

Such a view is a very undramatic one. It is typical of the Chinese theater to see to it that what is shown on stage is unsensational. If the dramatic, tragic view of fate is a thema in Greek theater, then the emphasis on the unsensational as a desirable ideal is its antithema.[7] As Plato recognized, and as has been repeatedly claimed ever since, the highest form of man, the *sage*, is of a very undramatic nature.[8] The sage is not a Stoic figure who has learned to overcome his involvement in the "momentary and the personal" but one who has eliminated it altogether and exists on a different level.

Einstein's world was not that of the Greek drama, nor was he a Stoic. Einstein was a genuine sage, his world an epic theater. There was no inevitability about the order of nature: ever again an unpredicted and unexpected new reality could emerge. That it was unexpected did not speak against it. If "good" theory brought the new picture, it constituted reality; the new picture was not accompanied by an inner conviction that the world had to be such and no other just because it turned out to be such.[9] So too in the moral sphere. Einstein did not believe in any predetermined moral order of the world that all humans were required to follow. With Spinozistic rigor and certainty he *chose* a moral code of his own, with his own priorities among its components, and he stuck to this code all his life. As a "free man" he freely chose his own binding principles, and no element of the divine or of an inexorable law of destiny entered into the choice. It was unsensational, relaxed, and permanent. Einstein would have repudiated Seneca, who "maintains that the Divinity has determined all things by an inexorable law of destiny, which he himself decreed but which Himself obeys."[10] His chosen moral code was all-embracing. It was reality in the same sense as his own unifying theory of the world—once formulated, it becomes reality. So also in music. Einstein often said that he felt a distaste for Beethoven's very dramatic orchestral works. He loved the architectonic majesty of Bach, but he was closest to Mozart. There the symmetry, the fundamental simplicity, the rational "inner perfection" created an epic distance from the dramatic and from the "merely personal."[11]

It is the main claim of this paper that Einstein thought in the epic manner here described. His is a very complex world-view, and, needless to say, I do not claim that Einstein expressed himself in the terms in which I shall describe his world. However, I believe that this description will emerge convincingly correct in the following arguments.

In epic theater the only historically meaningful question is: Why did it happen the way it did, since it could have happened otherwise? According to this view, science could have developed differently; other discoverers could have discovered different laws of nature; there is nothing inevitable in the uniqueness of Western science; a "comparative science" between different cultures is meaningful; lessons can be drawn from history for future use. All in all, this is an optimistic perspective.

Idealistic attitudes, whether reductionism, positivism, or behaviorism, share the Greek dramatic view of science. On the other hand, the attitude of historical sociology of scientific knowledge is that of the epic theater, and it is this world-view that underlies the following theses and the consequent study of the role of simplicity in Einstein's life and work.

In the following argument I shall use concepts and ideas that I developed and argued elsewhere. Since the type of argument here leads to a hermeneutic circle (not a vicious one!), I shall try to overcome this inherent difficulty by referring the reader to other publications.[12] At the same time, I offer here a series of theses or presuppositions that I shall not try to defend now. A brief explanation accompanies each.

1. A progressivist view of knowledge is adopted: knowledge grows.
2. Knowledge grows dialectically by way of ongoing critical dialogue—between competing metaphysics and theories in the body of knowledge; between competing images of knowledge; between competing normative tenets.

   a. The body of knowledge contains all statements about the subject of discourse—the physical world, life, society, the individual. It includes "true" as well as "false" knowledge; it includes metaphysical claims and testable theories; it includes both what is at a given point called theory and what is called fact.

   b. Images of knowledge are statements about knowledge, that is, sources of knowledge (senses, mind, revelation, tradition, authority, novelty), aims of knowledge (for use, for ornament, for finding Truth), methodologies, and so on.

   c. Normative claims are political ideologies, religious views, moral codes. Normative claims influence what is thought about knowledge, that is, our images of knowledge. The images serve as selectors among the possible problem-choices in the body of knowledge. The distinction between the body of knowledge and images of knowledge is not permanent but time- and context-dependent, that is, it itself depends on yet other images of knowledge.
3. Images of knowledge are held by groups of people—societies, members of a laboratory, practitioners of a discipline—and are thus shaped by sociopolitical factors.
4. One of the most important scientific activities, problem-choice, is made in terms of images of knowledge: for example, relevance, doability, beauty, worthiness, frontier of knowledge, newness, level of support, ease, difficulty, importance. In other words, the body of knowledge does not tell us on what problem to work.
5. All dimensions of culture—religion, ideology, common sense, art, music, and *science*—are *cultural systems* (a term borrowed from Clifford Geertz).[13] All are historically constructed, and their standards of comparison are time- and culture-dependent. All are constructed by man and yet serve as a scaffolding or web to which man holds.
6. All cultural processes, including science, are best described by way of *thick descriptions*—an iterative interactionist process between sense-experiences and man-invented theories.
7. Most people in most cultures treat the physical and the moral

world as real: for example, moral reality is the opposite of seeing moral choices as fine connections. That is, they attribute an existence to the world that is independent of our knowledge of it, although the description of that world is a human construct. When coming to compare different "worlds"[14] or wishing to translate from one world to another, people discover that there is no easily available external-to-all world (a context of contexts) with regard to which absolute criteria of translation and comparison can be developed. Then the choices are seen to be either to decide that "our world" is absolute reality—the external-to-all world (a basically religious view)—or to decide that, realism having failed, relativism holds. *Two-tier thinking* is the ability and readiness to admit that there is no absolute, external-to-all frame of reference with which we can compare any two frameworks in order to find out which is "true," or which translation is correct, or which moral dictum is "right." Thus, all frameworks are selected relativistically, but once a framework is selected, we relate to it realistically.

8. Realism is an interactionist view of nature and of the human observer; it assumes a wide spectrum of sources of knowledge (faith, intellect, senses) without specifying an absolute hierarchy among them; it is an epistemological refusal to accept rigid dichotomies. This type of realism is not an opposite of relativism, but it is the realism of the two-tier thinker. Though naturally neither Bacon nor Einstein thought in these terms, both Bacon and Einstein were realists of this type, and their philosophy of science was similar.
9. There is coherence, though no logical connection, among the views and life of every person. This is especially true of Einstein. There is strong coherence between his theories of the mind and his theory of our awareness of the world.
10. Simplicity and complexity are themata that have interacted dialectically as thema and antithema in Western culture ever since Greek times.
11. Themata such as simplicity and complexity, atomicity or continuum gain new content in a culture- and context-dependent way according to the prevalent images of knowledge (see next section).
12. Einstein was a two-tier-thinking realist as far as the contents of the world is concerned. However, in his epistemology the

methodological rule of the inner simplicity of a theoretical system is an absolute framework—an attribute of God.

THEMATA AND IMAGES OF KNOWLEDGE

Gerald Holton has introduced the concept of themata into the language of history and philosophy of science. It is a fruitful notion that underlies a research program: "Thematic analysis allows discernment of some constancies or continuities in the development of science, of relatively stable structures that extend across supposed revolutions and among apparently incommensurable rival theories."[15] Holton has traced the influence of themata on scientific thought and creativity in a classical and modern physics, with Kepler and Einstein being the central case studies. He has observed that a dialectic between thema and antithema occurs. In one case he analyzed part of a paper by a leading theoretical physicist and demonstrated the various themata being invoked.[16]

I would like to complicate matters. The concept of themata as a historical research tool is important, and no doubt it is of moment that themata actually recur with the same name again and again. Yet we must take note that although themata seem constant and continuous, they gain their meaning in a context-dependent way through the images of knowledge prevailing in a given culture, time, place, and discipline. Thus the historical question shifts. The problem is not whether the atomistic thema occurs in the thought of Democritus, Newton, Mach, Bohr, and Einstein but rather, what was understood by atomism in each case and why, in spite of differences in meaning, does the object nevertheless seem to represent the same thema. Each thema gains a context-dependent meaning; moreover, it is the meaning of other themata that lends a meaning to each. This interdependence makes a systematic, complete description very difficult. What we have is a cultural analysis that is *essentially incomplete,* and the only way to deal with it is by way of thick description.[17] For example, the thema of conservation in a physics with several disconnected conservation laws differs from the thema of conservation in a unitary physics where conservation laws tend to be reduced to as few as possible, and ideally to one.[18] In the first case, the thema of conservation goes well together with a dualistic or pluralistic thema or with atomism as a thema; in the second case, the thema of conservation will not tolerate any themata other than continuum and unitariness. Elegance or simplicity in a historical explanation rarely implies a monocausal explanation. Biological simplicity may mean geometrical symmetry or analogy between function and form, but it will seldom mean conceptual sparsity. Ernst Mach's quest for simplicity is antithetical to the thema of abstractness; Henri Poincaré's is neutral to it; Einstein's is contingent on it. Concrete sense experiences, as we shall see, are for Einstein the sign of disjointed, atomistic complexity. In idealistic world-views, certainty and completeness are coherent themata. For the epic theater world-view—that is, for two-tier-thinking realists—certainty and completeness are antithetical. These different views are not merely the personal views of a Mach, a Poincaré, or an Einstein, but their versions of socially determined images of knowledge: sources of knowledge, aims of knowledge, hierarchy among the sources, images of discipline or methodological principles.

There is an amusing illustration of what I have in mind in the Born-Einstein letters. Born tells of having received from Einstein a copy of his own book, *Natural Philosophy of Cause and Chance*, with Einstein's marginal comments. In the book Born had written: "With regard to simplicity, opinions will differ in many cases. Is Einstein's law of gravitation simpler than Newton's? Trained mathematicians will answer yes, meaning the logical simplicity of the foundations, while others will say emphatically no, because of the horrible complications of the formalism." Einstein added in his handwriting at the margin of this passage: "The only thing which matters is the *logical* simplicity of the *foundations*'." To this Born the empiricist remarked: "I agree with this, as a trained mathematician, but I am unable to condemn the other point of view entirely. After all, what really matters in the end is whose formulae do more justice to the observations, Newton's or Einstein's."[19]

So, clearly, Holton's emphasis on the role of themata is justified, and his view that thema and antithema interact dialectically is correct. Simplicity and complexity constitute such a pair, and they played an important role in the work of Einstein and his contemporaries. I have tried to show what diverse meanings "simplicity" had in different contexts. To relate these different meanings of simplicity to the various images of knowledge in the thinking of individuals is a major task still to be undertaken.

TWO-TIER THINKING IN EINSTEIN'S EPISTEMOLOGY

Einstein's dialectical treatment of the tension between simplicity and complexity as thema and antithema shows his two-tier thinking: he makes the relativistic choice of a world, which he then treats as reality. Needless to say, Einstein does not speak in these terms, and I am not at all sure that he would agree with my analysis, yet I hope to make a convincing case for recasting his epistemology in terms of images of knowledge, themata, and two-tier thinking, which form a hermeneutic circle.

Consider a passage from one of Einstein's most important and interesting philosophical writings: his address "Motiv des Forschens," delivered before the German Physical Society in Berlin in 1918 on the occasion of Max Planck's sixtieth birthday:

> one of the strongest motives that leads men to art and science is escape from everyday life. . . . A finely tempered nature longs to escape from personal life into the world of objective perception and thought. . . .
>
> . . . Man tries to make for himself in the fashion that suits him best a simplified and intelligible picture of the world; he then tries to some extent to substitute this cosmos of his for the world of experience. . . .
>
> . . . [T]he theoretical physicist's picture of the world . . . demands the highest possible standard of rigorous precision in the description of relations, such as only the use of mathematical language can give. . . . Supreme purity, clarity, and certainty at the cost of completeness. . . .
>
> The supreme task of the physicist is to arrive at those universal elementary laws from which the cosmos can be built up by pure deduction. There is no logical path to these laws; only intuition. . . . one might suppose that there were any number of possible systems of theoretical physics all equally well justified. . . . in practice the world of phenomena uniquely determines the theoretical system . . . this is what Leibnitz described so happily as a "pre-established harmony."[20]

Man is surrounded by uncertainty, and his daily life is dreary, crude, and hopeless. The aim of knowledge is to gain certainty by leaving personal life behind and escaping into the "world of objective perception and thought." Such an objective world constitutes reality. In order to give certainty and a semblance of objectivity, this reality must be simple and intelligible. What we observe and experience is neither simple nor intelligible, nor is it determined enough. We can construct many possible worlds out of our complex and unreliable direct experiences. The painter, the poet, the philosopher—all create their own distinctive "objective and certain world." And even physicists will construct different worlds. Clearly, every individual must make up his mind which of the "any number of possible systems of theoretical physics all equally well justified" is superior to the others. The superior one is that *theory* from which all phenomena can be deduced in the most rigorous, mathematical fashion. Had such a deduction been possible, there would have been one superior theory, one absolute reality. However, such a deduction is beyond the capacity of human intellect! Our description is and will remain necessarily incomplete. Moreover, no logical path leads even to our partial theoretical laws; we arrive at them after a leap of intuition. Thus, we have to choose the *simplest* and *most intelligible* description. Such a choice will be relative, depending on our images of knowledge and our "domain of experience." This is the relativistic tier of two-tier thinking. However, once we arrive at a self-constructed (relativistically selected) "simplified and intelligible picture of the world," we then "substitute this cosmos . . . for the world of experience and thus . . . overcome it." It becomes reality, in the form of theory, and this is the realistic tier of the Einsteinian epistemology.

To recapitulate: our experience is complex and diverging; we must *overcome* it by selecting from among the many world-descriptions, all incomplete, the one that best answers our criteria of simplicity and intelligibility and by substituting for our experiences this one picture, which then serves as reality itself. That we tend to find that this self-constructed world actually fits our incomplete, complex world of experience is due, according to Einstein, to preestablished harmony. Einstein cites Leibnizian preestablished harmony in order to try to escape from the silent unease that his own epistemology causes him. But there is no escaping from two-tier thinking. Leibniz would not have accepted Einstein's description of how we construct our world and would have rejected Einstein's admission that it is beyond the power of human intellect to find the one, overarching description of the world. Or, alternatively, had Einstein succeeded in convincing Leibniz of such severe limitations of the human intellect, then Leibniz would have claimed that preestablished harmony exists only between the facts and the ideal theory (not to be achieved by humans), and not between the world of experience and a relative, temporary, partial theory of the world.

The process of substitution of a simple cosmos for the world of experience is itself a very complex work of construction. Here the dialectic between simple and complex, thema and antithema, reaches its full expression. Logic too is simple, but as there is no logical bridge between our complex experiences and simple theories, we have to undertake the complex substitution in order to gain purity, clarity, certainty.

In the address quoted above, Einstein claimed that not only the natural scientist but also the painter, the poet, and the speculative philosopher creates his own cosmos, which he then substitutes for his world of experience. However, unlike Einstein, most natural scientists are unaware of this complex process and generally do not, in the course of their daily work, think in terms of such concepts as "many possible worlds" and the substitution of a self-created, simple, intelligible cosmos for a world of complex experiences.

Other areas of the cognitive quest have a much greater degree of self-awareness and understanding of the process by which man constructs reality: sociology and anthropology are at the forefront. The thick description that is the anthropologist's way of doing cultural analysis is a consciously complex process aware of its own limitations and essential incompleteness. *Cultural analysis* is the sum total of our attempt to understand the body of knowledge, the images of knowledge, and the ideologies acting in a culture; to seek out causal relations; to attempt to translate one culture into another in which we are at home; to comprehend the ways by which reality is constructed in a culture; to analyze the structure of the context on which the images of knowledge depend. Thus cultural analysis is "intrinsically incomplete. And worse than that, the more deeply it goes the less complete it is. It is a strange science whose most telling assertions are its most tremulously based, in which to get somewhere with the matter at hand is to intensify the suspicion, both your own and that of others, that you are quite getting it right."[21]

The parallel between the inevitability of the world order and our knowledge of it, as explained above in connection with the Greek dramatic view of the world, recurs here in connection with the completeness of our picture of the world and of our knowledge of theory construction. According to Einstein, our knowledge of the world is essentially incomplete, and we have seen that the same is true of our knowledge of the way in which we construct our knowledge of the world. We could say that description of the world in physics, when the body of knowledge is the physical world, is analogous to cultural analysis in anthropology, where the body of knowledge is a culture and the way it constructs its reality. On the other hand, we could say that a physical theory—that is, the way we describe our physical world—is part of our construction of reality in our culture—that is, part of our cultural analysis. In this case we have to admit essential incompleteness both in our knowledge of the world and in our epistemology. The deeper we go, the more complex the description becomes, the more we are in need of simplicity and intelligibility in our relativistically constructed cosmos, which we provisionally take as reality. The price we have to pay for gaining such simplicity and intelligibility—that is, one kind of certainty—is giving up another kind of certainty, namely, completeness.

INTERACTIONIST REALISM

In the eighth of my presuppositions I claimed that Einstein, like Bacon, was an interactionist realist. By this I meant that he was a two-tier thinker whose realism consisted in adhering to a world-view consciously chosen from among other possible world-views. As discussed above, the world-view chosen is the simplest and most intelligible, being the result of interaction between our diffuse and nonintelligible sense experiences and our speculations. Their interaction exhibits a preestablished harmony between them. This view requires elucidations.

Following Geertz and going beyond him, religion, ideology, common sense, ethics, art, and music are all cultural systems; moreover, science, too, is a cultural system.[22] Geertz, like Max Weber before him, considers culture as webs of significance spun by man himself, and yet man is suspended in them. We have seen above that our world-views are constructed by us, and yet we operate within their framework as if they constituted reality. Such world-views are part of our culture. In this view cultural analysis is "not an experimental science in search of law, but an interpretative one in search of meaning."[23] For Einstein, the poet, the philosopher, the painter, as well as the natural scientist, construct worlds of their own. Each such world is constructed in a sphere of culture, or as a dimension of it, and constitutes a cultural system. Talking of common sense, Geertz tells us about cultural systems: "if common sense is as much an interpretation of the immediacies of experience, a gloss on them, as are myths, painting, epistemology, or whatever, then it is like them *historically constructed, and like them subjected to historically defined standards of judgement. It can be questioned, disputed, affirmed, developed, formalized, contemplated, even taught,* and it *can vary dramatically from one people to the next. It is, in short, a cultural* system . . ." (italics added).[24]

Science, too, is historically constructed and is subjected to historically defined standards of judgment. Moreover, science can be questioned, disputed, affirmed, formalized, contemplated, and taught. Above all, science varies dramatically from one people to the next; in some aspects it may vary dramatically from one discipline to another, and it certainly does vary dramatically from one period to another. How, then, are we to proceed with a cultural analysis of science? There are several possibilities.

The *positivist bias* is to act as if complete objectivity in describing the world of experience were possible. Such a view is essentially religious: it presupposes the existence of an absolute external framework in which absolute criteria of truth and validity hold and in which results of translation from one language or conceptual framework to another can be compared in order to find out which is the *true* translation. In short, it assumes a third world where Truth resides. This view is religious because a framework independent of social and cultural contexts is above all other frameworks, and thus it amounts to a way of thinking of God. It is not an accident that all absolutist philosophical views were at one time admired or denounced, as the case may be, as religious. Einstein was not a positivist of this kind, and if religion is so understood, he was not religious either.[25]

The exact opposite to positivist objective certainty is the *subjective attitude*: we cannot claim completeness or certainty in our cultural analysis; therefore, whatever we tend to claim is equally reasonable. This view recently gained some popularity in Paul Feyerabend's version that "anything goes."[26] It amounts to relativism unhindered.

The solution is the *interactionist* view, *two-tier thinking,* the back-and-forth movement between sense experiences and speculation. Thinking, according to this view, involves the technique of thick description, and its main characteristic is that it escapes the artificial choice between "total objectivity" and complete relativism. The term "thick description" was introduced by Ryle and then developed by Geertz for ethnography.[27] Thick description is the most fundamental everyday activity of the ethnographer: he interviews informants, observes rituals, identifies kin terms, traces property lines, takes a census of households, and keeps a journal. For Ryle, thick description is a way of describing the complexity of thinking. He starts with the most elementary one-layer activity—for example, counting the number of cars on a street. Describing this activity involves a very "thin description." Then, layer by layer (or step by step on a ladder), the activity becomes more complex and its description thicker. Thus, the kind of description we have to give when relating what a person is doing is somewhere on a continuum between the very thin and the very thick, the thickness depending on the kind of activity we are describing. In ethnography any description consists of a translation; therefore, whatever the ethnographer relates can no longer be a thin description. This multiplicity of complex conceptual structures, many of them superimposed onto one another, is also a good description of what a scientist does: he formulates problems; chooses phenomena, that is, demarcates the seemingly self-evident from the seemingly puzzling; observes selected relevant motions, changes, processes, sizes, colors, and the like; traces interactions and connections between previously firmly or vaguely admitted unconnected phenomena; reduces one phenomenon to another and then changes frameworks, going in reverse order; counts the population of a newly determined independent unit; describes his experiments. No doubt, he must deal with a multiplicity of complex conceptual structures, most of them superimposed on one another.

Einstein's solution to cultural analysis of science appears to be different from that of thick descriptions. Einstein is willing to give up completeness for the sake of "supreme purity, clarity and certainty"—in other words, to substitute for complexity a *simplified cosmos*, a pale reconstruction, which, however, as we have seen, "can be brought within the domain of our experience." Had he actually escaped the complexity of thick description? Not at all.

The difference between thick description and Einstein's reconstruction of a simplified world is only apparent. Einstein did not actually believe that he could reduce a complex world of experiences to the simplified cosmos. Just as with the Gestalt psychologists or the cognitive psychologists of the Russian school (L. Vygotsky and A. Luria), Einstein never even attempted to account for complex processes or events in terms of simple ones. Just as for Vygotsky the adopted psychological theory "mediates" between complex experimental data and the speculative intuition to which we *leap*, so for Einstein the simplified cosmos that we adopt as reality is the result of the long and complicated process by which the experimental data interact with our speculative leaps into theory.

The process of "mediation" in Vygotsky's case, or of the interaction between what we intuit and what we experience, involves thick description. As Einstein remarked in "Motiv des Forschens," there is no methodological certainty about the road from experience to that simplified cosmos that can be built up by "pure deduction." This process, too, requires thick description: first coming to intuitive leaps and then deducing phenomena, comparing the sense experiences to another, "corrected" intuition, and so on. We have noted that the result of the thick description is a fit between the experience and the adopted cosmos, but a fit that is not a Leibnizian preestablished harmony. The intuitions to which the scientist leaps are individual. There is no Kuhnian mature science, not even theoretical physics, that is ever one-paradigmatic. Different intuitions, different thick descriptions, different simplified and intelligible pictures of the world are continuously competing with one another.

The type of reasoning involved in thick description is typical of cunning reason[28]—the same cunning reason that Einstein attributes to God in his famous dictum, "God is cunning but he is not malicious," or to nature when he says that nature's secrets must be wormed out of her. This type of reasoning and of knowing has always been with mankind, but it has often been overshadowed by the official epistemological reason that is typical of logic and of all rational reconstruction. Cunning reason can be seen in action as early as ancient Greece, alongside *episteme* and *metis*, only to be dismissed as merely practical and lacking in rigor. Yet it plays an important role in Homer, in Greek drama, and even in Plato. According to Marcel Detienne and Jean-Pierre Vernant, "*Metis* implies a complex but very coherent body of mental attitudes and intellectual behaviour which combines flair, wisdom, forethought, subtlety of mind, deception, resourcefulness, vigilance, opportunism, various skills and experience acquired over the years. It is applied to situations which are transient, shifting, disconcerting and ambiguous, situations which do not lend themselves to precise measurements, exact calculation or rigorous logic."[29]

The rational reconstructionist would like to disregard cunning reason, but it is an integral part of the scientific enterprise. Einstein at least, unlike others, admitted this freely and tried to take *metis* into account in his epistemology, in his interactionist realism.[30]

THE INTUITIVE LEAP

There is no logical bridge between the host of our fragmented sense experiences and our overarching world-view. Which, then, is our best approach to reality? Einstein answered: by *intuition*. That intuition works at all is owing to the essential simplicity of nature: God does not play dice. Nature's unfolding is inevitable, and we can intuit bits and pieces. If our knowledge is not statistically complex (for Einstein, all statistical averages are complex) but deterministically simple and symmetrical, we can then tease nature's secrets out of her. This is the meaning of Einsteinian intuition. It is an interactionist view of reality, that is, interaction between complicated sense-experiences and simple general theories. It is not a logical bridge. Nor is the interaction Kantian; it is not our preconceived ideas that shape reality: "Kant, thoroughly convinced of the indispensability of certain concepts, took them—just as they are selected—to be the necessary premises of every kind of thinking and differentiated them from concepts of empirical origin."[31] But instead of distinguishing the a priori from the empirical, Einstein rather conceived of a continuum between "experience-near" and "experience-far" concepts.[32] In his own words: "I am convinced, however, that this differentiation is erroneous, i.e., that it does not do justice to the problem in a natural way. All concepts, even those which are *closest to experience*, are from the point of view of logic freely chosen conventions . . ." (italics added).[33]

In my terms, it can be said that just as Einstein realized that all observation is theory-laden, so he also knew that all theory is observation-laden. One of the favorite games of twentieth-century philosophy, namely, the distinction between observation language and theoretical language, is thus seen as spurious, except in a well-defined, relativistically selected conceptual framework in which it is always possible to make such a distinction! (Again, two-tier thinking!)

Einstein held to an absolute criterion of simplicity: nature (which for him was identical with God) is itself simple and intelligible, even if only partially. As he says somewhere, "Nature conceals her mystery by her essential grandeur, not by cunning." The mystery is how simple ideas fit to produce a reasonable harmony between theory and experience. It is in this light that Einstein's oft-repeated remark on God's not playing dice must be understood. To Born he wrote: "You believe in the God who plays dice, and I in complete law and order in a world which objectively exists, and which I, in a wildly speculative way, am trying to capture. I firmly *believe*, but I hope that someone will discover a more realistic way, or rather a more tangible basis than it has been my lot to find."[34] According to Ronald Clark, Louis de Broglie tells of Einstein having explained to him "that all physical theories, their mathematical expressions apart, ought to lend themselves to so simple a description 'that even a child could understand them.' "[35] And for Einstein, simplicity was equivalent to beauty.

Einstein tried to grapple with his own concept of intuitive leap. He discussed it in his essay on Bertrand Russell and in a late letter to his friend Maurice Solovine.[36] Einstein called sense impressions the "raw materials" for knowledge. He distinguished the "aristocratic illusion concerning the unlimited penetrative power of thought" from the "plebeian illusion of naïve realism, according to which things 'are' as they are perceived by us through our senses."[37] Although he admits that such illusions or prejudices can lead to great achievements, he shows how they must finally break down. Russell had already argued that "Naïve realism leads to physics, and physics, if true, shows that naïve realism is false."[38] But Einstein claims that this argument works the other way, too: seemingly naive realism, or the physical mode of thought that is typical of the natural sciences, should contradict the skeptical thinking of Hume or Berkeley:

> If Berkeley relies upon the fact that we do not directly grasp the "things" of the external world through our senses, but that only events causally connected with the presence of "things" reach our sense-organs, then this is a consideration which gets its persuasive character from our confidence in the physical mode of thought. . . .
>
> It was, however, the very same physical mode of thought and its practical successes which have shaken the confidence in the possibility of understanding things and their relations by means of purely speculative thought.[39]

Concepts are free creations of our thought ("freely chosen convention" as we have seen) but must somehow be anchored in sensory experiences. They are not directly created by the senses, nor are they abstractions from them. The more primitive the concepts are, the more difficult it is to realize that they are independent creations of thinking.[40] Concepts are created by a free leap of our imagination from our sensory experiences. The nature of the leap is, and will remain, unknown:

> The concepts and propositions get "meaning," viz., "content," only through their connection with sense-experiences. The connection of the latter with the former is purely intuitive, not itself of a logical nature. The degree of certainty with which this connection, viz., intuitive combination, can be undertaken, and nothing else, differentiates empty phantasy from scientific "truth."[41]

Again, Einstein expresses his complicated *interactionist realism*. It is the strongest epistemological link between Francis Bacon and Albert Einstein. One of Einstein's letters to Maurice Solovine, written late in his life, on 7 May 1952, fully explains the intuitional leap.[42]

In the framework of such an interactionist point of view, the question about the reality of the external world simply does not arise. There is no tripartite distinction of external world, sense experiences, and world picture without a logical bridge from the first to the last and with no hope of reducing to logical steps the intuitive leap from one to the other. However, our theoretical framework, if constructed according to the absolute rule of simplicity (that is, sparsity of independent basic elements), will fit most experimental facts and is thus reality. This is a reality created by the mind, reached intuitively, and yet we tend to believe in it independent of our knowledge of it. It is the realism of two-tier thinking, not a convention-turned-Reality, as Eddington's Kantianism would have it. Though we cannot reduce the intuitive leap into logical steps,[43] we know that it involves organizing the complex of sense-experience. The organizing principles change with time and knowledge. As a result, not only are the intuitions of different individuals different, but those of the same individual change with time. Einstein emphasized his opinion that there are no final Kantian categories. According to Einstein's images of knowledge, the comprehensibility of nature is not put in danger by the inclusion of an unexplainable intuitive leap. Furthermore, simplicity in the foundations is not a statement about nature (that is, about a part of the body of knowledge) but becomes a source of knowledge. There are no presuppositions here about nature, only about our knowledge. Einstein does not ask questions about the reality of the external world; but theory, when constructed according to the rules of simplicity, touches on reality. It becomes reality in the interactionist two-tier-thinking way. Therefore, all the Einsteinian absolutes—symmetry, simplicity, harmony, ordering principles—are images of knowledge, that is, statements about our knowledge and not about the world. It is this unique epistemology that makes Einstein a revolutionary in conceiving possible worlds and a conservative determinist in conceiving his epistemology.[44]

Since one arrives at the leap to generalized theory by sheer intuition, we must now determine the absolute criterion of simplicity that assures reality.

THE IMAGES OF SIMPLICITY

For Einstein, the concept of simplicity is intimately connected with the concepts of harmony, beauty, and symmetry. There must be one, generalized, simple concept of field or law; these three concepts are organizing principles for the intuition, and phenomena will behave accordingly. Asymmetry is disruptive, unnatural, misleading.

There are many examples of this imagery in Einstein's thinking. The first sentence of the 1905 relativity paper refers to the fundamentally important concept of asymmetry: "It is known that Maxwell's electrodynamics—as usually understood at the present time—when applied to moving bodies, leads to asymmetries which do not appear to be inherent in the phenomena."[45] Here Einstein is referring to empirical asymmetries, but it is clear that for him symmetry is simple and good; asymmetry, complex and bad. As plain as that. However, symmetry can be of many sorts—conceptual, mathematical, geometrical, experimental or methodological. A fundamentally dualistic world-view has symmetries different from those in a unitary conception of the world. In a pluralistic universe, the most complicated probabilistic-statistical considerations may constitute the element of symmetry and thus a kind of simplification. Every type of dualism is actually a kind of asymmetry and is therefore undesirable. "It is my opinion therefore that the next stage in the development of theoretical physics will bring us a theory of light that can be interpreted as a kind of fusion of the wave and emission theories."[46]

After the successful completion of the special theory in 1905,

Einstein was deeply disturbed by the lack of symmetry between uniform and non-uniform motion. As he wrote, "We feel compelled at the present juncture to grant a kind of absolute reality to non-uniform motion."[47] But this was in opposition to symmetry in nature and thus to his new general principle of relativity. It could not be maintained, and indeed, the general theory did away with the asymmetry. As Dennis Sciama put it: "This was displeasing to Einstein who felt that the harmony of his theory of relativity required that all motion should be equally relative."

After completing the general theory of relativity, Einstein set himself the inhumanly complex task of simplifying our world-view one further and last step:

> The conceptual foundations of the General Theory have no relations with the electromagnetic field. These facts suggest the following questions: Is it not possible to generalize the mathematical foundations of the theory in such a way that we can derive from them not only the properties of the gravitational field, but also those of the electromagnetic field?"[48]

As Einstein said to Pauli in 1916, "For the rest of my life I want to reflect on what light is."[49] Light combined mechanical and electromagnetic characteristics and thus offered him hope for a solution. Once he thought for a moment that he had succeeded, and he gave an interview:

> For years it has been my greatest ambition to resolve the duality of natural laws into unity. This duality lies in the fact that physicists have hitherto been compelled to postulate two sets of laws—those which control gravitation and those which control the phenomena of electricity and magnetism. Many physicists have suspected that two sets of laws must be based upon one general law, but neither experiment nor theory has, until now, succeeded in formulating this law. I believe now that I have found a proper form.... The purpose of my work is to further this simplification, and particularly to reduce to one formula the explanation of the field of gravity and of the field of electromagnetism. For this reason I call it a contribution to "a unified field theory." . . . Now, but only now, we know that the force which moves electrons in their ellipses about the nuclei of atoms is the same force which moves our earth in its annual course about the sun, and is the same force which brings us the rays of light and heat which make life possible upon this planet.[50]

When speaking of constants of nature, Einstein identified *simplicity* with *intelligibility*[51] of nature:

> I would like to state a theorem which at present can not be based upon

anything more than upon a faith in the simplicity, i.e., intelligibility of nature: there are no *arbitrary* constants of this kind; that is to say, nature is so constituted that it is possible logically to lay down such strongly determined laws that within these laws only rationally completely determined constants occur (not constants, therefore, whose numerical value could be changed without destroying the theory).[52]

A final example:

it is impossible to deduce from the field equations alone an interaction between bodies, which can be described separately by means of solutions of the system. For this reason all theories up to now required, in addition to the field equations, special equations for the motion of material bodies under the influence of the fields.[53]

This situation, according to Einstein is "genuinely complicated." The simplification would come if "one had the field-equation of the total field, [then] one would be compelled to demand that the particles themselves would *everywhere* be describable as singularity-free solutions of the completed field-equations. Only then would the general theory of relativity be a *complete* theory.[54]

If it is presupposed that there are two different force fields, electromagnetic and gravitational, and that the one needs an ether to be transmitted, then it is symmetrical to expect the other field also to need a substratum of transmission. If some kind of unity is presupposed, it will be symmetrical to demand that the same ether transmit both fields. If, on the other hand, a basic dualism is acceptable, why not look for a different substratum for each field? If the concept of ether is abolished—that is, if it stops being "the substratum of thought in physics"[55]—then both fields are transmitted through the void. And Einsteinian symmetry demanded here the reduction of both fields to one unified field and progression in form of waves.

Einstein identified simplicity with naturalness and inner perfection in his "Autobiographical Notes":

The second point of view . . . may briefly but vaguely be characterized as the "naturalness" or "logical simplicity" of the premises. . . . This point of view, an exact formulation of which meets with great difficulties, has played an important rôle in the selection and evaluation of theories since time immemorial. . . . among theories of equally "simple" foundation that one is to be taken as superior which most sharply delimits the qualities of systems in the abstract. . . . The second point of view may briefly be characterized as concerning itself with the "inner perfection" of a theory. . . .[56]

However, Einstein thought that "among the 'augurs' there usually is agreement in judging the 'inner perfection' of the theories. . . ."[57] He was wrong. What to him seemed "greater inner perfection," that is, simplicity, seemed to others unnecessary epistemological complexity that did not pay off as against the conceptual structure of the probabilistic interpretation of the microscopic world, which seemed much simpler to most of his eminent colleagues. Clearly, simplicity is an image of knowledge. Since Einstein thought of simplicity this way, it will not be so much of a surprise that in 1905 he shattered the structure of classical physics by rethinking the concepts of space and time. The world "happens" in terms of events, and for events, space and time can be expected to be symmetrical. But then action-at-a-distance becomes a distorting feature. Moreover, if an action-at-a-distance were admitted, then duality again creeps in, "Since then there exist two types of conceptual elements, on the one hand, material points with forces at a distance between them, and on the other hand, the continuous field."[58] This state of affairs must be an intermediate state in physics, and it is "unsatisfactory."

There is a third type of asymmetry that is also unnatural and must also disappear. Even energy must become conceptually unique. There may seemingly be different kinds of energies, but not *essentially* different ones: "the division of energy into two essentially different parts, kinetic and potential energy, must be felt as unnatural. . . ."[59] As Einstein saw it in his later years, the only part of physics that satisfied his demand for simplicity—and thus, in his opinion, was here to stay—was classical thermodynamics:

> A theory is the more impressive the greater the simplicity of its premises is, the more different kinds of things it relates, and the more extended is its area of applicability. Therefore the deep impression which classical thermodynamics made upon me. It is the only physical theory of a universal content concerning which I am convinced that within the framework of the applicability of its basic concepts, it will never be overthrown. . . .[60]

Simplicity is also basic for the architecture of matter. Nature (= reality) is simple and cannot be fooled by devices. If our experiential world is complex and points to an atomistic structure of the world while our intuition tells us that reality as experienced in theory is simple and intelligible and consists of a continuum, this continuum must be essential and cannot be introduced through statistics, which is a mere mathematical device. Speaking of contemporary quantum theorists, Einstein says: "Above everything else, . . . they believe that the apparently discontinuous character of elementary events can be described only by means of an essentially statistical theory, in which the discontinuous changes of the systems are taken into account by the way of the continuous changes of the probabilities of the possible states."[61] For these quantum theorists, Arnold Sommerfeld among them, statistical theory was simple, and it was more than a mere mathematical device. Einstein, on the other hand, according to Sommerfeld, on the issue of continuum versus discontinuity most decisively took a position on the side of the continuum.[62]

∞

The role of mathematics plays a special part in the dialectic of simplicity and complexity. For Einstein, God is a God of abstract concepts, not of mathematical cunning. Mathematics is complex, a cunning device; it cannot replace simple reality. However, when properly used, it can help in creating conceptual simplicity: "the approach to a more profound knowledge of the basic principles of physics is tied up with the most intricate mathematical methods."[63] Einstein provided a classical formulation of the dialectic: "Equations of such complexity as are the equations of the gravitational field can be found only through the discovery of a logically simple mathematical condition which determines the equations completely or [at least] almost completely."[64]

Lorentz may have agreed later that Einstein had simplified Lorentz's own theory: "Einstein's theory . . . gains a simplicity that I had not been able to attain."[65] However, what Lorentz called simplicity was not identical with Einstein's concept. Lorentz himself had remarked of the way the younger man proceeded, "Einstein simply postulates what we have deduced, with some difficulty and not altogether satisfactorily, from the fundamental equations of the electromagnetic field."[66] Adding further postulates, simple as this seemed to Einstein, was not *simplicity* to Lorentz, but the opposite. Like most of his contemporaries, Lorentz did not give up the necessity of the ether. If any new postulate were added, the number of fundamental assumptions would not diminish but grow!

Einstein's remarkable cosmological formulation that the world is finite but not limited can also be better understood in light of the simplicity-complexity tension. Infinity is complicated—much too complicated—for reality: "The hypothesis that the universe is infinite and Euclidean at infinity, is, from the relativistic point of view, a complicated hypothesis."[67] That the fundamental guiding concepts are constantly being reinterpreted in accordance with what I have called images of knowledge was clearly stated by Einstein himself (though not in the same terms) in one of the few responses he made to attacks on him:

> what seems obvious to people and what does not seem obvious, has changed. Opinions about obviousness are to a certain extent a function of time. I believe that physics is abstract and not obvious, and as an example of the changing views of what is clear, and what is not, I recommend you to consider the clarity with which Galilean mechanics has been interpreted at different times.[68]

COHERENCE OF THOUGHT AND LIFE

Presupposing that the human mind is coherent in the different areas of its expression, we must now ask what form these images of simplicity and complexity take in Einstein's work, in his daily life, and in the moral fabric of his personality.

An astonishing coherence characterizes Einstein's psychology (that is, his attitude to people), his theory of mind, and his theory of the world. (We shall come to the moral sphere later.) In all three areas, two dimensions coexist: the dimension of immediate sensations and the dimension of theoretical generalization. No logical bridge links the two. The certainty about the correctness of the theoretical dimension is assumed, since simplicity of the conceptual foundations has been secured. Then the fit with the experimental evidence is sought.

As we have seen, this is a very complex process. It proceeds as follows. In all three spheres—namely, in our awareness of ourselves, in our "thinking," and in our accounting of the world—we directly experience only disjointed, partial, and elementary fragments of experience! Once more, the similarity with Bacon's epistemology is very great: Bacon repeatedly emphasized the disjoint, superficial, and complex character of misleading commonsensical sense-data. For Bacon, as for Einstein, only theory can lend certainty. In all three spheres we leap *intuitively* to overarching generalizations, which constitute our fundamental theories, whether these are of our subjective selves or the mind, or of the world. At best, these theories will comply with a metatheoretical rule of conceptual simplicity. If achieved, this very simplicity is the proof that we are describing *reality*. If we have succeeded, then the partial, elementary, disjointed fragments of experience must follow, that is, they can be "deduced" from our fundamental theories. If it has been ascertained that there was conceptual simplicity, we have *reached* reality. If the experimental fragments somehow do not follow, there is something amiss on the experimental side, and sooner or later a proper solution will be found.

It should be noted that such expressions as "intuitive leap," "nonlogical step," and "deducing" experience from theories are Einstein's own very suggestive, but never fully explained, conceptions. That intuition is the opposite of logic seems to have been self-evident for Einstein. Also, for him, logic meant deductive steps, because whenever intuition is described at all (as above), what is meant is that no logical steps, that is, deductions, lead from our complex, disjointed world of experience to our simple and intelligible theories. On the other hand, once such a theory (≡ reality) is formulated, the world of experience can be "deduced" from it. Let us look among Einstein's writings for instances in all three areas.

Awareness of ourselves. Einstein explained that "the actual experience takes place in kaleidoscopic particular situations. The manifoldness of the external situations and the narrowness of the momentary content of consciousness brings about a sort of atomizing of the life of every human being."[69] As we have seen, sense experience is atomistic, theory is continuous. Einstein sought to introduce into his own life a continuum-type generalization that would lend him the perspective of overview: "In a man of my type, the turning-point of the development lies in the fact that gradually the major interest disengages itself to a far-reaching degree from the momentary and the merely personal and turns towards the striving for a mental grasp of things."[70] The resulting "mental grasp of things" creates a simple generalization conceiving of a continuum of being; the continuum is from God, through man, to every creature in Nature.

Lest this remark be taken as mere romanticism, let us connect it with Einstein's attitude to problems of life and death. It has often been remarked, both critically and incredulously, that his copious correspondence with his best and closest friends, just like the "Autobiographical Notes," is almost devoid of the "momentary" and of the "merely personal." The death of his second wife, Elsa, his true and helpful companion for so many years, is almost ignored in letters written in the very weeks and days after her death.[71] A clue lies in his answer to a visitor who asked him what he would think of the meaning of his own life when on his deathbed: "I would not be interested in such a question, either on my deathbed or at any time. After all, I am only a tiny particle of nature."[72] And when he was seriously ill in 1916, Hedwig Born, astonished that he showed no sign of fear, asked whether he did not fear death. He responded, "No. I feel myself so much a part of everything living that I am not in the least concerned with the beginning or ending of the concrete existence of any one person in this eternal flow."[73]

This was the great, simple continuum of life. The feeling of unity of all *nature* helped him to eliminate the disturbing "momentary and personal." Once again, this was not Stoicism, not an enormous effort to overcome human frailty, not a subjugation of the emotional by the rational, but a genuine, painless, epic distance from all that is "here and now."[74] In his maturity Einstein used to say that loneliness stopped hurting and became rather a liberating force. Or, as he put it in the beginning of the "Autobiographical Notes":

> It is quite clear to me that the religious paradise of youth, which was thus lost, was a first attempt to free myself from the chains of the "merely-personal," from an existence which is dominated by wishes, hopes and primitive feelings. Out yonder there was this huge world, which exists independently of us human beings and which stands before us like a great, eternal riddle, at least partially accessible to our inspection and thinking. The contemplation of this world beckoned like a liberation, and I soon noticed that many a man whom I had learned to esteem and to admire had found inner freedom and security in devoted occupation with it. The mental grasp of this extra-personal world within the frame of the given possibilities swam as highest aim half consciously and half unconsciously before my mind's eye.[75]

Thus, the unity with nature, the getting away from the merely personal, is rooted in a new awareness of reality, and this is what gave Einstein freedom and certainty.

Our epistemological awareness. Upon receiving sense impressions, the cognitive process starts with the emergence of a memory picture, that is, unordered, fragmentary images. These images are atomistic and complicated, and their emergence does not yet constitute thinking. "When, however, a certain picture turns up in many such series, then—precisely through such return—it becomes an ordering element for such series, in that it connects series which in themselves are unconnected. Such an element becomes an instrument, a concept. . . . all our thinking is of this nature of a free play with concepts; the justification for this play lies in the measure of survey over the experience of the senses which we are able to achieve with its aid."[76]

As before, here the immediate level of experience is discontinuous and complex. With the help of simplifying order, we connect the fragments and create a continuum. The ordering and creation of a continuum is *thinking*. We now go to what Einstein called his epistemological credo. This is best expressed in "Physics and Reality" (1936):

> I believe that the first step in the setting of a "real external world" is the formation of the concept of bodily objects and of bodily objects of various kinds. Out of the multitude of our sense experiences we take, mentally and arbitrarily, certain repeatedly occurring complexes of sense impressions (partly in conjunction with sense impressions which are interpreted as signs for sense experiences of others), and we correlate to them a concept—the concept of the bodily object. . . .
>
> The second step is to be found in the fact that, in our thinking (which determines our expectation), we attribute to this concept of the bodily object a significance, which is to a high degree independent of the sense impressions which originally give rise to it. This is what we mean when we attribute to the bodily object "a real existence."[77]

Clearly, the significance we attribute to objects, which is "independent of the sense impressions" is arrived at intuitively. Our intuition, simple and intelligible, has "real existence."

How we construct the world. We are flooded by complexes of atomistic sense impressions. From there we *leap* to great theoretical generalizations characterized by overall applicability, simplicity, and great unifying power:

> The aim of science is, on the one hand, a comprehension, as *complete* as possible, of the connection between the sense experiences in their totality, and, on the other hand, the accomplishment of this aim *by the use of a minimum of primary concepts and relations*. . . .
>
> . . . Thus the story goes on until we have arrived at a system of the greatest conceivable unity, and of the greatest poverty of concepts of the logical foundations, which is still compatible with the observations made by our senses.[78]

In all three processes we can observe the dialectical interplay between thema and antithema: simplicity versus complexity, and continuum versus atomicity. That which we can be aware of is fragmental, partial, and irreducible to more basic elements. That is, our world of awareness is *atomistic and complex.* The theoretical generalizations, on the other hand, are overarching, symmetrical, and expressible in very simple conceptual structures. The theoretical level is *simple, intelligible and continuous.* These generalized theories tolerate no asymmetries among natural phenomena that should be symmetrical; no typically atomistic explanations, such as action-at-a-distance, are admitted; no statistical explanations are accepted as ultimate descriptions of reality.

It does not follow that atomistic experience is necessarily microscopic or that theoretical overview is macroscopic. We experience in broken-up details. These can be macroscopic chunks of the world of atomic particles, just as there are continuum theories that apply to the microscopic world. The formulation should rather be the following: the harmony of the continuum must be such that it does not ignore that individual constituent particles.

Einstein's rational realism is not of an absolutist kind. This is not a world of Platonic ideas or a Popperian third world. It is a reality accessible to us through our process of thinking, and our only certainty that we have indeed reached reality is that our theoretical generalizations are of the simplest kind. How do we know that we have achieved such simplicity and intelligibility? A facetious answer would be that we have to be an Einstein. A more serious answer, and one in Einstein's terms, will be our conviction that we are on the right track, a conviction that comes from the interaction between our intuition and our sense experiences. If they reinforce each other, then we are all right. If they clash, but we still feel certain about our intuition, we must rely on it and not on the experience. The world need not be the way it is: it could have been different. In the final account, reality is chosen relativistically and the criterion of reality is the simplicity of our theories. This, as we have seen, is Einstein's two-tier thinking. However, he is an absolutist of a special kind: the criterion of simplicity is a methodological absolute. This is what I mean when I say that Einstein was a revolutionary as far as the world is concerned but that his epistemology was conservative, and his commitment to simplicity in the conceptual foundations absolute.

We can follow this line of thought in Einstein's daily life.[79] He was well known for his disregard of social conventions and for the resulting peculiarities of his behavior—wearing no socks, wearing one brown and one black shoe, using regular soap for shaving, and many other such storied patterns. There are two points to be made here. First, it is usually considered that this behavior expressed Einstein's quest for simplicity; second, it is usually claimed that these actions were done naively or unawares.

With regard to simplicity-symmetry, certainly it seems simple enough not to wear socks or not to use two soaps, one for shaving and one for washing, and it also may be a simplicification of life not to have to heed colors of shoes—if, according to one's images, the social habits and conventions, the momentary and the merely personal are unimportant. But if someone's basic attitude to life is aesthetic in the realm of the momentary and the personal, what could be more disharmonious, disjointed, complex, and asymmetrical than wearing a black and a brown shoe? (One might actually aim at such a color scheme in a specially sophisticated aesthetic-sartorial context.) It is this coherence that, in addition to Einstein's greatness as a physicist, added a further dimension to his personality and secured for him such a unique place in our culture.

There are serious Einstein scholars who claim that even if Einstein had not been one of the greatest physicists of all times, he would still have been a great humanist philosopher and an elemental moral force. Whether we accept this judgment or not, there is no question that Einstein carried an unprecedented moral weight and took carefully considered, courageous, and often original stands on a host of issues. Among these, the most important clusters were: the philosophy of science; questions of nationalism; problems of war and peace; and considerations of individual freedom and dignity. Above all, he fought tirelessly on behalf of the underdog.

Einstein's views on all of these issues were so clear-cut, so seemingly simple-mindedly idealistic, so consistent in considering human frailties that many called him naive and dismissed his views with a condescending smile. To them, Einstein was a simplistic philosopher, a confused professor, naive in matters political and social, oblivious to the multiple forces aligned against him. I do not find this view acceptable. On the contrary, Einstein's various political and humanistic activities can be seen as wise and common-sensical, stemming from a world-view that was fully in line with his scientific metaphysics and the images of knowledge that guided him in his physical researches.

Einstein was a realist insofar as he took the existence of an external world for granted—a world that did not depend on or behave according to our knowledge of it. He was a realist in the same sense with respect to human society and the individual human nature as given. Yet in both these realms he knew that we do not have objective access to this reality and that our description of the world is a result of the interaction of that world with our individual minds. He therefore expected to reach the clearest, simplest conclusion of how the world—physical and human—is supposed to behave. Having once reached that conclusion, no seeming observational or experimental counter-evidence could divert him. As C. P. Snow said of him, Einstein was "unbudgeable." In physics, this stubbornness took the form of one or another of the famous sayings about experimental evidence having to yield to the theory. It was the same clear thinking, reducing all phenomena to what he thought to be fundamental, that made him appear to others to be naive or ignorant of human nature. He believed that he knew what was important and what was basic; accordingly, he chose consciously to disregard all contrary evidence, claiming that in the final account the disturbing phenomenon—be it physical or human—would change or disappear. This is not naiveté but exactly the source of his towering moral strength.

THE SIMPLICITY OF THE MORAL CODE

In the preceding sections we have examined the coherence in the various aspects of Einstein's cognitive world-view. But this coherence ran even deeper. Einstein's preoccupation with moral issues was urgent and intense, but quite different from his untiring quest for understanding the world and conceptualizing reality. Yet here, too, there is coherence between these two mainsprings of life. Just as immediate awareness through the senses of the self, of the mind, and of the world—all these in the cognitive domain—was seen to be fragmentary and partial, so our immediate moral judgments were seen as tending to be too personal, too involved in the momentary, too atomistic and disjointed. It is therefore our supreme task to reduce all these intuitions to a few, simple, general moral principles that apply everywhere and forever. We are not to allow our impressions, our judgment of what people are and how they do or ought to behave, to stand in the way of the moral imperative. This was the attitude that created among those surrounding Einstein the misleading image of his naiveté or unworldliness. Just as in our awareness of the world we can know that our most general theories touch reality once the demand for conceptual simplicity has been satisfied, so in the moral area we are expected to apply "unbudgeable" standards and principles and think away the distorting particulars in order to be able to judge ourselves and the world in the light of such principles.

One difference between the cognitive and the moral realm is that our theories of the world are descriptive, whereas our moral principles are normative. A second difference is that in the physical-descriptive sphere, once a discrepancy occurred between overall continuous world pictures and the mosaic of atomist experiences, Einstein blamed the sense experiences but considered it his duty to deal with the discrepancy. Thus, the repeated preoccupation with suggested experiments and thought-experiments as tests of his speculations.[80] Although Einstein never tired of searching for the distortions in our sensory world in order to eliminate the "mistakes," so that what remains should follow deductively from the most general theories, in the moral-political sphere he was satisfied with the general principles and did not invest time and energy in trying to change the unprincipled moral reaction of his fellow men. He never made an effort to suggest practical ways of dealing with human frailty in order to make people behave according to the general principles. To complete the analogy to the cognitive realm, he behaved as if the simplicity and beauty of theory were all and as if he could afford not to care whether experimental evidence clashed with theory. Einstein stuck to theory. But he did so because, being an interactionist-realist, he *knew* that sooner or later experiments would fit the theory. In the moral sphere, this interaction was outside his main interest.

One exception to this aloofness from the "momentary" and the "personal" in this sphere were his repeated attempts to give practical advice on Zionist issues and Arab-Jewish affairs. After the anti-Jewish riots in Hebron in 1929 he wrote:

> what has to be done to obviate any possibility of a recurrence of such horrors?
>
> The first and most important necessity is the creation of a *modus vivendi* with the Arab people. . . . We Jews must show above all that our own history of suffering has given us sufficient understanding and psychological insight to know how to cope with this problem of psychology and organisation: the more so as no irreconcilable differences stand in the way of peace between Jews and Arabs in Palestine. Let us therefore

above all be on our guard against blind chauvinism of any kind, and let us not imagine that reason and common-sense can be replaced by British bayonets.[81]

To Chaim Weizmann he wrote:

Unless we find the way to honest cooperation and honest dealings with the Arabs, we have not learned anything on our way of two thousand years' suffering and deserve the fate that is in store for us. In particular, I feel we should beware of relying too much on the English. For unless we succeed in arriving at a genuine cooperation with the leading Arabs, the English will drop us, though not formally, but de facto. And the latter will commiserate with our debacle, piously casting their eyes to heaven and proclaiming their innocence, without lifting a finger to help us. . . .[82]

And after an exchange of letters with the editor of the Arab newspaper *Falastin*, Einstein came to an almost ridiculously detailed proposal:

Your letter has given me great pleasure. It shows me that there is good will available on your side, too, for solving the present difficulties in a manner worthy of both our nations. I believe that . . . they can be got over if both sides bring honesty and good will to the task. . . .

A Privy Council is to be formed, to which the Jews and Arabs shall each send four representatives, who must be independent of all political parties:—

Each group to be composed as follows:—

A doctor, elected by the Medical Association.

A lawyer, elected by the lawyers.

A working men's representative, elected by the trade unions.

An ecclesiastic, elected by the ecclesiastics.

These eight people are to meet once a week. They undertake not to espouse the sectional interests of their profession or nation but conscientiously and to the best of their power to aim at the welfare of the whole population of the country. Their deliberations shall be secret and they are strictly forbidden to give any information about them, even in private. . . .[83]

Excellent analyses of Einstein's role in Jewish affairs and in the Zionist cause have been provided by others (including some in this volume), and it is not my main task here to go into details. I want to make only two points. First, Einstein's model of complex, atomistic, disjointed, immediate experiences and reactions are organized by simple, continuum-type, intelligible principles or theories in the moral sphere as well as in the cognitive sphere, and there is an astonishing coherence between Einstein's epistemology in various spheres and his ethical world-view. Second, Einstein's Jewish-Zionist views and activities are in direct contrast to his usual attempt to distance himself from the "momentary and merely personal." What he refused to do for himself in order to find his peace of mind, he unhesitatingly agreed to do for the suffering others and for the underdog—in this case, his people. This particular inconsistency raised Einstein to an even higher level of humanity.

COULD GOD HAVE CREATED A DIFFERENT WORLD?

In an oft-repeated but undocumented story, Einstein one day turned to his assistant, Ernst Straus, and said, "What really interests me is whether God had any choice in the creation of the world."[84] That is, could God have created a different world? Einstein's answer was that, in principle, yes, God could have done so. We have seen that different conceptual frameworks can be taken as different worlds. In any case, if our reality is arrived at by an intuitive leap, even if "anchored" in sense experiences, and if our reality is essentially incomplete, many worlds are certainly possible.[85] There is nothing logically binding in the way the world is constituted. In physics itself, a strong case can be made that the less comprehensive a physical theory is, the greater the number of possible worlds it allows. Thus, Steven Weinberg writes:

> Quantum mechanics without realitivity would allow us to conceive of a great many possible physical systems. . . . However when you put quantum mechanics together with relativity, you find that it is nearly impossible to conceive of any possible physical system at all. Nature somehow manages to be both relativistic and quantum mechanical; but these two requirements restrict it so much that it has only a limited choice of how to be—hopefully a very limited choice.[86]

Einstein would not have agreed; it is not this kind of multiplicity of possible worlds that he had in mind. For him this type of difficulty is only a sign of unsatisfactory theory, not a proof of many possible worlds. Nature does *not* manage to be both relativistic and quantum mechanical. Indeed, the noncorrespondence, or even the contradiction, between these two great physical theories was, according to Einstein, and remains today, the great scandal of physics.[87] Einstein would have agreed that once these two are comprehended in a single more general theory, we shall then have reality. But, unlike most of his colleagues then and now, he would not have considered this reality as the only possible one. He would only have claimed that this reality is the one we get if we stick to the methodological absolute demand of simplicity. Although Einstein never expressed his thoughts in these terms, he left enough ruminations on the subject to allow us this interpretation. To James Franck he once wrote, "I can imagine that God created a world without any laws: a chaos in short. But the notion that statistical laws are final and that God draws a lot, is highly unsympathetic to me."[88]

Not just *any* world is possible, but there is more than one possible world! Actually, the entire famous discussion about God's not playing dice can be recast in terms of logically possible worlds. Different worlds are possible, and, according to Einstein, the one called reality is the one arrived at intuitively when the scientist is guided only by simplicity and intelligibility. That there is only one such intuition or that all great scientists would necessarily leap to the same intuition, Einstein never claimed.

One way of understanding the question is to look at the natural constants and ask whether they are uniquely determined by the universal scheme. In the "Autobiographical Notes," as already quoted in another context, Einstein says:

> Concerning such [dimensionless constants] I would like to state a theorem which at present can not be based upon anything more than upon a faith in the simplicity, i.e., intelligibility, of nature: there are no *arbitrary* constants of this kind; that is to say, nature is so constituted that it is possible logically to lay down such strongly determined laws that within these laws only rationally completely determined constants occur (not constants, therefore, whose numerical value could be changed without destroying the theory).[89]

But every theory is essentially incomplete; therefore the circle is not closed. A different network of theories (again partial) could be developed with a different set of constants.

Einstein often deliberated on the possibility of worlds other than the one that constituted his reality. Explaining to Solovine why the world could not be infinite, he wrote: "An infinite world is only then possible if the average density of matter in the world vanished. Such a supposition is logically possible, but less probable than the assumption that there is a finite average density of matter in the world."[90]

The question of whether a different world is possible is intimately connected with Einstein's religiousness. As we shall see, this kind of religiousness takes recourse to the mysterious in the cosmos. However, the mystery offers a wonderful fit between our atomistic, untrustworthy experiences and our continuous, simple, intelligible speculations. These and the subsequent "fit" are not uniquely determined. In the last paragraph of his 1932 credo Einstein affirmed:

> The most beautiful and deepest experience a man can have is the sense of the mysterious. It is the underlying principle of religion as well as of all serious endeavour in art and in science. . . . He who never had this experience seems to me, if not dead, then at least blind. The sense that behind anything that can be experienced there is a something that our mind cannot grasp and whose beauty and sublimity reaches us only indirectly and as feeble reflexion, *this* is *religiousness.* In *this* sense I am religious. To me it suffices to wonder at these secrets and to attempt humbly to grasp with my mind a mere image of the lofty structure of all that there is.[91]

This religiousness is certainly not objective or simplified, even though at the end Einstein echoes his reconstruction of a simplified world where what he can grasp with his mind is "a mere image" of what there is. Elsewhere Einstein explains that the conviction of the rationality or intelligibility of the world that lies behind all scientific work of a higher order is akin to religious feeling.[92] Thus religiousness is a sense of the mysterious. But the mysterious is the inner perfection, the symmetry, the simplicity of the universe and thus of God.

I have argued that two-tier thinking is an essentially nonreligious view and that Einstein adopted two-tier thinking with regard to reality. However, I have also argued that any admission of an external context of all contexts is an essentially religious view. This is exactly the view that Einstein adopted when he claimed absolute validity for his principle of simplicity.

Thus, the intuitive leap from our senses of atomistic, disjoined, disorganized complexity to generalized, unitary, continuous, simple theories and their fit with experience is called reality by us, and it is this that brings us closer to an understanding of God. Einstein once said, "I want to know how God created this world. I am not interested in this or that phenomenon, in the spectrum of this or that element. I want to know His thoughts, the rest are details."[93]

THE MYTH OF SIMPLICITY

The theory of the growth of knowledge that has guided our argument holds that knowledge grows through a continuous critical dialogue among the various components of the dialectical process of progress: ongoing debates among opposing scientific metaphysics, competing images of knowledge, and rival ideologies. An important role is thus given to metaphysics in the growth of knowledge.

Einstein saw an even greater role for metaphysics. He often declared that science *is* metaphysics. He fiercely criticized the logical empiricists and tried to argue the case for metaphysics in all areas of thinking. In his "Remarks on Bertrand Russell's Theory of Knowledge" he happily noted that Russell is almost free from "fear of metaphysics." Yet, even in Russell's *Meaning and Truth*, "the spectre of the metaphysical fear has caused some damage."[94]

In such a world-view there was not much place for the impossible in nature. True, there always would remain an "ignorabimus" in the mysterious fit of human intuition with the world of the senses,[95] but whatever the mind set itself to do would become a mere "ignoramus": "With the 'impossible' the case is as follows: it changes into possible when people *really want* it, even if this happens only out of anxiety in a state of unbearable uncertainty. In order to bring about such a state of wanting we should try with all our might."[96]

This thought is more complex than it seems. There are two issues here: what about the "impossible," and what is the will? Einstein meant that the impossible, or "ignorabimus," depends on our intellectual will, and although our reality emerges from our intuition, the unbearable uncertainty may influence us to will an intuitive solution. Needless to say, Einstein did not express this thought in so many words, but it seems a highly plausible interpretation. With regard to the will, Einstein often said, following Schopenhauer, that man cannot simply will what he wills.[97] No contradiction here. The meaning of Einstein's words follows from his epistemology: our speculations about the world, that is, what we really will, are severely constrained by the absolute demand for simplicity; however, when we have satisfied this methodological absolute, reality will be, so to say, in our hands. The "fitting" is the inherent mystery in nature, that is, the way we leap to our theories, which are capable of worming out, teasing out, prying out nature's secrets. All this is to some extent a result of our will. Einstein, like Bacon, actually thought of nature in such terms: "The scientist has to worm these general principles out of nature by perceiving in comprehensive complexes of empirical facts, certain general features which permit of precise formulation."[98]

Not only is the fit between theory and sense experience a sign of "magic" in nature, but scientific theory itself carries magic. The introduction to the 1915 paper on general relativity closes with these words: "No one who genuinely understood this theory will be able to escape its inherent magic."[99] In the introduction to his son-in-law's biography of him, Einstein wrote: "What has perhaps been overlooked is the irrational, the inconsistent, the droll, even the insane, which nature, inexhaustively operative, implants in an individual, seemingly for her own amusement. But these things are singled out only in the crucible of one's own mind."[100] And again, a magnificent passage of a letter to Solovine, written late in life:

> Now I am coming to the most interesting point of your letter. You find it strange that I regard the comprehensibility of the world (in so far as we may be justified in speaking of such) as a miracle or as an eternal secret. Now, a priori, one ought to expect the world to be chaotic, in no way comprehensible through thinking. One could (even *should*) expect the world to be governed by law only to the extent that we intervene by introducing some order. This would be a kind of order similar to the alphabetical order of the words in a language. The kind of order created, for example, by Newton's theory of gravitation, is of an entirely different character. Even though the axioms of the theory are posited by man, yet the success of such an undertaking presupposes a high degree of orderliness of the objective world, which was not to be expected a priori. Herein lies the "miracle" which is becoming increasingly deep with the development of our knowledge.[101]

Because our sense experiences are disjoint, chaotic and our theories purely speculative, we ought to expect nature to be incomprehensible. But nature is comprehensible. The incomprehensibility, that is, the mystery, thus shifts to the "fit." In short, "The most incomprehensible thing about the universe is that it is comprehensible."[102] Wherein lies simplicity? As we have seen, simplicity is a perennial thema in science, dialectically related to its antithema, complexity. The contents of this thema is given by context-dependent, socially woven images of knowledge. The contents of Einstein's simplicity was the inner perfection, the beauty, and the sparsity of fundamental concepts, namely, symmetry. For this simplicity the leaving aside of all mathematical, historical, and epistemological complexity was not too heavy a price. But the last of these was the most difficult one to let go. It is difficult to conceive of a more complex intellectual activity than the Einsteinian method of thinking away all that is disturbing, atomistic, asymmetrical, or disharmonious.

How unfashionable Einstein was became somewhat hidden by the seeming similarity between Einstein's demand for his simplicity and the very fashionable demand for simplicity by the various idealistic schools, such as positivism, behaviorism, and reductionism. But for these schools, simplicity consisted of reducing all metaphysics to sense experiences. For them, "simple" is whatever is directly available to the senses; all so-called theoretical statements are reduced to so-called observational ones.

From the philosophy of science (Quine's above all), from sophisticated historical sociology of scientific knowledge, from comparative anthropology (Cole, Horton, Gellner), and from studies on chimpanzees (Premack's and others), we have learned that the simpler, more readily available items of awareness are the ones that presuppose theoretical terms and mental attributes such as motives, wishes, and some holistic traits. Even chimpanzees have a theory of mind; that is, they attribute knowledge and intention to others. It is the thinking away of these terms that involves such an enormously complex effort in order to reduce the world to a so-called experimental immediacy. The basic difference between the branches of positivism and Einstein's view is that Einstein was aware of the complexity of thinking away asymmetries, superfluous entities, experimental noise, or atomistic ripples on the underlying continuum.[103] A quest for simplicity, while fully aware of the complexity of the process leading up to it, is *the myth of simplicity.*

As it has been repeatedly pointed out, Einstein was a gigantic counter-trend figure.[104] Such is the power of consensus in science that even Einstein did not escape the dictatorship of the majority. When he stopped being the leader, when he continued on his lonely path in his quest for determinism, he, like Kant, was "forgiven" for senility. Einstein, as usual, was aware of this attitude. He wrote to Born in 1944: "Even the great initial success of the quantum theory does not make me believe in the fundamental dice-game, although I am well aware that our younger colleagues interpret this as a consequence of senility." Yet he ended that letter, symbolic of his last years, on an optimistic note: "No doubt the day will come when we will see whose instinctive attitude was the correct one."[105] The world of ordinary man is seen as a Greek drama: God is reality, and the world of the sage is an epic theater.

I wanted to give this essay a different title, but it might have been misinterpreted, leading to an accusation of irreverence. Now, having presented my arguments, I dare to disclose the intended title, a paraphrase of Keynes's famous phrase about Newton: "Einstein, the Last of the Scientists, the First of the Magicians."

ACKNOWLEDGMENTS

Anyone who writes today on any problem connected with Einstein will have to rely heavily on the work of the many Einstein scholars but especially on the biographies of Einstein by Philipp Frank and Banesh Hoffmann and on the work of Gerald Holton. Indebtedness to their work is gratefully acknowledged.

I am also deeply grateful to my friends Robert S. Cohen, Robert K. Merton, and Ady Ophir for their criticism and editorial help in rewriting this paper.

NOTES

1. Epicurus, "The Menoecus," in Epicurus, *The Extant Remains*, Oxford, 1926, pt. 134. Quoted and adapted by John Passmore, *Science and Its Critics*, New Brunswick, N.J., Rutgers University Press, 1978, p. 29.

2. Alfred North Whitehead, *Science and the Modern World* (Lowell Lectures, 1925), New York, Free Press, 1969, p. 10.

3. Ibid.

4. Ibid.

5. The quotation is from a manuscript by Ilse Rosenthal-Schneider, "Reminiscences of Conversation with Einstein," dated 23 July 1957 and quoted by Gerald Holton in *The Thematic Origins of Scientific Thought: Kepler to Einstein*, Cambridge, Mass., Harvard University Press, 1973, pp. 236-237. In note 52, pp. 255-256, Holton also quotes a story by P. A. M. Dirac, who concludes: "I think there is a moral to this story, namely, that it is more important to have beauty in one's equations than to have them fit experiment."

6. Walter Benjamin, *Understanding Brecht*, New York, New Left Books, 1973, p. 8.

7. This is another illustration of Holton's insight that themata and

their antithemata relate dialectically in the history of culture. This is the central theme of Holton's *Thematic Origins of Scientific Thought.* See also Gerald Holton, "On The Role of Themata in Scientific Thought," and R. K. Merton, "Thematic Analysis in Science: Notes on Holton's Concept," *Science* **188** (1975), pp. 328-338.

8. Many Brechtian "heroes" are such, as pointed out by Benjamin: for example, Galy Gay in *A Man Is a Man* and Azdak in *The Caucasian Chalk Circle.*

9. If there was dramatic inevitability about anything in Einstein's conceptual world, it was about his epistemology. Human knowledge had to be deterministic, had to be predictable.

10. Quoted from Seneca (De Provident, V) by W.E.H. Lecky, *History of European Morals* (1869), New York, Arno Press, 1975, p. 163n.

11. Einstein's expression, in his "Autobiographical Notes," trans. Paul Arthur Schilpp, in Schilpp, ed., *Albert Einstein: Philosopher-Scientist,* Evanston, Ill., Library of Living Philosophers, 1949, p. 7.

12. Yehuda Elkana: "The Distinctiveness and Universality of Science: Reflections on the Work of Professor Robin Horton," *Minerva,* **15** (1977), pp. 155-173; "Two-Tier-Thinking: Philosophical Realism and Historical Relativism," *Social Studies of Science* **8** (1978), pp. 309-326; "Transformations in Realist Philosophy of Science from Victorian Baconianism to the Present Day," in E. Mendelsohn, ed., *I. B. Cohen Festschrift,* forthcoming; "A Programmatic Attempt at an Anthropology of Knowledge," in E. Mendelsohn and Y. Elkana, eds., *Science and Culture,* Yearbook of the Sociology of Sciences, vol. 5, Dordrecht, Reidel Publishing Co., 1981, pp. 1-76; "Of Cunning Reason," in Thomas Gieryn, ed., *Robert K. Merton Festschrift,* New York Academy of Sciences, forthcoming; "Science as a Cultural System: An Anthropological Approach," in V. Mathieu and P. Rossi, eds., *Scientific Culture in the Contemporary World,* Milan, Scientia, 1979, pp. 269-290.

13. See Clifford Geertz, *The Interpretation of Culture,* New York, Basic Books, 1973.

14. Nelson Goodman's "worlds" in his *Ways of Worldmaking,* Sussex, Harvester Press Ltd., 1978, are very close to my relativistically selected frameworks, inside of which we are realists. However, his arguments are purely philosophical and do not branch into sociology or anthropology.

15. Introduction to Gerald Holton's *The Scientific Imagination: Case Studies,* Cambridge, Cambridge University Press, 1978, p. ix. See also Merton, "Thematic Analysis in Science."

16. Holton, "Themata in Scientific Thought," in *Scientific Imagination,* pp. 3-14. The paper analyzed is Steven Weinberg's "Unified Theories of Elementary Particle Interactions," *Scientific American* **231**, no. 1 (1974).

17. This is the view of Geertz, and I fully share it. See Geertz, "Thick Description: Toward an Interpretative Theory of Culture," in Geertz, *Interpretation of Culture.*

18. Holton is aware of this: "There is of course much room for personal disagreement in what 'simplicity' or 'restrictiveness' consists. Thus Einstein and Planck debated strongly in 1914 whether the simplest physics is one that regards as basic *accelerated* motion (as Einstein had come to believe) or *unaccelerated* motion (as Planck insisted)." Holton, *Scientific Imagination*, p. 299, n. 8.

19. *The Born-Einstein Letters: Correspondence between Albert Einstein and Max and Hedwig Born from 1916 to 1955 with Commentaries by Max Born*, trans. Irene Born, New York, Walker and Co., 1971, pp. 163-164.

20. From "Motiv des Forschens," translated as "Principles of Research" in Albert Einstein, *Ideas and Opinions* (based on *Mein Weltbild*, ed. Carl Seelig), new trans. and rev. Sonja Bargmann, New York, Crown Publishers, 1954, pp. 225-226.

21. Geertz, *Interpretation of Culture*, p. 29.

22. See Elkana, "Science as a Cultural System."

23. Geertz, *Interpretation of Culture*, p. 5.

24. Clifford Geertz, "Common Sense as a Cultural System," *Antioch Review* **33** (1975), p. 8.

25. Yet with Einstein, epistemological matters are never simple or unambiguous. Gerald Holton rightly notes "that positivism and its antithesis were also dialectically intertwined in Einstein's writings." Moreover, referring to the 1905 relativity paper, Holton adds: "we find there both positivism of the instrumentalist and operationalist variety, which Einstein uses in defining the concept, and on the other hand, the rational realism inherent in the *a priori* declaration of the two basic principles of relativity." Holton, "Finding Favor with the Angel of the Lord," in Y. Elkana, ed., *The Interaction between Science and Philosophy*, Atlantic Highlands, N.J., Humanities Press, 1974, p. 367.

26. P. Feyerabend *Against Method*, London, New Left Books, 1975.

27. Ryle introduced "thick description" in his "Thinking and Reflecting" (1966), reprinted in his *Collected Papers*, London, Hutchinson, 1971, vol. 1, pp. 465-479, and "The Thinking of Thoughts: What Is 'Le Penseur' Doing?" (1968), reprinted in ibid., pp. 480-486.

28. This of course has nothing to do with the Hegelian "cunning reason" (*List der Vernunft*).

29. Marcel Detienne and Jean-Pierre Vernant, *Cunning Intelligence in Greek Culture and Society*, Sussex, Harvester Press Ltd., 1978, p. 3.

30. See Elkana, "Of Cunning Reason."

31. Einstein, "Autobiographical Notes," p. 13.

32. On experience-near and experience-distant concepts in physics see my "The Distinctiveness and Universality of Science."

33. "Autobiographical Notes," p. 13.

34. *Born-Einstein Letters*, p. 149.

35. Ronald W. Clark, *Einstein: The Life and Times*, New York and

Cleveland, World Publishing Co., 1971, p. 344. No exact source of the story is given.

36. A. Einstein, "Remarks on Bertrand Russell's Theory of Knowledge," trans. Paul Arthur Schilpp, in Schilpp, ed., *The Philosophy of Bertrand Russell*, Evanston, Ill., Library of Living Philosophers, 1946, pp. 277-291; letter of 7 May 1952, in Einstein, *Lettres à Maurice Solovine*, Paris, Gauthier-Villars, 1956, p. 121.

37. Einstein, "Russell's Theory of Knowledge," p. 281.

38. Russell, *An Inquiry Into Meaning and Truth*, quoted in ibid., p. 283.

39. Ibid., p. 283.

40. In the "Autobiographical Notes" Einstein says that "one does not easily become aware of the free choice of such concepts, which, through verification and long usage, appear to be immediately connected with the empirical material" (p. 49).

41. Ibid., p. 13. Robert S. Cohen has correctly noted the similarity of this explanation to the verification = meaning model.

42. The letter is fully quoted and beautifully analyzed in Gerald Holton, "Einstein's Model for Constructing a Scientific Theory," in P. C. Aichelburg and R. Sexl, eds., *Albert Einstein: His Influence on Physics, Philosophy and Politics*, Braunschweig/Wiesbaden, Friedrich Vieweg & Sohn, 1979, pp. 109-136.

43. "That the totality of our sense-experience is such that they can be arranged in an order by means of thinking . . . is a fact which strikes us with amazement, but which we shall never be able to comprehend." A. Einstein, "Physik und Realität," *Journal of the Franklin Institute* **221** (1936), p. 315; translated by Ilse Rosenthal-Schneider in her penetrating discussion comparing the epistemologies of Kant, Planck, Einstein, and Eddington. In terms of classical philosophy, her analysis is the most illuminating I know. The argument needs to be repeated here, since I want to cast the discussion in terms of images of knowledge and the interactionist view of realism. See Ilse Rosenthal-Schneider, "Presuppositions and Anticipations in Einstein's Physics," in Schilpp, ed., *Einstein: Philosopher-Scientist*, pp. 129-146.

44. Wolfgang Pauli, in his "Impressionen über A. Einstein," notes that in each of his revolutionary works Einstein stuck conservatively to some earlier result or idea: in his early quantum theory he held to Boltzmann's statistical interpretation of entropy; in the special theory of relativity he kept the foundations of Lorentz's electrodynamics; and in the general theory he stuck to the concept of field as developed in the continuum physics of Faraday and Maxwell. Pauli is correct, but only up to a point. Einstein applied this conservative trend not to the world but to our knowledge of it. Continuum at this stage, just like the conservation of energy, is no longer part of scientific metaphysics in the body of knowledge but is a basic image of knowledge. Our knowledge of the world

must be such that a methodological rule of simplicity guides it. And as we have seen, simplicity and continuum are inseparable. See W. Pauli, *Aufsätze and Vorträge über Physik und Erkenntnistheorie,* Braunschweig/Wiesbaden, Friedrich Vieweg & Sohn, 1961, p. 82.

45. A. Einstein, "Zur Electrodynamik bewegter Körper," *Annalen der Physik* **17** (1905), p. 891.

46. A. Einstein, "Entwicklung unserer Anschauungen über das Wesen und die Konstitution der Strahlung [Development of our views on the nature and constitution of radiation]," a paper read at the Salzburg meeting of the Society of German Scientists in 1909 and published in *Physikalische Zeitschrift* **10** (1909).

47. A. Einstein, *Relativity, the Special and the General Theory: A Popular Exposition* (1916; first English ed., 1920), London, Methuen, 1976, p. 62.

48. A. Einstein, "Theory of the Affine Field," trans. R. W. Lawson, *Nature* **112** (1923), pp. 448-449.

49. W. Pauli, *Theory of Relativity* (1920), London, 1958.

50. *London Daily Chronicle,* 26 January 1929; reprinted in *Nature* **123** (1929), p. 175.

51. See also R. S. Cohen and Yehuda Elkana, eds., *Hermann von Helmholtz: Epistemological Writings,* Boston Studies in the Philosophy of Science, vol. 37, Dordrecht, Reidel Publishing Co., 1977.

52. "Autobiographical Notes," p. 63.

53. Ibid., p. 79.

54. Ibid., p. 81.

55. Gerald Holton reminds us of this phrase by René Dugas in his "Einstein, Michelson and the 'Crucial' Experiment," *Isis* **60** (1969), pp. 133-197.

56. "Autobiographical Notes," p. 23.

57. Ibid.

58. Ibid., p. 27.

59. Ibid., p. 31.

60. Ibid., p. 33.

61. Ibid., p. 87.

62. Arnold Sommerfeld, "To Albert Einstein's Seventieth Birthday," trans. Paul Arthur Schilpp, in Schilpp, ed., *Einstein: Philosopher-Scientist,* p. 105.

63. "Autobiographical Notes," p. 17.

64. Ibid., p. 89.

65. Note added to the 1915 edition of Lorentz's *The Theory of Electrons* (1909), quoted by Holton, *Thematic Origins,* p. 304.

66. Ibid., p. 304.

67. A. Einstein, *The Meaning of Relativity* (Stafford Little Lectures of Princeton University, May 1921), 5th ed., Princeton, N.J., Princeton University Press, 1955, p. 99.

68. In response to Philipp Lenard, cited in Clark, *Einstein: The Life and Times*, p. 264.

69. "Autobiographical Notes," p. 7.

70. Ibid., p. 7.

71. In an undated letter to Max Born, Einstein wrote: "I have settled down splendidly here: I hibernate like a bear in its cave, and really feel more at home than ever before in all my varied existence. This bearishness has been accentuated still further by the death of my mate [*Kameradin*] who was more attached to human beings than I." *Born-Einstein Letters*, p. 128.

72. Reported in Carl Seelig, ed., *Helle Zeit—Dunkle Zeit: In Memoriam Albert Einstein*, Zurich, 1956, p. 87.

73. Ibid., p. 36.

74. Of course, such an epic attitude to life is analyzable in psychological and psychoanalytic terms, as is the Stoic subjugation of the emotional in favor of the rational. However, in this discussion of coherence it is enough to relate to what Einstein says and take it at face value.

75. "Autobiographical Notes," p. 5.

76. Ibid., p. 7.

77. In *Ideas and Opinions*, p. 291.

78. Ibid., pp. 293-294.

79. Gerald Holton has pointed this out in his essays, especially in his "Finding Favor with the Angel of the Lord."

80. It was Galileo's interest in thinking up experiments to smooth the deductive road from theory to implications that made Einstein such an admirer of him.

81. A. Einstein, *About Zionism: Speeches and Letters*, ed. and trans. Leon Simon, New York, Macmillan, 1930, pp. 69-70.

82. Einstein to Weizmann, 1930, The Weizmann Archives, Yad Chaim Weizmann, Rehovoth.

83. Letter to the editor of *Falastin*, 15 March 1930; reprinted in *Ideas and Opinions*, pp. 172-173.

84. One of the many versions of this story is given by B. Kuznetzov in *Einstein and Dostoyevsky*, Hutchinson Educational, 1972, p. 39. There we find: "What interests me is: could God have made the world differently? Does the requirement of logical simplicity leave any latitude?" Ernst Straus was Einstein's assistant at Princeton from 1944 to 1948. In a short "Memoir" printed in A. P. French, ed., *Einstein: A Centenary Volume*, Cambridge, Mass., Harvard University Press, 1979, pp. 31-32, he relates several other stories of great intellectual charm. It is typical that in A. Moszkowski, *Conversations with Einstein*, London, 1970, there is a whole chapter called "Of Different Worlds" (pp. 125-142).

85. I refer once more to Goodman, *Ways of Worldmaking*. See also Richard Rorty, "The World Well Lost," *Journal of Philosophy* **69** (1976), pp. 649-665.

86. Steven Weinberg "Where We Are Now," *Science* **180** (1973), pp. 276-278.

87. See the Abstract of Eugene Wigner's talk at the Special Session to Celebrate the Hundredth Anniversary of Dr. Albert Einstein's Birth on 25 April 1979, "Basic Conflict Between the Concepts of General Relativity and of Quantum Mechanics," printed in *Bulletin of the American Physical Society* **4** (1979), p. 633. (I owe this reference to G. Holton.)

88. In Carl Seelig, "Albert Einstein: Leben und Werk eines Genies," in *Unsere Zeit*, Zurich, 1960, p. 396.

89. "Autobiographical Notes," p. 63.

90. *Lettres à Solovine*, p. 32.

91. From "My Credo," recorded at the "initiative of the German League for Human Rights" in Berlin, autumn 1932, a few days before Einstein left Europe forever. A copy of the tape is kept in the Phonotèque of the Hebrew University of Jerusalem. The German text was published by Friedrich Harneck in *Naturwissenschaften* **1** (1965), p. 98. The translation is from the catalogue of the National and Hebrew University Einstein Centennial exhibition.

92. Einstein's answers to questions of a Japanese scholar in 1929; reprinted as "On Scientific Truth," in *Ideas and Opinions*, pp. 261-262.

93. Quoted without indication of source (I could not locate it either) by Silvio Bergia, "Einstein and the Birth of Special Relativity," in French, ed., *Einstein*, p. 67.

94. Einstein, "Russell's Theory of Knowledge," p. 289. At the end of that article he said: "In view of these endeavours I am particularly pleased to note that, in the last chapter of the book, it finally crops out that one can, after all, not get along without 'metaphysics.' The only thing to which I take exception there is the bad intellectual conscience which shines through between the lines" (p. 291).

95. Emil du Bois-Reymond in a famous paper, "Über die Grenzen des Naturerkennens," introduced the distinction between "ignoramus" (what we do not know) and "ignorabimus" (what we cannot possibly know). See du Bois-Reymond, *Reden*, Leipzig, Verlag von Veit & Co., 1912, vol. 1, pp. 441-473.

96. Einstein to Solovine, 7 May 1952, *Lettres à Solovine*, p. 119.

97. Einstein's position is beautifully represented by Quine. In an interview, the question was posed: "Does this mean that you deny the existence of the age-old problem about whether or not we have free will?" Quine answered: "Clearly we have free will. The supposed problem comes of a confusion, indeed a confusing turn of phrase. Freedom of the will means that *we* are free to *do* as we will; not that our will is free to will as it will, which would be nonsense." In Bryan Magee, *Men of Ideas: Some Creators of Contemporary Philosophy*, London, 1978, p. 43.

98. From an address to the Prussian Academy of Sciences, 1914, in *Ideas and Opinions*, p. 221.

99. Quoted by Cornelius Lanczos in *The Einstein Decade (1905-1915)*, London, Elek Science, 1974, p. 12.

100. Rudolf Kayser's biography of Einstein, written in 1929 under the pseudonym Anton Reiser, quoted by Clark, *Einstein: The Life and Times*, p. 409.

101. Einstein to Solovine, 30 March 1952, *Lettres à Solovine*, p. 115.

102. A. Einstein, "Physik und Realität," *Journal of the Franklin Institute* **221** (1936), p. 315.

103. "Herein lies the weakness of the positivists and the professional atheists who feel smug in their conviction that they have been successful in divesting the world not only of the gods but also of miracles. The beauty of it is that we have to content ourselves with the recognition of the 'miracle,' beyond which there is no legitimate way out." Einstein to Solovine, 30 March 1952, *Lettres à Solovine*, p. 115.

104. See Yaron Ezrahi's paper in this volume.

105. *Born-Einstein Letters*, p. 149.

Yaron Ezrahi

EINSTEIN AND THE LIGHT OF REASON

THE OPEN ACADEMY AND THE PUBLIC

AROUND 1660, the Fellows of the Accademia del Cimento in Florence assembled to perform an experiment that would test Christiaan Huygens's theory that Saturn does not really consist of separate globes and that the changing appearances of the planet that had baffled earlier observers were produced by the ring around it. Committed to "finding out the truth by means of experimental tests,"[1] the Fellows constructed a small model of Saturn according to the proportions stated by Huygens. On the basis of the memorandum written of the experiment, we know that "the model was set up at the end of a gallery . . . 75 meters long—no doubt in Pitti Palace—and illuminated by four torches, located so that they were hidden from the eyes of the observers. The results were what Huygens predicted. . . . Then the observations were repeated in daylight from 23 meters with the naked eye. Fearing that they might be deceiving themselves, the Fellows called on 'many people, among them illiterate persons, who had not seen from nearby the arrangement of the machine that they were to observe' and had them make sketches, one at a time, of what they saw from 23 meters. Almost all drew the disk of Saturn between two round bodies separated from it by a short distance."[2]

Like Galileo, who broke the monopoly of Latin by publishing his works in the vernacular, the experimenters of the Accademia del Cimento projected an image of science that would open the academy's doors before the layman and invite the illiterate to add the testimony of their senses to that of the experts. Although the laymen who entered the sanctuary of science were often baffled by the instruments and the mathematics that seemed to carry science beyond the domain of the natural eye, the ideal of science as a form of public knowledge drawn from the "facts" of common perception persisted in both the ethos of science and the popular mind. Even though the inevitable incongruities between scientific and common-sense constructions of reality have reappeared throughout the social history of science, the belief that science and common sense refer ultimately to the same objective reality remained powerful enough to support the persistence of Enlightenment ideas about the role of scientific knowledge in reforming public affairs.

Nearly three centuries after the experiment at Florence, Albert Einstein, the living symbol of twentieth-century physics, invited laymen to recognize that, with respect to the science whose logical base "departs more and more from the facts of experience,"[3] the value of the testimonies of their senses is limited, that what ordinary men perceive and discover by means of their standardized clocks and measuring rods are not the "objective facts" that stand for a reality common to the layman and the physicist. Instead, what the layman discovers depends upon perceptions that tend to prejudice scientifically valid concepts of the world.

In coupling the most spectacular scientific advance of the age with an image of science that questions the continuity between common-sense and scientific definitions of truth and reality, and in positing a theory of knowledge that affirms the role of "intuition and deductive thought,"[4] Einstein emerged willy-nilly as the messenger of a dramatic break with the social notions of science and of its place in culture that prevailed within the liberal-democratic tradition. In order to examine the significance of this development, one should first consider briefly some of the main elements of the common conceptions of science that Einstein's physics appeared to undermine.

The notion that in the search for truth the sense of the natural philosopher is not privileged as against that of the layman can already be found in the work of Francis Bacon, who held that the inductive method is not only a preferred method for the advancement of knowledge but also, by elevating "examining" over "telling," a superior method of presentation and persuasion.[5] Bishop Thomas Sprat, who in the first history of the Royal Society (1667) largely subscribed to Bacon's concept of knowledge, asserted that "unlike the magical glass used by the astrologers to deceive the ignorant, the mind of man is like a glass which is able to represent to itself all the works of nature which have been brought before it."[6] Consistent with their trust in the "glass" of the human mind, the Fellows of the Royal Society are reported to have invited the "merchants" and other "citizens who are men of traffic" to assist in making observations in distant countries and to have welcomed the nobility and the gentry "to propose their doubts."[7] According to this description, the pursuit of knowledge was based on a "union of eyes" extending across separate callings, faiths, and classes.[8] From the British colonies in the New World, Corresponding Fellows of the Royal Society sent specimens of "diverse curiosities of nature," which the London Fellows carefully examined and classified.[9] On some occasions, when a Correspondent of the Royal Society would send written and illustrated accounts of observations that appeared to mix the facts of nature with theological speculations or an appeal to other "authorities," the Royal Society responded with criticism, reiterating the standards that would render such reports acceptable.[10]

Such early attempts at setting standards for reliable testimonies indicate the kinds of processes that were eventually to separate the observations of the scientists from those of the laymen. But the persisting commitment to the role of observations, trials, and experiments in restraining the claims of scientists continued to feed a widespread popular image of scientific knowledge as merely a refinement and a systematic classification of publicly shared experience. Viewed in this way, science seemed to have the authority to set standards for guiding the observations and interpretations of laymen.

As part of the attempt to bring the ordinary man to see nature with the eyes of science, the collections of objects that the nobility had locked up in private curiosity cabinets to use for social entertainment were opened to the public. But the objects in these private collections—which could mix shells from foreign beaches with the Gospel of St. John written in microscopic letters on a piece of parchment[11]—were reclassified, and the objects of nature separated from other curios.[12] Thus, even as mathematical reasoning and technical professional standards seemed repeatedly to challenge the image of science as a form of publicly accessible knowledge and leave the amateur outside the closed doors of the academy, the rise of modern public museums, especially after the French Revolution and intensifying toward the second half of the nineteenth century, helped to sustain the precarious accountability of science to the world of common-sense experience. Technical inventions that opened the way for mass diffusion of standardized pictorial representations of "natural facts" in books and, later, the invention of photography further enhanced pictorial images of knowledge as a record of facts and thus continued to project science as a publicly accessible enterprise.[13] To be sure, the distance induced by progressive professionalization increasingly excluded the layman as an equal participant in correcting and assessing the evidence of nature. But scientists like Humphry Davy and Michael Faraday lectured before laymen at the Royal Institution during the first half of the nineteenth century and encouraged the lay public to remain, if not equal participants, at least an audience of science.

Assuming that observable reality is both the shared record and the theater of a public meeting of many individual minds, the participatory social image of science has come over the years to be identified with wider social and political ideals of democratization. As against the more radical populistic strains of democratic movements, which coupled demands for wider political participation with the zeal of reform and heightened popular sentiment, the example of scientists reaching agreements through careful observations appeared to exemplify a more moderate, even conservative, method of democratic participation that could secure discipline and authority within freedom of discourse and deliberation. It seemed to confirm the feasibility of freedom based on individualism and restrained by empiricism, freedom that did not lead to endless conflicts or isolations. No wonder that in the second half of the seventeenth century the notion of science as a mode of tolerant and restrained discourse could be integrated into the world-view of religious groups like the Latitudinarians.[14] The harmonizing cultural import of science was stressed by people like Thomas Sprat, who, in referring to the Scotch, Irish, and English gentry of mid-seventeenth-century England, observed that "in their assemblies for making laws [these estates] are separated and in their customs and manners of life, they differ," but that "in the Royal Society they meet and communicate without any distinction of countries or affections." In matters of science their "universal agreement is almost the only thing wherein the nobility of all three kingdoms are united."[15]

The persistent belief that this "universal agreement" of men as scientists could be extended to encompass laymen as well was, of course, strained by the periodic manifestations of discrepancies between scientific and lay notions of reality. That Einsteinian physics appeared to have unsettling effects on this faith was not a new situation. What was novel was the dimensions of this challenge, which derived from a combination of factors, including Einstein's prominence in twentieth-century science, his explicit criticism of reliance upon perceptions as a source of scientific knowledge, and the historical circumstances of the period, especially in the second quarter of the twentieth century, which revived widespread commitments to the liberal-democratic mission of science to enlighten and rationally restrain public discourse.

Contrary to the entrenched faith in the ultimate continuity between the common perception and the scientific construction of reality, Einstein insisted that the very concept of reality that permits universal agreement between scientists and laymen is a constraint on the advancement of knowledge. In his effort to pave the way for public understanding of the relativity theory, Einstein tried to discard as a prescientific prejudice the very "reality" impressed upon the "glass" of the mind by common perceptions. But in this effort he also furnished a suggestive description of the compelling process through which different persons develop their common-sense notions of an externally independent, objective world and come to forge a "universal agreement." Einstein suggested that we consider, for example, the experience of lightning. He observed that at first the experience "it is lightning" appears as a personal "experience." When the person who has had the experience of lightning also experiences other persons in a way that brings their behavior "into relation with his own experience 'it is lightning,' " that experience is "no longer interpreted as an exclusively personal experience, but as an experience of other persons (or eventually only as a 'potential experience')." "In this way," concludes Einstein, "the interpretation that 'it is lightning,' which originally entered into the consciousness as an 'experience,' is now also interpreted as an (objective) 'event.' It is just the sum total of all events that we mean when we speak of the 'real external world.' "[16] What concerns us here is that when Einstein attempted to modify this procedure of constructing the world ("In order to arrive at the idea of an objective world, an additional constructive concept still is necessary"), when his physics induced renewed strain in the presumed continuity between what is "seen" and what is "happening,"[17] he cast doubt not only on the validity of some accepted scientific theories but also on the grounds of a host of well-entrenched cultural attitudes and social commitments that had rested on the idea that the truth is accountable, as it were, to the sphere of commonly accessible perceptions.

Already in the seventeenth century, this idea had been associated with the elevation of the eye not only as an instrument for emancipating knowledge from its bondage to texts or the spoken word of the sages but also as a factor in new notions of discourse and persuasive communication in the civil society.[18] According to John Locke, in a polity that rejects violence or transcendence as principles of order, to eliminate the "imperfections of language" by reference to "settled standards in nature" is a way to advance not only knowledge but also peace.[19] The "language, which was given us for the improvement of knowledge and bond of society, should not be employed to darken truth and unsettle people's rights. . . . Or . . . at least, if this will happen *it should not be thought learning or knowledge to do so.*"[20] Locke warned the philosophers who "had learning and subtlety enough to prove that snow was black; i.e., to prove that white was black," that such proofs "destroy the instruments and means of discourse, conversation, instruction and society."[21]

Nonetheless, what has given philosophers the power to damage "the instruments and means of discourse" is not so much the cultivation of false ideas (the common appearance of snow may not indicate its composition). It is rather the cultural attitude that both defines knowledge as derived from experience and elevates the surface of things as the external-impersonal standard for fixing the true meaning of words. When experienced reality is made to function like a "gold standard," determining the value of the words that enter the currency of civil discourse, radical changes in the cultural perceptions of reality are bound to precipitate a crisis. At such moments, science may appear to license a floating, instead of a fixed, standard of truth, thus opening the civil discourse to the abuses that Locke feared. As "the chief end of language in communication is to be understood," observed Locke, "words serve not well for that end, neither the civil nor philosophical discourse, when any word does not excite in the hearer the same idea which it stands for in the mind of the speaker."[22]

The academy that opened its doors before the layman gave rise to a faith in the existence of a "reality" that could ultimately excite the "same ideas" in the discursive universes of both science and the civil society. But a democratic concept of knowledge binding both scientists and citizens was also a liability. The science that had been authorized to set standards for civil discourse could not radically shift its grounds without appearing to threaten the very "instruments and means of discourse, conversation, instruction and society."

THE NEW PHYSICS AND THE BROKEN MIRROR OF REALITY

During the second quarter of the twentieth century, against a background of escalating ideological confrontations and political conflicts, the attraction of a language controlled by direct references to experience inspired a renewed faith in the remedial effects of science on the "disease of discourse."[23] John Dewey was one of the leading exponents of the idea of reforming public discourse along the supposed principles of the language of empirical science. Like Locke, he celebrated the "development of political forms that promote the use of symbols in place of arbitrary power." Nevertheless, he insisted, the symbols "are significant only in connection with realities behind them."[24] Using words as a substitute for realities instead of as a means of contact with them leads inevitably to propaganda and deception, thought Dewey.[25]

By contrast, the "constructive speculations"[26] of an Einstein disengaged commonly observable "facts" from the grounds of scientific truth. In the climate fostered by Dewey and others, which demanded that the uses of words and the claims of speakers be restrained by references to the publicly accessible facts of "reality," such "constructive speculations" inevitably induced profound anxieties. Einstein himself resisted the conclusion prevalent among his peers "that the experimentally assured duality of nature (corpuscular and wave structure) can be realised only by . . . a weakening of the concept of reality."[27] But in subscribing, even temporarily, to the idea that light is both corpuscular and wave, he seemed to have committed the very sin against which Locke had warned the philosophers whose subtle proof that snow is both black and white threatened to "destroy the instruments and means of discourse."[28]

Einstein was, of course, well aware of the predicament of a science that, in failing to mirror common-sense reality, gives license to the abuses of discourse that Locke associated with civil disorder and Dewey with deception and propaganda. "When language becomes thus partially independent from the background of impressions," he observed, "a greater inner coherence is gained. . . . Only at this further development where frequent use is made of so-called abstract concepts, language becomes an instrument of reasoning in the true sense of the word. *But it is also this development which turns language into a dangerous source of error and deception.*"[29] Einstein's concerns here reflect the dilemma of science: in pursuing the advancement of knowledge, science uses and develops language as an instrument of thought, while in the larger society it is expected to help to improve language as a tool of disciplined social discourse and a means for promoting consensus and peace.

In a political context where any radical transformations of the prevailing conceptions of empirical reality would appear to license the return of the ghosts of metaphysics and political deception, the spectacular victory of the new physics was inevitably linked with a sense of cultural crisis. Einstein, who as we saw, sensed this anxiety, often compared the losses with the gains. "The general public," he admitted, "may be able to follow the details of scientific research to only a modest degree; but it can register at least one great and important gain: confidence that human thought is dependable and natural law universal."[30]

But how can the physics that attempts "to grasp reality as it is thought independently of its being observed"[31] gain the confidence of the layman except by means of the very trust that he had formerly denied to the priest? Neither Einstein's physics nor any other science can alter the conditions that render "empirical realism" a useful and convenient attitude in the context of ordinary life. Yet the new physics appeared to weaken the perceived validity of the concept of an objective and universally accessible external reality as an effective check on the claims of grand metaphysical and ideological visions. Reassurances about the persisting invariance of natural laws anchored beyond the domain of "native" sense experience did not appear sufficiently distant from earlier assurances concerning the existence of an invisible God. Science could not distinguish itself as a domain safe from religion and mysticism. When moving relative to another inertial observer, the clocks that are retarded in high velocities and the measuring rods that are contracted with accelerated movement are still sufficiently reliable in our world of sensations to ensure that we meet on time or that our maps are accurate. But the very idea that clocks and measuring rods can change their behavior in a universe of high velocities seemed unsettling to the notion that science, as a mirror of the external reality that we commonly experience, culturally and epistemologically validates the public realm of the civil society. The idea of an infinite number of equally objective "realities" erodes the social authority of science as a source of fixed universal standards that commensurate our cognitive constructions of reality in everyday experiences and ensure a domain in which the social and the political enterprise can rest on the solid basis of impersonal standards rather than on the more fragile basis of personal trust.

In a society in which the only type of knowledge that can bind the public is public knowledge, the perceived flight of the new physics beyond the realm of appearances generated a profound disillusionment. Hannah Arendt observed in her posthumously published book that in its "relentless search for the base underneath mere appearances . . . modern science has indeed forced the ground of appearances into the open so that man, a creature fitted for and dependent on appearances, can catch it." But ultimately, she continued, "the results have been rather perplexing. No man, it has turned out, can live among 'causes' or give full account in normal human language of a Being whose truth can be scientifically demonstrated in the laboratory and tested practically in the real world through technology. It does look as though Being, once made manifest, overruled appearances except that nobody so far has succeeded in living in a world that does not manifest itself of its own accord."[32]

Internally, this development was not a loss for science: Einstein's physics exposed the extent to which the habits of commonsense perception have prejudiced the formation of scientific concepts. But the classical alliance between the culture of science and the liberal-democratic ideals of politics has been irreparably damaged. One of the fundamental tenets of this alliance, as Leonard Bloomfield observed, was the premise that "linguistically, as well as in handling, science is a *public* activity . . . science deals with the phases of response that are alike for all normal persons. Its observations and predictions can be tested by everyone. . . . Unique personal or communal behavior figures in science as an object, which may be observed like any other, but it does not figure as part of scientific procedure."[33]

The new physics, to be sure, could not be viewed as a negation of science as a suprasubjective or public enterprise "in principle." In the sense in which the findings of science are subject to the tests of the scientific community, Einsteinian physics remained, of course, dependent upon public assessment. Only distortions and misunderstandings could lead to the view that relativity theory gives license to a relativism that grants recognition only to private experience and particular dispositions. Nevertheless, even though the growing accent on mathematics and on experience mediated by secluded laboratories has not made science inaccessible to any person in principle, it has progressively excluded the layman in fact. Simultaneously, in the context of the larger society, convulsed by two world wars and the threat of widespread dictatorship, there was a renewed sense of urgency in the public mission of science as a messenger of light. Just when the new physics dramatically exposed the limitations of common-sense experience and its tendency to prejudice our scientific construction of the world, the social and political upheaval of the time seemed to urge liberal-democratic spokesmen to renew the commitment to the role of science in securing the solidity, communicability, and objectivity of a sphere of facts resting on publicly shared perceptions. The very factors that appeared to the physicist to lead to a false, yet socially shared, concept of objective and external reality were precisely the factors that appeared to recommend communal testimony to the facts of experience and the "method of scientific discourse" as constraints on the prejudices and conflicts engendered by the contemporary abuses of language. As formulated by one of the leading popular writers of the time, Stuart Chase, it was only by extending into the sphere of social discourse the scientists' insistence on the clear correspondence between language and the facts of experience that the "disease of discourse" could be remedied or contained.[34] "Dictators can force a kind of duress agreement, primitive men cannot reach agreement in their magical realms of good and evil spirits. Even the rationalists cannot achieve it for they hold reason alone uncoordinated by the outside world. The scientists above all others have won agreement."[35] The "ills of the age," diagnosed as "verbalism," "metaphysics" and "ideological abstractions," seemed just the kind of problems that science, as a discourse disciplined and coordinated by the "outside world," could help to remedy.

EINSTEIN AND THE CRISIS OF PUBLIC CULTURE

"A culture which permits science to destroy traditional values but which distrusts its power to create new ones is a culture which is destroying itself," according to John Dewey.[36] It is precisely with respect to this affirmative-constructive task assigned to science—that is, to reconstruct civil discourse in a free polity—that Einstein's physics had its unsettling effects. The paradoxes and elusive referents associated with the new physics in the popular mind appeared to negate the belief in the continuity of the "publicness" of scientific and common-sense knowledge.

If, for Bishop Sprat, science represented the cooperative enterprise of many minds, which are "like a glass which is able to represent to itself all the works of nature which have been brought before it,"[37] Einsteinian science appears to the popular mind rather like the product of what Sprat called the "magical glass used to deceive the ignorant."[38] Instead of the science of the open academy resting on the "union of eyes" and the cooperation of the learned and the lay, Einsteinian science emerges in the popular mind as the inexplicable work of the brain of a single genius, a product that bears all the marks of a contact with secrets of nature inaccessible to the ordinary man. Instead of the science that advances bit by bit through the cumulative contributions of many industrious men, Einsteinian science, "like the lock which suddenly opens after a thousand unsuccessful attempts," suggests the idea of total knowledge discovered at once by a single act of divine intuition.[39]

In a culture that has celebrated the death of genius by inventing the "average man" and calibrating the "glasses" of all minds, Einstein's mythical brain was commended to medical research after his death to become a singular curio.[40] And it is only fitting that, following the desire of the saintly scientist, the manner of disposal of his ashes was kept secret from the world so that there would be no place, however humble, that might become a shrine.[41] As an anachronistic symbol of the scientist as hero, "Einstein" is, of course, the most fitting iconography of his own physics. The mystique surrounding Einstein's figure in the popular mind seems to correspond to the astonishing message of his physics.

Ironically, the uneasiness and anxiety engendered by the growing difficulties of sustaining a scientifically certified concept of reality as a public check on civil discourse was aggravated by Einstein's refusal to depart entirely from the classical commitment to the idea of representative physics as a notation of reality. On the one hand, Einstein pointed to the illusions generated by the failure to distinguish between what is "seen" and what is "happening"[42] and insisted that "we must make our mind to accept the fact that the logical basis [of physical theory] departs more and more from the facts of experience. . . ."[43] On the other hand, he did not give up the idea that "the whole of science is nothing more than a refinement of every day thinking."[44] But it is just this latter claim that inevitably empowers scientific revolutions to have disruptive effects on the world of the ordinary man and the social order. Cardinal Bellarmine had urged Galileo to speak "hypothetically, and not absolutely."[45] Similarly, Henri Bergson proposed to Einstein to eliminate the paradoxical appearance of his theory and "reconcile it with men as men" by "a distinction between physical truth and truth," or between the physicist's time and the layman's time, not reserving the attributes of "real" or "truth" only for the former.[46] Maurice Merleau-Ponty, who reported this encounter between Einstein and Bergson, is entirely sympathetic to the latter's attempt to obtain Einstein's cooperation in reconciling the "concrete world of our perception" with science and in allowing physics to "freely develop its paradoxes without authorizing unreason."[47]

In the cases of Galileo and Einstein, the ultimate refusal to concede the notion of physics as a representation of reality was sufficiently strong to resist the pressures to save the world of appearances from the advances of science. Einstein would probably have subscribed to Galileo's words when he asked rhetorically, "Who indeed will set bounds to human ingenuity?"[48] It is reported that Einstein's response to Bergson was equally uncompromising. Already in his popular exposition of relativity, published about a year and a half before the meeting with Bergson, Einstein insisted when discussing simultaneity that unless the requirement of a definition of simultaneity is satisfied "such that this definition supplies us with the method by means of which . . . [it can be decided] by experiment whether or not both the lightning strokes occurred simultaneously . . . I allow myself to be deceived as a physicist (and of course the same applies if I am not a physicist). . . ."[49]

In subsequent meetings with Bergson, Einstein indeed made a distinction between "psychological time" and the "time of the physicist."[50] But he did not seem to give his blessing to Bergson's suggestion that the attributes "truth" or "real" would be allowed also with reference to notions that evolve in the context of ordinary experience. "Psychological time," observed Einstein, "is different from the time of the physicist. . . . Originally individuals have the notion of simultaneity of perception; they can hence understand each other and agree about certain things they perceive; this is a first step toward objective reality. But there are objective events independent of individuals, and, from the simultaneity of perceptions one passes to the events themselves."[51] This leap from simultaneity of perceptions to simultaneity of events in the world of the objects themselves is, according to Einstein, an instinctual step not permissible in physics. "Nothing in our minds permits us to conclude to the simultaneity of events for the latter are only mental constructions, logical beings."[52]

This objection to extending notions evolved in the field of ordinary perceptions to the field of physics does not necessarily mean that the physicist's concepts of time or space can or should replace the conventional or intuitive notions that are accepted in the sphere of ordinary human interaction. What is important to understand here is that even if Einstein did not explicitly claim that the truths of physics are superior to those of common sense, the authority and the cultural influences of physics in Western society could, and in fact did, lend such import to his distinction between psychological and physical time. Einsteinian physics was widely construed as invalidating pervasive common-sense "psychological" concepts of time as well as ordinary notions of objects and events. Merleau-Ponty, for one, interpreted Einstein's position in the exchange with Bergson as suggesting the idea that "it is science alone we must go to for the truth . . . and the experience of the perceived world is no more than a stutter which precedes the clear speech of science."[53] Merleau-Ponty is led, therefore, to conclude that Einstein's position puts us "face to face with the crisis of reason. . . ."[54]

A science that appears to weaken man's confidence in the ordinary construction of reality is not congenial for sociopolitical order, which rests on the commitment to the value and reasonableness of the common human judgment. Einstein's own commitment to a notion of science that progressively departs from the facts of experience in achieving "greater unity in the foundations"[55] and at the same time does not abandon "the idea of direct representation of physical reality in space and time"[56] could only accentuate the cultural tensions generated by the new physics in the relations between science and ordinary experience. There persisted a widespread commitment to the classical role of science in correcting common-sense notions and substituting verifiable facts for "metaphysical" ideas about the world. But the new developments in the theoretical and experimental bases of physics appeared progressively to weaken science as an instrument for solidifying the foundations of an intersubjective realm of publicly shared "truths" and "realities." A science that generates knowledge within esoteric mathematical formulae and in which images of reality refer to laboratory-bound observations accessible only to a few can still, to be sure, remain a potent tool for criticizing common-sense notions of the world. Einstein, refusing the cultural compartmentalization of physics, considered it an important task for the physicist to approach critically "the problem of analyzing the nature of everyday thinking."[57] But the Enlightenment idea about the corrective and educational role of science required that scientific criticism of common-sense thinking not undermine the trust in the reality that evolves from shared perceptions.

During the second quarter of the twentieth century, when the idea of social discourse accountable to reality by references to "public facts" was invoked by liberal-democratic spokesmen in their confrontation with what they saw as the "metaphysical" and "propagandistic" language of fascist or communist leaders, the authority of science in backing responsible everyday discourse was widely regarded as essential. Science was associated with the task of exposing deception and ideological fictions by authorizing the force of "real facts." Hence, precisely during this period any dramatic evidence to the effect that science does not confirm a reality that is publicly accessible at the level of commonly shared perceptions was bound to be unsettling. Despite considerable differences in the theories of such people as the philosopher John Dewey, the linguist C. K. Ogden, the literary critic I. A. Richards, American and immigrant affiliates of the Vienna Circle, and a publicist like Stuart Chase, these and other intellectuals shared a widely held belief in the potential remedial effects of science on discourse diseased by the proliferation of fantasies, metaphysical ideas, and abstractions. They held that a language disciplined by direct and unambiguous references to concrete objects of perception would be congenial to responsible discourse and would enhance accord rather than conflict. Despite Einstein's frequent insistence that the use of words should be disciplined by experiments,[58] his physics did not appear to support these hopes for rendering civil discourse more responsible by reference to scientifically certifiable facts.

Despite repeated and ingenious attempts to popularize Einstein's physics, its empirical and conceptual components remain elusive to the ordinary man.[59] The strains induced by the remoteness of Einsteinian science in the movement to enlist the method of scientific discourse to rid society of metaphysical speech and unresolvable ideological conflicts are clearly detectable in Stuart Chase's effort to minimize the damage. In his *Tyranny of Words* (1938), Chase, inspired by Bridgman's operationalism, tried to save the notion of "public science" from the shocking effects of a physics that melts the solid, external *objects* of our experience by resting science on the publicness of the scientist's *operations*. Considering Einstein's relativity as responsible for a "crisis in meaning" and assessing its cultural implications, Chase noted that beyond the sphere of science, Einstein "also revolutionized certain aspects of human communication. In the long run, relativity may prove a more important factor in language than in physics. Its impact has caused thoughtful men everywhere to look to their words, to question the validity of their concepts."[60]

In light of the popular reflections of Einstein's physics, the dilemma before Chase was how to continue to defend science as a weapon against the outbursts of "irrationalism" and what the eighteenth-century thinkers called "enthusiasm," how to persuade his audience that Einstein's physics had not invalidated the notion that "the scientific method is concerned with how things do happen" and that "most of us are amateur scientists today, though we are seldom aware of it," how to impress upon the public "that Einstein has not shunted science into ghostly realms where 'everything is electricity—electricity is unknown—therefore everything is unknown.' "[61] Chase wanted instead to convince his audience that "Einstein only brought us closer to the world outside."[62] By invoking Bridgman's operationalism, Chase clearly hoped to substitute operations that are the same for all actors in place of objects that are the same for all observers as a public external check on the uses of words. By insisting, with Bridgman, that "concepts not subject to operations are meaningless,"[63] Chase attempted to secure the continued role of science as discourse accountable to objective facts in the construction of nonarbitrary civil discourse controlled by reference to public, repeatable actions. As one might expect, Chase tended to buttress his revised definition of public science by reassuring his readers that actually nothing had changed and that when "the experimental proof comes," the scientists who squabble over the right theory "will cease their bickering and remove their hats. . . . The fact has *always been* for the physicist the one ultimate thing from which there is no appeal."[64]

THE NIGHTMARE OF THE SAINTLY PHYSICIST IN THE LIGHTHOUSE

The notion of the scientist removing his hat before the facts of experience brings to mind the scientists who went to the artisan's workshop or who, like Galileo, mingled with the shipbuilders of Venice. In a speech delivered in 1933 at Albert Hall in London, Einstein presented his very different notion of the proper place for a scientist to think out his problems. "I lived in solitude in the country," he said, "and noticed how the monotony of quiet life stimulates the creative mind. There are certain callings in our modern organization which entail such an isolated life without making a great claim on bodily and intellectual effort. I think of such occupations as the service in lighthouses and lightships."[65] As against the bustling workshop or laboratory, where science evolved from "the union of eyes and hands,"[66] Einstein idealized the solitude and the isolation of the country or the lighthouse. There is, of course, a fit between the image of the solitary scientist in the lighthouse and Einstein's physics as the inexplicable product of the "constructive speculations" of a genius. But in a liberal-democratic culture that had come to celebrate the progress of science as a testimony to "civilization" and the industry of the many, the image of science as the product of solitary genius was an oddity that had to be countered. It threatened to discredit the hard-won gains that had come with the emancipation of science from the stigma of aristocratic "high culture."

The association of Einstein's physics with an epistemology in which, as Gerald Holton has observed, "reality does not need to be validated by the individual's sensorium" and with Einstein's own more frequent reference in his nonscientific writings to religious questions[67] was not reassuring for those who sought in science the remedies for the proliferation of opposing political theologies and for "verbalism" unrestrained by empiricism. For some theologians, the new physics suggested a sense of the limits of human knowledge and offered a belated acknowledgment on the part of science of a place for religion and belief. Such notions appeared incompatible with popular concepts of science as the classification of the visible and the systematic refinement of common-sense perceptions. Moreover, as Holton has shown, they appeared inconsistent with what had evolved within the folklore of the scientific community as the established image of the growth of knowledge.[68]

The force of this image is manifest in the persistent tendency within the scientific community to exaggerate the role of the Michelson experiments in the evolution of Einstein's special relativity theory. In numerous accounts of the history of the special theory of relativity, especially in science textbooks, the Michelson-Morley experiment is described as the "crucial" experiment leading directly to the birth of the theory.[69] On the basis of the extensive evidence he collected, Holton concluded that "the role of the Michelson experiment in the genesis of Einstein's theory appears to have been so small and indirect that one may speculate that it would have made no difference to Einstein's work if the experiment had never been made at all."[70] At the same time, Holton, supported by Einstein's own observations, ascribes to the Michelson experiments a more significant role in facilitating the reception of Einstein's theory within both the scientific community and the larger society.

Reflecting on this consistent deviation from the historical record, Holton suggests that the emphasis on inductive reasoning and the tendency to depersonalize the scientist's research activities are meant to socialize the student to "accepted public norms of professional behavior."[71] These norms tend to belittle the role of speculations and other nonpublic factors in scientific research and the evolution of knowledge. Viewed from this perspective, a theory that describes knowledge as growing out of facts and observations minimizes the dependence of students upon teachers. As a public norm, a theory that renders teachers accountable to what students can witness, implicitly also renders leaders accountable to what citizens as autonomous and independent observers can confirm. The idea that "hearing causes us to believe, since we believe through teachers while . . . we cannot experience what we learn except by means of vision"[72] is, of course, an ancient wisdom. But in the modern democratic society, the shift from teachers who persuade by telling to teachers who convince by presenting and demonstrating has been more than a development in the realm of education and pedagogy.[73]

On the more fundamental level, where shifting cultural attitudes forge new models of authority and change the canons of persuasion in the polity, such developments have reflected the belief that "democracy [has] to be a government by judgment rather than by arbitrary decision [and] scientific method and research [are] necessary to guarantee the validity of judgment."[74] In such a cultural context, theories that claim that knowledge grows independently of the accumulation of public facts validated by shared perceptions and common experience are bound to appear as "subversive genealogies" of knowledge.[75] They discredit both the origins of the social authority of science and the type of political authority that rests on the premise that citizens secure the accountability of leaders by universally shared standards of "facts" and "reality." That the layman remained a humiliated or bewildered outsider to the great debate between Einstein and his peers, as testified by the repeated failures to popularize Einstein's physics,[76] dramatized the unintended cultural repercussions of modern physics.

Again, it was not the gulf between what science asserts and what the layman understands that was the novel element in the situation. What was significant was the combined impact of a scientific advance that could stimulate the *New York Times* to publish six editorials within one month[77] and the insistence that understanding this theory requires "getting rid of the notions which are useful in ordinary life."[78] Particularly offensive to the layman was the notion that the new physics discards commonsense constructions of reality *in principle.* The *New York Times* protested that "the most disturbing feature of the situation is the assumption that only men of wonderful learning have the ability, and therefore the right, to see what meaning there is in the fact that light, being subject to a turning from the straight path by a mass of matter like the sun, must itself have of matter at least the quality of weight."[79] Einstein's humorous remark that only "twelve persons understood the new theories made the *New York Times* only a little happier."[80] Einstein's physics stands as the primary example not only of the repeated failures to popularize science but also of a much deeper challenge to the realizability of the very ideal of popularization.[81]

Although the scientist whose "free conceptual constructions" revolutionized modern physics had a dream of the lonely physicist in the lighthouse, his physics, insulting to common sense, failed to sustain the popular image of Newton's physics as a symbol of the guiding light of reason.[82] Many thinkers and publicists who have reflected upon the impact of Einstein's physics have noted this alienation of light from truth. Milic Çapek observed that the most revolutionary aspect of the new physics is "that the words 'picture' and 'view' entirely lose their etymological meaning. As the so-called primary qualities of matter now join the secondary qualities in their exit from the objective physical world, it is clear that the future conception of matter ought to be devoid of all sensory qualities."[83] John Dewey referred to the same development as the end of "the old spectator theory of knowledge."[84] Seeming to strain the old bond forged by the Enlightenment between the advancement and the diffusion of knowledge, Einstein's physics appeared to set the success of science on a collision course with the objective of civilizing and restraining the rapidly expanding body of rightful political participants by democratizing the knowledge of firm truths. These apparent effects of Einstein's physics, dimming the light of scientific reason, did not escape the observer who described the shift from Newtonian to Einsteinian physics as a shift from light to night.[85]

If the Enlightenment crowned science as a messenger of light, rendering reality transparent to all through its visible surface, Einstein's physics seemed to drive science back again to the hidden dimensions of Being. Contrary to Einstein's own aspirations, his physics, inevitably vulgarized in popular culture, suggested to the mystified layman a universe bearing some of the characteristics ascribed to it by the prescientific mind. In the feudal universe, observed Marc Bloch, "the material world was scarcely more than a sort of mask, behind which took place all the really important things."[86]

The tensions induced by the new physics between the "culture of science" and the ideals of democratization disciplined and stabilized by enlightenment led thoughtful scientists to rethink the relations between science and the world of the ordinary man. J. Robert Oppenheimer's *Science and Common Understanding* (1954) and James B. Conant's *Modern Science and Modern Man* (1953) are typical examples. Oppenheimer protested against the philosophers and popularizers who mistake relativity for the doctrine of relativism and who construe "Einstein's great works as reducing the objectivity, firmness, and consonance to law, of the physical world."[87] At the same time, he had to acknowledge that by altering our ideas on causality and the objectivity of the physical world, the new physics reminded us "in a quite unexpected way, of the nature and the limitations, as well as the power, of human knowledge itself."[88] Juxtaposed with the example of Newton, the contemporary physicist has not brought to his reflections on culture and society the optimism that has traditionally come with a faith in reason confirmed by the advancement of knowledge. His tone is more pessimistic, and his message is not the power of knowledge, but its limits. Oppenheimer thought that "if we err today . . . it is in expecting too much of knowledge."[89] In a similar tone, James Conant addressed himself to the "deeply troubled" modern man, mindful of the fact that "the attitude of the American citizen towards science and scientists has undergone a profound change."[90] In order to mitigate the apparent disenchantment of the layman, Conant points to what he regards as the fallacies underlying the popular notions of science—fallacies that appear, on the one hand, to encourage unrealistic faith in the power of science to spread the light of reason in the larger society and, on the other, to nourish the disappointments that come with the disconfirming images projected by the new physics.

What Conant had in mind was, first of all, the harm caused by the myth of "the scientific method," which "requires the habit of facing reality" and evolves scientific theories as "pictures."[91] But in a society in which reality functions as an escape from subjectivity and metaphysical isolation, in a culture in which reality is empowered to uphold the public realm, science as a picture of reality can easily become the first article of faith. This is why Conant, disconcerted by the cultural resonance of the new physics, recommends that we withdraw our belief in scientific theory as a creed and regard it merely as a policy.[92] Instead of conceiving of scientific theories as maps of reality, Conant suggests that they be reduced to mere "guides to human action."[93] Such changes, he acknowledges, are required in part because "the basic assumptions of those who once popularized science have been challenged."[94] Instead, he would now subscribe to the position stated in the Book of Job, that "the universe is essentially inexplicable,"[95] although he expresses some concern that the withdrawal of the claim that science is a map of the universe may be abused by the superstitious.[96]

Both Oppenheimer and Conant recognized that the science that fails to bring salvation through illumination cannot continue to be thought of as a source of light. Both reflect the recognition that the cultural role of science is transformed by the influence and implications of the new scientific developments. The science that no longer furnishes a map of reality is not the science that can, by correcting the imperfections and abuses of language, mitigate controversies or, as Locke wished, lead to the advancement of both knowledge and peace.

But in the second half of Einstein's century, the science that does not decipher the riddle of the universe for the layman is culturally compatible with the politics that abandon the zeal for a total, comprehensive creed. In both science and politics the earlier aspiration for the cognitive construction of reality has been tempered by an increasing preoccupation with operations and techniques that work. Although Einsteinian physics has not constituted a retreat from public to private science or from objective to subjective experience, its reflections in the popular mind have helped to undermine science as a hallmark of public culture. Today the ideal of science as a mirror of reality is just as remote as the ideals of perfect international order and world government. The affirmative roles of science in constructing a common consciousness and in sustaining the public realm have been eroded by the reassertion of doubt, private interest, and pessimism. The new physics has not given support to the liberal-democratic propagandists who sought to elevate what they saw as the canons of scientific method and discourse to the level of a formula for breeding discipline and agreement in a society of freethinkers.

Since World War II, "works," to paraphrase Bacon, "have increasingly come to be esteemed more for the comforts of life they afford than as pledges of truth."[97] But the unexpected technological consequences of the science that was supposed to convert mental discipline into social moderation have come to symbolize excess. Engulfed in the clash of interests, science can no longer represent the "objective" social space of the public realm. It becomes instead a part of the politics that has substituted compromise for consensus and has come to legitimate decisions not by reference to "truth" but by reference to "fair representation."

Einstein's physics, by advancing truth beyond the grasp of the ordinary man, did not serve the aspirations of Einstein the man and the citizen of the world. These cultural effects were not consistent with his hopes that "intellectual workers will spread clear ideas and contribute through enlightenment to prevent able statesmen from being hampered in their work by antiquated opinions and prejudice."[98] The Einstein who exemplifies the shifting cultural role of the scientist, whose preoccupation with the truth and the riddle of the universe is eclipsed by the pressing problems of responsibility and the ethical uses of power, is a tragic figure. The conversion of the light of reason into the excesses of power was not the dream, but the nightmare, of the lonely, saintly "physicist in the lighthouse."

NOTES

1. Christiaan Huygens, *Oeuvres*, publieés par la Société Hollandaise des Sciences, The Hague, 1888-1950, vol. 3, pp. 152-158, quoted in W. E. Knowles Middleton, *The Experimenters: A Study of the Accademia del Cimento*, Baltimore, Johns Hopkins University Press, 1971, p. 260.
2. Ibid.
3. Albert Einstein, "Physics and Reality" (1936), reprinted in Einstein, *Out of My Later Years*, New York, Philosophical Library, 1950, p. 96.
4. Albert Einstein, "The Experimental Confirmation of the General Theory of Relativity," in Einstein, *Relativity, The Special and the General Theory: A Popular Exposition* (1916; first English trans., 1920), trans. Robert W. Lawson, 15th ed., New York, Crown Publishers, 1961, p. 123.
5. See Bacon's distinction in *De Augmentis*, bk. 6, chap. 2, *The Works of Francis Bacon*, ed. J. Spedding and R. L. Ellis, London, 1875, vol. 4, p. 449.
6. Thomas Sprat, *History of the Royal Society* (1667), ed. Jackson I. Cope and Harold Whitmore Jones, St. Louis, Washington University Press, London, Routledge and Kegan Paul Ltd., 1958, p. 97.
7. Ibid., pp. 129-131.
8. Ibid., p. 85.
9. "John Winthrop, Jr.'s Major Gift to the Royal Society," an excerpt from the *Journal Book* of the Royal Society of London, 10 Feb. 1669/70. Cited in Raymond Phineas Stearns, *Science in the British Colonies of America*, Urbana, University of Illinois Press, 1970, p. 691. Stearns's discussion illuminates many sides of the participation of laymen in the enterprise of the Royal Society of London.
10. Ibid., pp. 414-424.
11. The reference is to the collection of Jean de Berry, duke of Burgundy, described in Alma S. Wittlin, *The Museum: Its History and Its Tasks in Education*, London, Routledge and Kegan Paul Ltd., 1949, p. 68.
12. See, for instance, ibid., and Silvio A. Bedini, "The Evolution of Science Museums," *Technology and Culture* **6**, no. 1 (1965), esp. pp. 18, 25.
13. On the evolution of standardized pictorial communication, see William M. Ivins Jr., *Prints and Visual Communication*, Cambridge, Mass., MIT Press, 1953.
14. See, for instance, in Margaret C. Jacob, *The Newtonians and the English Revolution 1689-1720*, Hassocks, The Harvester Press, 1976, esp. pp. 22-71, 162-200.
15. Sprat, *History of the Royal Society*, p. 131.
16. Einstein, "Relativity and the Problem of Space," in *Relativity*, pp. 139-140.
17. Ibid., p. 140.
18. The role of vision in new modes of discourse is discussed by Walter

J. Ong, S.J. See, for instance, his *Ramus, Method, and the Decay of Dialogue*, New York, Octagon Books, 1974; *The Presence of the Word*, New Haven, Yale University Press, 1967; and "I See What You Say," in Ong, *Interfaces of the Word*, Ithaca, Cornell University Press, 1978.

19. John Locke, *An Essay Concerning Human Understanding*, collated and annotated by Cambell Fraser, New York, Dover Publications, 1959, vol. 2, pp. 106-120 (bk. 3, chap. 9).

20. Ibid., p. 131 (emphasis added).

21. Ibid., p. 129.

22. Ibid., p. 105.

23. The tendency to regard the sociopolitical upheaval of the time as, at least in part, a crisis of communication and the belief that this crisis can be resolved or mitigated by a reform of language was shared by an impressive group of linguistic scholars. Among the most influential were Alfred Korzybski, C. K. Ogden, and I. A. Richards.

24. John Dewey, *Liberalism and Social Action* (1935), New York, Putnam, Capricorn Editions, 1963, pp. 71-72.

25. Ibid., p. 92.

26. Einstein, "Physics and Reality," p. 76.

27. Einstein, "Relativity and Space," p. 157.

28. Locke, *Human Understanding*, p. 129.

29. Albert Einstein, "The Common Language of Science," in *Out of My Later Years*, pp. 111-112 (emphasis added).

30. Einstein, "Science and Society," in ibid., p. 137.

31. Albert Einstein, "Autobiographical Notes," trans. Paul Arthur Schilpp, in Schilpp, ed., *Albert Einstein: Philosopher-Scientist*, Evanston, Ill., Library of Living Philosophers, 1949, p. 81.

32. Hannah Arendt, *The Life of the Mind*, vol. 1: *Thinking*, London, Secker and Warburg, 1978, pp. 25-26.

33. L. Bloomfield, "Linguistic Aspects of Science," in *International Encyclopedia of Unified Science*, vol. 1-2: *Foundations of the Unity of Science*, Chicago, University of Chicago Press, 1939, vol. 1, pp. 46-47.

34. Stuart Chase, *The Tyranny of Words*, New York, Harcourt, Brace, 1939, p. 171.

35. Ibid., pp. 360-361.

36. John Dewey, *Freedom and Culture*, New York, Putnam, 1939, pp. 148-149.

37. Sprat, *History of the Royal Society*, p. 97.

38. Ibid.

39. Roland Barthes, "The Brain of Einstein," in Barthes, *Mythologies* (1957), trans. A. Lavers, London, Jonathan Cape, 1977, p. 69.

40. Ibid., p. 68.

41. Banesh Hoffmann, with the collaboration of Helen Dukas, *Albert Einstein: Creator and Rebel*, New York, Viking Press, 1972, p. 261.

42. Einstein, "Physics and Reality," p. 69.

43. Ibid., p. 96.
44. Ibid., p. 59.
45. Cardinal Bellarmine's reply to Foscarini, 12 April 1615, cited in Jerome J. Langford, *Galileo, Science and the Church*, Ann Arbor, University of Michigan Press, 1971, p. 60.
46. Maurice Merleau-Ponty reported on this meeting in his "Einstein and the Crisis of Reason," in *Signs*, trans. Richard C. McCleary, Evanston, Ill., Northwestern University Press, 1964, pp. 192-197.
47. Ibid., p. 196.
48. From Galileo's letter to the grand duchess of Tuscany. The English translation of the letter is published in *Discoveries and Opinions of Galileo*, trans. and ed. Stillman Drake, New York, 1957, p. 187.
49. Einstein, "On the Idea of Time in Physics," in *Relativity*, p. 22.
50. See in *Bergson and the Evolution of Physics*, trans. and ed. P.A.Y. Gunter, Knoxville, University of Tennessee Press, 1969, p. 133.
51. Ibid.
52. Ibid.
53. Merleau-Ponty, "Einstein and the Crisis of Reason," pp. 196-197.
54. Ibid.
55. Einstein, "Physics and Reality," p. 73.
56. Einstein, "The Fundaments of Theoretical Physics," in *Out of My Later Years*, p. 110.
57. Einstein, "Physics and Reality," p. 59.
58. Einstein, "Idea of Time," p. 22.
59. On the failure of popularization, see for instance, Ronald C. Tobey, "The Einstein Controversy, 1919-1924," in Tobey, *The American Ideology of National Science, 1919-1930*, Pittsburgh, University of Pittsburgh Press, 1971, pp. 96-132.
60. Chase, *The Tyranny of Words*, pp. 47, 117.
61. Ibid., p. 138.
62. Ibid., p. 139.
63. Ibid., p. 128.
64. Ibid., p. 125.
65. Albert Einstein, "Science and Civilization," in *Out of My Later Years*, pp. 149-150.
66. Sprat, *History of the Royal Society*, p. 85.
67. Gerald Holton, "Mach, Einstein, and the Search for Reality," in Holton, *Thematic Origins of Scientific Thought: Kepler to Einstein*, Cambridge, Mass., Harvard University Press, 1973, p. 243.
68. Gerald Holton, "Einstein, Michelson, and the 'Crucial' Experiment," in ibid., p. 263.
69. Ibid., pp. 261-352.
70. Ibid., p. 327.
71. Ibid., p. 271.
72. Roger Bacon in his discussion of the science of optics in *The Opus*

Majus of Roger Bacon, ed. J. H. Bridges, vol. 2, pp. 2-3. Cited in English by David C. Lindberg and Nicholas H. Steneck, "The Sense of Vision and the Origins of Modern Science," in Allan G. Debus, ed., *Science, Medicine and Society in the Renaissance: Essays to Honor Walter Pagel*, Science and History Publications, vol. 1, New York, 1972, pp. 36-37.

73. See on this aspect, the works of Walter Ong cited in note 18 above.

74. Tobey, *Ideology of National Science*, p. 169.

75. On subversive genealogies in the political sphere see Judith N. Shklar, "Subversive Genealogies," *Daedalus*, Winter 1972, pp. 129-154.

76. Tobey, *Ideology of National Science*, pp. 96-132.

77. Ibid., p. 105.

78. Russell, *The ABC of Relativity*, London, Kegan Paul, 1941 p. 5.

79. Tobey, *Ideology of National Science*, pp. 105-106.

80. Ibid., p. 106.

81. Ibid., esp. pp. 114-115.

82. This image was sustained even though the association of Newtonian science with the concept of the cosmos as something *seen* was in some respects inconsistent with such Newtonian concepts as "actions-at-a distance" and with other phenomenologically less dependent elements of his theory. Einstein observed that "physics (at least part of it), originally more phenomenologically constructed, was reduced, by being founded upon Newton's mechanics for atoms and molecules, to a basis further removed from direct experiment, but more uniform in character." Einstein, "Physics and Reality," p. 74.

83. Milic Çapek, *The Philosophical Impact of Contemporary Physics*, Princeton, N.J., D. van Nostrand Co., 1961, p. 379.

84. John Dewey, *The Quest for Certainty* (1929), New York, Putnam, Capricorn Books, p. 204.

85. This observer added a new ending to Pope's famous epitaph celebrating the light of Newton:

> Nature and nature's law lay hid in night.
> God said "let Newton be" and all was light.
> It did not last: the Devil howling "Ho!
> Let Einstein be!" restored the status quo.

Cited in Ronald W. Clark, *Einstein: The Life and Times*, New York, Avon Books, 1971, p. 336.

86. Marc Bloch, *Feudal Society*, trans. L. A. Maynon, Chicago, University of Chicago Press, 1964, p. 83.

87. J. Robert Oppenheimer, *Science and the Common Understanding* (1954), New York, Simon and Schuster, 1966, p. 4.

88. Ibid., pp. 33-34.

89. Ibid., p. 93.

90. James B. Conant, *Modern Science and Modern Man*, Garden City, N.Y., Doubleday, Anchor Books, 1953, p. 12.

91. Ibid., pp. 35-36.

92. Ibid., p. 91.
93. Ibid., p. 101.
94. Ibid., p. 104.
95. Ibid., p. 157.
96. Ibid., p. 172.
97. Bacon's original statement is in "The New Organon," bk. 124, in *The Works of Francis Bacon*, ed. Spedding and Ellis, vol. 4, p. 110.
98. *New York Times*, 18 November 1946, p. 25, col. 1.

Einstein and Developments in the Jewish World

Isaiah Berlin

EINSTEIN AND ISRAEL

WHEN THE MAYOR of Jerusalem, Teddy Kollek, and Professor Aryeh Dvoretsky, President of the Israel Academy, invited me to speak about Albert Einstein, my instinctive (and some might think, right) reaction was to decline this great and wholly unexpected honor. My reason was obvious enough: Einstein's chief title to immortal fame is his transcendent scientific genius, about which, like the vast majority of mankind, I am totally incompetent to speak. Einstein was universally revered as the most revolutionary innovator in the field of physics since Newton. The exceptional respect and attention that were everywhere paid to his person and to his opinions on other topics sprang from this fact. He knew this himself; and although he was a genuinely modest man, embarrassed by the adulation which he excited, and disliked publicity, he expressed pleasure at the thought that, if homage was to be paid to individuals at all, it should go to those who could claim achievement in fields of intellect and culture rather than of power and conquest. Indeed, that a mathematical physicist should have become a great world figure is a remarkable fact and a credit to mankind.

If the impact of Einstein's ideas outside the realms of theoretical physics and, perhaps, of the philosophy of physics is compared to that made by the ideas of other great scientific pioneers, an odd conclusion seems to emerge. Galileo's method, to go no further back, and his naturalism, played a crucial role in the development of seventeenth-century thought, and extended far beyond technical philosophy. The impact of Newton's ideas was immense: whether they were correctly understood or not, the entire programme of the Enlightenment, especially in France, was consciously founded on Newton's principles and methods, and derived its confidence and its vast influence from his spectacular achievements. And this, in due course, transformed—indeed, largely created—some of the central concepts and directions of modern culture in the west, moral, political, technological, historical, social—no sphere of thought or life escaped the consequences of this cultural mutation.

This is true to a lesser extent of Darwin—the concept of evolution affected many fields of thought outside biology: it upset the theologians, it influenced the historical sciences, ethics, politics, sociology, anthropology. Social Darwinism, founded on a misapplication of Darwin's and Huxley's views, with its eugenic and sometimes racist implications, did social and political harm. I should perhaps hesitate to refer to Freud as a natural scientist; but there is no doubt that his teaching, too, affected fields far outside psychology—history, biography, aesthetics, sociology, education.

But Einstein? His scientific achievement touched on the philosophy of science; his own views—his early acceptance of Mach's phenomenalism, and his subsequent abandonment of that view—show that he possessed the gifts of a philosopher, and so, indeed, did his views of the central doctrines of Spinoza, Hume, Kant, Russell. In this respect, Einstein and Planck were virtually unique among the outstanding physicists of our century. But his influence on the general ideas of his time? On educated opinion? Certainly he presented a heroic image of a man of pure heart, noble mind, unusual moral and political courage, engaged in unswerving pursuit of the truth, who believed in individual liberty and social equality, a man sympathetic to socialism, who hated nationalism, militarism, oppression, violence, the materialistic view of life. But apart from embodying a combination of human goodness with a passion for social justice and unique intellectual power, in a society in which many seemed to live by the opposite values—apart, that is, from his exemplary life, from being, and being seen to be, one of the most civilised, honourable and humane men of his time—what impact did Einstein have?

It is true that the word 'relativity' has been, to this day, widely misinterpreted as relativism, the denial of, or doubt about, the objectivity of truth or of moral and other values. But this is a very old and familiar heresy. Relativism in the sense in which Greek sophists, Roman sceptics, French and British subjectivists, German romantics and nationalists professed it, and in which theologians and historians and ordinary men have, in modern times, been tormented by it—this is the opposite of what Einstein believed. He was a man of simple and absolute moral convictions, which were expressed in all he was and did. His conception of external nature was that of a scientifically analysable, rational order or system; the goal of the sciences was objective knowledge of an independently existent reality, even though the concepts in which it was to be analysed and described were free, arbitrary human creations.

What general impact did his doctrines have? Modern theoretical physics cannot, has not, even in its most general outlines, thus far been successfully rendered in popular language as Newton's central doctrines were, for example, by Voltaire. High-minded public men in England like Haldane and Herbert Samuel tried to derive general metaphysical or theological truths, usually somewhat trite ones, from the general theory of relativity, but this only showed that their gifts lay in other spheres.

But if the impact of Einstein's scientific thought on the general ideas of his time is in some doubt, there can be none about the relevance of his non-scientific views to one of the most positive political phenomena of our time. Einstein lent the *prestige mondial* of his great name, and in fact gave his heart, to the movement which created the state of Israel. Men and nations owe a debt to those who help to transform their realistic self-image for the better. No Zionist with the least degree of self-esteem can refuse to pay him homage if the opportunity of doing so is offered to him. Einstein's support of the Zionist movement and his interest in the Hebrew University were lifelong. He quarrelled with Weizmann more than once; he was highly critical of the Hebrew University and, in particular, of its first President; he deplored the shortcomings of Zionist policy towards the Arabs; but he never abandoned his belief in the central principles of Zionism. If young people (or others) today, whether Jews or gentiles, who, like the young Einstein, abhor nationalism and sectarianism and seek social justice and believe in universal human values—if such people wish to know why he, a child of assimilated German Jews, supported the return of the Jews to Palestine, Zionism, and the Jewish state, not uncritically nor without the anguish which any decent and sensitive man cannot but feel about acts done in the name of his people which seem to him wrong or unwise, but nevertheless steadily, to the end of his life—if they wish to understand this, then they should read his writings on the subject. With his customary lucidity and gift for penetrating to the central core of any issue, whether in science or in life, Einstein said what had to be said with simplicity and truth. Let me recall some of the things he said and did, and in particular the path which led towards them.

He was born in Ulm, the child of irreligious parents. He was educated in Munich, where he seems to have encountered no discrimination; if he reacted strongly against his school and suffered something approaching a nervous breakdown, this does not seem to have been due to anti-Jewish feeling. What he reacted against was, perhaps, the quasi-military discipline and nationalist fervour of German education in the 1890s. He studied intermittently in Milan and Zurich, taught in Zurich, obtained a post in the Patent Office in Bern, then held university chairs in Prague and Zurich, and in 1913 was persuaded by Nernst and Haber, as well as Planck, whose reputations were then at their peak, to accept a research post in Berlin.

I do not need to describe the atmosphere of Prussia on the eve of the First World War. In a letter written in 1929 to a German Minister of State, Einstein said, 'When I came to Germany fifteen years ago [that is, in 1914] I discovered for the first time that I was a Jew. I owe this discovery more to gentiles than Jews.' Nevertheless, the influence of some early German Zionists, in particular Kurt Blumenfeld, the apostle to the German Jews, played a significant part in this—and Einstein remained on terms of warm friendship with him for the rest of his life. But, as in the case of Herzl, the decisive factor in his awakening as a Jew was not so much encounter with an unfamiliar doctrine (he had met adherents of it in Prague but apparently took no interest in it then) as the chauvinism and xenophobia of leading circles, in this case in Berlin, which led him to a realisation of the precarious predicament of the Jewish community even in the civilised west. 'The best in man can flourish', he declared, 'only when he loses himself in a community. Hence the moral danger of the Jew who has lost touch with his own people and is regarded as a foreigner by the people of his adoption.' 'The tragedy of the Jews is that they . . . lack the support of a community to keep them together. The result is a want of solid foundations in the individual which in its extreme form amounts to moral instability.'

The only remedy, he argued, is to develop a close connection with a living society which will enable individual Jews to bear the hatred and humiliation to which they are often exposed by the rest of mankind. Herzl is to be admired, Einstein tells us, for saying 'at the top of his voice' that only the establishment of a national home in Palestine can cure this evil. It cannot be removed by assimilation. The Jews of the old German ghettos were poor, deprived of civic and political rights, insulated from European progress. Yet

> these obscure, humble people had one great advantage over us—each of them belonged in every fibre of his being to a community in which he was wholly absorbed, in which he felt himself a fully privileged member, which asked nothing of him that was contrary to his natural habits of thought. Our forefathers of those days were pretty poor specimens intellectually and physically, but socially they enjoyed an enviable spiritual equilibrium.

Then came emancipation: rapid adaptation to the new open world: eager efforts to don clothes made to fit others, involving loss of identity, the prospect of disappearance as a group. But this was not to be:

> However much the Jews adapted themselves, in language, manners, and to a large extent even in the forms of religion, to the European peoples among whom they lived, the feeling of strangeness between them and their hosts never vanished. This is the ultimate cause of anti-Semitism, which is, therefore, not to be got rid of by well-meaning propaganda. Nationalities want to pursue their own goals, not to blend.

To ignore, or argue against, emotional prejudice or open hostility, Einstein declared, is wholly futile; the baptised Jewish *Geheimrat* was to him merely pathetic. National frontiers, armies, he regarded as evil, but not national existence as such: the life of peaceful nations, with reciprocal respect for one another and toleration of each other's differences, was civilised and just. There follows a statement of Zionism not unlike the reaction to a similar predicament of another internationalist and socialist, Moses Hess, in the 1860s. Let me quote Einstein's words in 1933: 'It is not enough for us to play a part as individuals in the cultural development of the human race, we must also attempt tasks which only nations as a whole can perform. Only so can the Jews regain social health.' Consequently: 'Palestine is not primarily a place of refuge for the Jews of eastern Europe, but the embodiment of the re-awakening of the corporate spirit of the entire Jewish nation.'

This seems to me a classical formulation of the Zionist creed, with an affinity to the unpolitical cultural nationalism of Ahad Ha'am: what Einstein was advocating was, in essence, the creation of a social and spiritual centre. But when British policy and Arab resistance, in his judgement, made the state inevitable, he accepted it, and the use of force to avoid annihilation, as being, perhaps, something of a necessary evil, but nevertheless as a burden and a duty to be borne with dignity and tact, without arrogance. Like all decent Zionists he was increasingly worried about the relationship with the Arabs of Palestine. He wished for a state in which Jews and Arabs could fully cooperate. But he realised, sadly, that events made this unlikely for the time being. He remained a consistent supporter of the Jewish state of Israel: here Jewish ideals must be pursued, especially three among them: 'The pursuit of knowledge for its own sake; an almost fanatical love of justice; desire for personal independence.'

I need hardly say how sharply this differed from the general attitude of the educated German Jews of his milieu, not to speak of men of similar origin and social and intellectual formation elsewhere in western Europe. When one remembers Einstein's earlier life, remote from Jewish affairs, his lifelong idealistic internationalism, his hatred of all that divided men, it seems to me to argue a remarkable degree of insight, realism and moral courage, of which his fellow Jews today have good reason to feel proud. After all, other eminent German-Jewish scientists, honourable men of unimpeachable personal integrity, Fritz Haber, Max Born, James Franck, reacted very differently. So did writers and artists like Schnitzler, Stefan Zweig, Mahler, Karl Kraus or Werfel, who were all too familiar with anti-Semitism in Vienna.

I do not wish to imply that Einstein necessarily condemned assimilation to the culture of the majority as always ignoble or doomed to failure. It was plainly possible for children of Jewish parents to find themselves so remote from their community and its traditions that, even if they considered it, they were unable psychologically to re-establish genuine links with it. He was clear that in a civilised society every man must be free to pursue his own path in the manner that seemed to him best, provided that this did not do positive harm to others. He did not accuse these scientists and writers and artists of dishonourable or craven motives; their human dignity was not, for him, in question, only their degree of self-understanding.

It was his incapacity for self-deception or evasion, his readiness to face the truth, and—if the facts demanded it—to go against the current of received ideas, that marked Einstein's bold rejection of the central elements in the Newtonian system, and it was this independence that characterised his behaviour in other spheres. He rejected conventional wisdom: 'Common sense', he once said, 'is the deposit of prejudice laid down in the mind before the age of eighteen.' If something did not seem to him to fit, morally or politically, no less than mathematically, he would not ignore, escape, forget it; adjust, arrange, add a patch or two in the hope that it would last his time; he would not wait for the Messiah—the world revolution—the universal reign of reason and justice—to dissolve the difficulty. If the shoe does not fit, it is no use saying that time and wear will make it less uncomfortable, or that the shape of the foot should be altered, or that the pain is an illusion—that reality is harmonious, and that therefore conflict, injustice, barbarism belong to the order of appearances, which superior spirits should rise above. If his philosophical mentors, Hume and Mach, were right, there was only one world, the world of human experience; it alone was real: beyond it there might be mystery; indeed, he regarded the fact, of which he was totally convinced, that the universe was comprehensible as the greatest of mysteries; yet no theory was valid which ignored any part of direct human experience, in which he included imaginative insight, arrived at by paths often far from conscious.

It was this sense of reality that saved him, despite his deep convictions, from being doctrinaire. When what he knew, understood directly, was in conflict with doctrinal orthodoxy, he did not ignore the immediate evidence of his moral, social or political sense. He was a convinced pacifist; during the First World War he made himself unpopular in Germany by denouncing it. But in 1933 he accepted the necessity of resisting Hitler and the Nazis, if need be by force, which horrified his pacifist allies. He was an egalitarian, a democrat, with an inclination towards socialism. Yet his sense of the need to protect individuals from the state was so strong that he believed that Bills of Rights would be trampled on unless an élite of educated and experienced persons in authority at times effectively resisted the wishes of majorities. He praised the American Constitution, and in particular the balance of power between the President, Congress, and public opinion (his early political mentor, the Austrian socialist Fritz Adler, would scarcely have approved). He hated walls between human beings, exclusiveness. But when Jewish students were being hounded by nationalist students in German or Polish universities, he declared that Weizmann was right; liberal and socialist resolutions were useless; the Jews must act, and create their own university in Jerusalem.

He hated nationalism all his life. But he recognised the acute need of the Jews for some form of national existence; above all, he did not regard a sense of national identity and nationalism as being one and the same thing. It is clear that he took political allegiance seriously. He renounced his German nationality twice. He would not, as a young man, have chosen to adopt Swiss, or, after Hitler, American citizenship, had he not felt that he could give his full allegiance to these democratic countries when, for obvious reasons, he found it unbearable to retain his German passport. It was this combination of social sensitiveness and concrete insight into what it is that men live by that saved him from doctrinaire fanaticism; it was this that made him morally convincing.

He was an innocent man, and sometimes, I should think, taken in by fools and knaves. But innocence has its own modes of perception: it sometimes sees through its own eyes, not those of the spectacles provided by conventional wisdom or some uncriticised dogma. The very same independence which caused him to reject the accepted notions of physical space-time, and boldly offer the hypothesis of gravitational waves and light quanta against the resistance of physicists and philosophers, also liberated him morally and politically.

Consequently this man who sought privacy, who remained wholly uncorrupted by adulation and unparalleled fame in five continents, who believed in salvation by work and more work to unravel the secrets of nature—secrets miraculously amenable to analysis and solution by human reason—this gentle, shy and modest man displeased many establishments: German nationalists, Germanophobe Frenchmen, absolute pacifists, Jewish assimilationists, Orthodox rabbis, Soviet Marxists, as well as defenders of absolute moral values in which, in fact, he firmly believed.

He was neither a subjectivist nor a sceptic. He believed that the concepts and theories of science are free creations of the human imagination, not, as Bacon or Mill or Mach thought, themselves abstracted from the data of experience; but what the scientist seeks to analyse or describe by means of these theories and concepts is itself an objective structure of which men, viewed scientifically, are themselves a part. Moral and aesthetic values, rules, standards, principles, cannot be derived from the sciences, which deal with what is, not with what should be; but neither are they, for Einstein, generated or conditioned by differences of class or culture or race. No less than the laws of nature, from which they cannot be derived, they are universal, true for all men at all times, discovered by moral or aesthetic insight common to all men, and embodied in the basic principles (not the mythology) of the great world religions.

Like Spinoza, he thought that those who deny this are merely blinded by the passions; indeed, he felt Spinoza to be a kindred spirit. Like Spinoza, he conceived God as reason embodied in nature, as being, in a literal sense, a divine harmony, *Deus sive Natura*; and, again like Spinoza, he showed no bitterness towards his detractors, nor did he compromise with them—he remained serene and reasonable, humane, tolerant, undogmatic. He did not wish to dominate, and did not demand blind fidelity from his followers. He supported any movement—say, the League of Nations or left-wing groups in America—if he thought that on the whole it did good, or at least more good than harm.

So with Jewish Palestine. He hated the chauvinists; he was critical, at times to an unrealistic degree, of the attitude of the Zionist leadership towards the Arabs, but this did not make him lean over backward occasionally as it did others; he denounced the Eisenhower Administration for seeking to please the Arab states at the expense of Israel, a policy which he attributed to American imperialism. He was critical of some of the Hebrew University's policies: for instance, he thought that, among the academic refugees from Fascist Europe, young scholars, not the old and famous, should be offered appointments. But his loyalties remained unimpaired. He was not prepared to abandon the Zionist movement because of the deficiencies of some of its leaders. His Zionism was grounded in the belief that basic human needs create a right to their satisfaction: men have an inalienable right to freedom from hunger, nakedness, insecurity, injustice, and from homelessness too.

He was somewhat homeless himself. In a letter to his friend Max Born he wrote that he had no roots; that he was a stranger everywhere. He was, on his own admission, a lonely man who instinctively avoided intimacy. He was a solitary thinker, not easy to know as a human being. His deep humanity and sympathy with the victims of political oppression, social discrimination, economic exploitation, were central to his outlook and need no special explanation; they were in part, perhaps, a compensation for his difficulty in forming close personal relationships.

Like many physicists connected in some way with the production of the atom bomb, he was, in his later years, oppressed by a sense of the responsibility of scientists for introducing a terrible new means of destruction into the world; and he condemned the use of it made by his adopted country, which seemed to him bent on a dangerously imperialist course. His hatred of the cruelty and barbarity of reactionaries and Fascists at times led him to believe that there were no enemies on the left—an illusion of many decent and generous people, some of whom have paid for it with their lives.

Perhaps his very gifts as a scientist led him to schematise, to oversimplify practical problems, including complex political and cultural ones, which allow of no clear-cut solutions, to be too sweeping and to ignore the wrinkles and unevennesses of daily life, insusceptible as they are to exact quantitative analysis. For it seems to me that there may exist a certain difference between the gifts of scientists and humanists. It has often been pointed out that major discoveries and inventions—as opposed to demonstrations of their validity—require great imaginative power and an intuitive sense—not rationally analysable—of where the right solution must lie, and that this is not dissimilar from the vision of artists or the sympathetic insight into the past of gifted historians or scholars. This may well be true. Yet those who deal with human beings and their affairs need some awareness of the essential nature of all human experience and activity, a sense of the limits of what it is possible for men and women to be or to do; without some such awareness of the limits imposed by nature there is no criterion for dismissing an infinity of logically possible but wildly improbable or absurd historical or psychological hypotheses.

About what makes men rational Aristotle and Kant and Voltaire and Hume may well be right: on this sense of what can, and what clearly cannot, be the case in human affairs, on the normal association of ideas, on such basic concepts as those of past, future, things, persons, causal sequence, logical relations—a closely woven network of categories and concepts—human rationality, perhaps even sanity, in practice, depends. Departure from these, as attempted, for example, by surrealist painters or poets, or aleatory composers, may be interesting, but it is deliberately counter-rational.

But in mathematics or theoretical physics this sense of reality does not necessarily seem to be required. Indeed, something close to the opposite may, at times, be needed. In the case of seminal discoveries—say, of imaginary numbers, or non-Euclidean geometry, or the quantum theory—it is precisely dissociation of commonly associated ideas, that is, departure from some categories indispensable to normal human experience, that seems to be required, namely a gift for conceiving what cannot in principle be imagined, nor expressed in ordinary language which is concerned with day-to-day communication, with the facts and needs of human life. It is this detachment from, even flouting of, everyday reality that leads to the popular image of the abstract thinker—Thales who falls into a well, the absent-minded professor who boils his watch in place of an egg.

This kind of escape into abstractions—an ideal world of pure forms expressed in a specially invented symbolism free from the irregularities and untidiness, or even the basic assumptions, of ordinary experience—may possibly, at times, be connected with a psychic disturbance, some kind of displacement in early life. Einstein's breakdown as a schoolboy in Munich is paralleled by similar childhood experiences of Newton and Darwin, who also remained somewhat inaccessible emotionally. These thinkers, too, spoke of a type of experience which Einstein described as a deeply religious feeling before a vision of the divinity revealed in the all-embracing unity and rational harmony of the rigorously causal structure of nature. This was a vision of reality which nothing could shake: consequently, Einstein remained an unyielding determinist, and never accepted the uncertainty principle as an ultimate category of natural knowledge, or as an attribute of objective nature—only as part of our provisional and incomplete analysis of it.

Such addiction to pure abstraction and generalisation may, at times, be connected with an incapacity for close personal relationships with others, a full social life; this appears to me to be a plausible hypothesis. It may well have been so with Albert Einstein. What he withheld from private life he gave to the world. Not only the fame of his achievement, but his figure, his face, are known to millions of men and women. His appearance became a visible symbol, a stereotype, of what people supposed a scientist of genius should look like, much as an idealised Beethoven became a commercialised image of the inspired artist. How many people know what other scientists of genius—Planck, Bohr, Rutherford—looked like? Or, for that matter, Newton or Galileo, or even Darwin? Einstein's features, with their simple, kindly, bemused, melancholy expression, moved men's hearts everywhere. He was very famous, virtually a folk hero, and his appearance was as familiar and as widely loved as Charlie Chaplin's, long before he was portrayed on American stamps or Israeli banknotes.

Let me return briefly, in conclusion, to the state of Israel. The Zionist movement, like the state of Israel, has often been attacked, today more than ever, both by countries outside its borders and from within; sometimes with, more often without, reason or justice. That Einstein, who tolerated no deviation from human decency, above all on the part of his own people—that he believed in this movement and this state and stood by it through thick and thin, to the end of his life, however critical he was at times of particular men or policies—this fact is perhaps among the highest moral testimonials on which any state or any movement in this century can pride itself. Unswerving public support by an utterly good (and reasonably well-informed) man, against a virtually complete lack of sympathy for it on the part of the members of his social and intellectual milieu (whose general moral and political views he largely shared), may not by itself be enough to justify a doctrine or a policy, but neither can it be dismissed; it counts for something; in this case for a great deal.

Yitzhak Navon

ON EINSTEIN AND THE PRESIDENCY OF ISRAEL

THIS SYMPOSIUM is under the auspices of the President of Israel, but not because of his knowledge of Einstein's philosophy and theories. I almost would say, "On the contrary." I remember Prime Minister David Ben-Gurion asking Supreme Court Justice Felix Frankfurter for his opinion of a certain leading personality in the United States. "Well, he has some gaps in his ignorance," said the Justice. Perhaps I can join that club.

Actually, we in Israel are much concerned with gaps. My experience in the political field has shown me that there are three principal gaps. One is the social gap between the rich and the poor. It is obvious; it is understood but not always effectively dealt with. Then there is the gap between the natural sciences and the humanities. Those who deal with science can understand and read history, philosophy, and literature, but one rarely finds a literary man capable of understanding physics and mathematics. How do you bridge that gap? I can tell you from experience that it is a most urgent problem, and it would be a great service to humanity if leaders of countries were enabled to understand the magnitude of scientific research.

The third gap is that between those who know and those who act. There are ever so many people who are most knowledgeable but do not undertake the responsibilities of action. On the other hand, there are ever so many people with executive responsibilities whose knowledge is, let us say, hardly complete, as is proved by experience over many years in this country and, I presume, in other countries, as well. Surely this is a gap that should be bridged.

Addressing you here, in the House of the President, I would

like to speak of Einstein's relation to the Presidency of Israel. I was serving as political secretary to Prime Minister Ben-Gurion when our first President, Dr. Chaim Weizmann, died. That morning in the office, Ben-Gurion said, "There is only one man whom we should ask to become the President of the State of Israel. He is the greatest Jew on earth. Maybe the greatest human being on earth. Einstein. What do you think?"

I said, "President of the State of Israel? Why not head of scientific research? I'm afraid he may stumble on the way. . . ." But Ben-Gurion insisted that he felt duty bound to offer the post to Einstein. If Einstein refused, the decision would be his.

Here I have the urgent cable Ben-Gurion sent on 16 November 1952 to the Israeli Ambassador in Washington, Abba Eban. It reads in my translation: "Please inquire immediately of Einstein whether he is prepared to become President of Israel if elected (by the Parliament). Immediately after election he would have to come to Israel and become an Israeli citizen. He could continue with his scientific work without interference. Please cable his answer immediately. Ben-Gurion."

The answering cable was dated 18 November 1952. The Ambassador had tried unsuccessfully to contact Einstein by phone and had finally wired him that he had instructions from the Israeli Government to convey a very urgent message to him. It was a message of historic importance that could not be transmitted over the phone; the Ambassador therefore requested that Einstein receive the second in command at the Embassy, Minister Goitein, who would bring a letter from the Ambassador. After Einstein had thought the matter over, the Ambassador would like to visit him and continue the discussion.

That night, the cable continued, Einstein phoned Eban at home and, in trepidation and with much emotion, asserted that he believed he knew what the Ambassador wanted to tell him. He could not even think of discussing a position that he was not free to take and for which he was completely unsuited. He never accepted any position unless he felt he could conscientiously fulfill its obligations. He had no talent at all for human relations; he only had some understanding in the world of matter and nature, and he would like to dedicate himself to study of that world without any interruption.

"I answered," Eban wrote,

> that this was no way to discuss the matter, and I hoped that he would not reach his decision till after assessment of the problem in historical

perspective. I repeatedly pleaded with him to receive Goitein and then me. He said it was a waste of time. His mind was made up. He begged me not to impose discussion of the matter upon him. I told him that for me such a discussion seemed of the utmost importance. Indeed, I found it unthinkable to cable my Government that its proposition had been rejected out of hand.

Einstein then agreed to receive my letter from Goitein and to meet me two days later in New York. He changed his mind because he did not want to hurt the feelings of those who had proposed him for the post. I said that I would ask him in the meantime to give no publicity to the matter. He started to laugh, telling me that newspapermen kept phoning him all morning to ask for confirmation of the rumors from Israel that the Prime Minister had cabled him about his accepting the Presidency. He told all of them that he never got such a cable—a denial which, to his regret, was true only from a purely formal point of view. We agreed that he should refer all questions to me at the Embassy. . . . I have the impression that the offer has not caused him any elation or pride, but rather sorrow.

The story ends of course with Abba Eban's meeting Einstein and being told again that the Presidential office requires understanding of human relations—something he, Einstein, was deficient in. It was his wish and intention to deal only with matter and nature.

When Ben-Gurion sent that cable to invite Einstein to become President, he asked me to come to his office and have coffee with him. He said, "Tell me what to do if he says yes! I've had to offer the post to him because it's impossible not to. But if he accepts, we are in for trouble." Two days later the Ambassador met Einstein at a reception to which he had come with black tie and without stockings. . . .

I know that what Ben-Gurion did was really a way of expressing the deep admiration and gratitude he felt for Einstein the humanist, and of proclaiming the pride of the Jewish people in a man who was a great universal figure and at the same time did not deny, but on the contrary was proud, that he belonged to this persecuted people, a people who indeed saw in science and culture the climax of achievement.

It is perhaps not without relevance that on this very day we are debating in Parliament the text of the peace treaty with Egypt. It seems evident that by tomorrow the debate will be concluded with a clear majority for confirmation. I must stress that this is no easy matter, for the concessions and sacrifices made by Israel are over and beyond what should have been expected or demanded, and there are still forces in the Arab world and outside of it that will do their utmost to sabotage peace. We rejoice soberly: in our hearts we are happy, in our minds we know that there are difficult times ahead. Propaganda over the air waves and acts of sabotage will attempt to destroy the peace treaty, but all of us are united in our hope that the forces of peace will prevail. We are prepared to make concessions and look forward to the beginning of a new era for the benefit of all nations in the Middle East, with, I think, helpful implications for the world at large.

Uriel Tal

JEWISH AND UNIVERSAL SOCIAL ETHICS IN THE LIFE AND THOUGHT OF ALBERT EINSTEIN

FORMS OF SOCIO-ETHICAL THOUGHT

TO THE QUESTION "Just what is a Jew?" Einstein in November 1938 replied that two features have been characteristic of Jewish life:

> the democratic ideal of social justice, coupled with the ideal of mutual aid and tolerance among all men. Even the most ancient religious scriptures of the Jews are steeped in these social ideals, which have powerfully affected Christianity and Mohammedanism and have had a benign influence upon the social structure of a great part of mankind. . . .
>
> The second characteristic trait of Jewish tradition is the high regard in which it holds every form of intellectual aspiration and spiritual effort. I am convinced that this great respect for intellectual striving is . . . responsible for the contributions that the Jews have made toward the progress of knowledge, in the broadest sense of the term.[1]

These two elements, social justice and intellectual aspiration, which in Einstein's opinion characterize Judaism, exemplify the two planes on which Jewish and human morality exist, that of the individual and that of society. For Einstein, perhaps the highest of all moral values is the individual. Of all the phenomena in the empirical world, the individual is the only one with the status of a subject and the ability to reason, the only one by nature capable of consciousness, of self-awareness, of self-criticism, of self-restraint, and thus also of moral responsibility: "*The true value of a human being* is determined primarily by the measure and the sense in which he has attained liberation from the self."[2]

The second plane, society, attains the status of a lofty value because of the individuals composing it. Consequently, society is superior to the state. It is the function of society to enable the individual to develop his particularity, his intellectual and moral potential, while the role of the state is to serve society, allowing it to function in freedom and autonomy: "the state is made for man, not man for the state. . . . The state should be our servant; we should not be slaves of the state"; "the only justifiable purpose of political institutions is to assure the unhindered development of the individual and his capacities."[3]

These moral values—that is, the individual as sovereign, society as a framework for the implementation of that sovereignty, and the state as their servant—are, according to Einstein, among the chief components of the legacy of Judaism. Together with the heritage of classicism, they represent the essence of Western civilization, or, as he termed it, of the "European spirit." Thus dictatorships, like Nazism in his day, are the extreme opposite of what Einstein termed the Judeo-Christian tradition and likewise of the "European spirit," for the "essence of despotism lies not only in the fact that one man with virtually unlimited authority holds power, but that society itself becomes an instrument for enslaving the individual. This is why I consider the servitude to the state as the main enemy of the European spirit."[4] And in that vein, in the famous speech at the Princeton Theological Seminary in May 1939, Einstein concluded:

> The most fundamental principles of our aspirations and valuations are given to us in the Judeo-Christian religious tradition. It is a lofty goal. . . . When one divests this goal of its religious form and regards only this purely human side, it may be expressed as follows:
>
> Free and self-responsible development of the individual so that he will freely and joyfully put his energies at the service of the community of man. . . . the same words may be considered as the expression of the fundamental democratic principle. The true democrat deifies his nation just as little as the religious person in our sense does.[5]

These principles, as formulated in the late 1930s, were for Einstein the essence of Jewish and universal social ethics. Here we shall analyze them from two aspects: the form of Einstein's manner of thinking and arguing in this area of Jewish and universal morality, and the content of his thought and argument in this area.

∞

One of the characteristic features of the form of Einstein's manner of thinking and argument on our present subject is its inner consistency throughout an extended period, in fact from the beginning of the First World War until his last days in the spring of 1955. Those years can be divided into five distinct segments. First, Einstein's support of Zionism begins in 1919 with his first public appearances, especially his first visit to the United States in 1921 with Chaim Weizmann for the Zionist movement and the Hebrew University. Next, his activity continued between the two world wars in international institutions for peace, among them the League of Nations Committee on Intellectual Co-operation in 1924, 1925, and 1929, and the International Labor Organization's Advisory Committee of Intellectual Workers in 1927. Then, in the 1930s, and especially in the immediate prewar years, Einstein warned of the imminent danger to Jews, to democracy, and to civilization presented by Nazism, Fascism, and dictatorial regimes in general. Following World War II, Einstein emphasized the moral and political responsibility that scientists on the one hand and political leaders on the other were charged with, in view of the atomic energy they controlled. As Einstein wrote in 1948, in his contribution to a college debaters' handbook: "In the shadow of the atomic bomb it has become even more apparent that all men are, indeed, brothers. If we recognize and act upon this simple truth, mankind may proceed to a higher level of human development. But should the angry passions of a nationalistic world engulf us any further, we are doomed."[6] In the last years of Einstein's life, after the Holocaust and as the State of Israel was established and struggled for its existence, the main thrust of his political reaction continued to be moral. As he said in a letter to one of the heads of the Jewish Agency a short time before his death: "The most important aspect of our policy must be our everpresent, manifest, desire to institute complete equality for the Arab citizen living in our midst . . . the attitude we adopt toward the Arab minority will provide the real test of our moral standards as a people."[7]

The consistent nature of Einstein's Judeo-ethical conception over forty years was apparent as well in cases where he felt obliged to take an exceptional stand. Thus, for example, a few days after the murder on 24 June 1922 of German Foreign Minister Walter Rathenau, Einstein wrote to a colleague that in this particular instance he was in favor of the proposal that the University of Berlin should hold a memorial service for Rathenau, for, although "In general, the intervention by a cultural institution into political affairs is to be condemned. . . . here, it is a question of *affirming a broad moral position*. . . . The university must . . . affirm that any society which does not insist upon respect for all life must necessarily decay."[8] Subsequently, from 1933 on, as the Nazis rose to power, as despotism spread throughout Europe, and as the freedom of the world was endangered, Einstein came to the conclusion that the pacifist principles he still believed in dared not to stand in the way of armed struggle against Nazism, for, as he wrote in a letter to a student who was a conscientious objector: "Organized power can be opposed only by organized power. Much as I regret this, there is no other way."[9]

Understanding the structural form of Einstein's argument is essential for understanding his actual stand. On the one hand, Einstein refrained from developing a formal or systematic creed of social or Jewish philosophy. Most of his statements on such problems were brief, concise, sometimes even fragmentary, often resembling formulas and employing such expressions as "God" in a symbolic manner. On the other hand, he remained consistent in his cognitive approach, which demanded that a position regarding social morality, in Jewry or humanity as a whole, must be based on the inseparable relationship between two components of knowledge: the rational and the empirical, that is, theory and experience, imagination and sophistication, or, as he once said when explaining his Zionist motivations, "not merely by my spontaneous feelings but on rational grounds. . . ."[10] On the one hand, general laws in matters of Jewish and universal social ethics cannot be inferred from personal or social experience but must originate in tradition, faith, or perhaps even more in the inventive and aesthetic faculties of the human mind. Belief in the possibility that the rules governing society, history, art, man's ethical responsibility, and man's religiosity are amenable to reason is a theoretical assumption that precedes experience and empirical observation. On the other hand, logical thinking, theoretical formulation, and ethical consciousness cannot in themselves provide us with an adequate knowledge and evaluation of the empirical reality of history, of actual social traditions, or of the ways of life. In short, a principle of Jewish or universal ethics cannot simply be inferred from experience, nor can social reality and its critical and constructive assessment be experienced without factual observation or perhaps even actual participation.

Finally, although logical and conceptual generalizations about Judaism and universalism are a cognitive necessity for rational attitudes on ethics, these generalizations should not be pushed too far. Reality includes a multiplicity of individualities, be they human beings or social values, and no generalization can fully represent this plurality. In social reality, for example, a specific, individual sequence of cause and effect, though derived from a theorem, can never be fully identical with it. One should remember, says Einstein, that ever since David Hume, causal interrelationships can be verified and substantiated theoretically, in thought and in logic, but not necessarily empirically. Sensuous experiences (*sinnliche Erlebnisse*) in themselves, Einstein wrote to Ernest B. Zeisler, cannot establish an adequate cognitive framework for social ethics, human rights, or Jewish values. Hence, Judaism must be understood as an empirical individual reality rooted in universal theoretical ethics. The individuality of each phenomenon, person, group, nation, or tradition must be safeguarded, but within the universality of mankind and as an inseparable part of it.[11]

ETHICS OF RESTRAINT

These forms of thought in the area of social ethics provided a framework within which Einstein reacted to some of the problems of his time, and we shall consider a number of these according to their substance and chronology.

A

In 1920-1921 and later in 1929, Einstein testified that the anti-Semitism he encountered in Germany upon his arrival there in 1914 had had a considerable influence on his development of his Jewish identity. What worried him from the outset was how anti-Semitism had damaged the Jew's self-image, "how . . . innumerable . . . forces of the Gentile majority undermined the confidence even of the best of my fellow-Jews. . . ."[12] The devastating effect of anti-Semitism was greatest, Einstein felt, during the Emancipation period, for at that time most Western Jews lacked consciousness of their identity as Jews and, to an extent, of their dignity as human beings. As a result, their moral resistance weakened, in contrast to earlier times, when the Jewish community was less amorphous. But with the emergence of Zionism, oppression and antagonism could be transformed; Jewish existence would be reawakened and flourish.[13]

With regard to the nature of anti-Semitism, why and how it persisted, Einstein proffered a number of answers. Hatred of the Jews developed because the Jews were a weak and defenseless minority and because individual Jews nevertheless achieved prominence in gentile society. "The Jews as a group may be powerless, but the sum of the achievements of their individual members is everywhere considerable and telling. . . . The forces dormant in the individual are mobilized, and the individual himself is stimulated to self-sacrificing effort, by the spirit that is alive in the group."[14] This intellectual vitality and power concentrated in a weak, scattered minority aroused fear and suspicion, and thus resentment, opposition, and even hatred, evident especially in the Nazi movement: "More than anything else in the world, they fear the influence of men of intellectual independence. I see in this the essential cause for the savage hatred of Jews raging in present-day Germany. . . . they see the Jews as a nonassimilable element that cannot be driven into uncritical acceptance of dogma. . . ."[15]

The motives for hating Jews also formed the mechanisms of anti-Semitism, primarily justification and projection. In the political sphere, for example, anti-Semitism was used to justify the interests of the privileged classes: in the case of Nazism, this anti-Semitism "enabled a small, unscrupulous, and insolent group to place the German people in a state of complete bondage."[16] The second mechanism operating in this case was projection. In the long history of Jew-hating,

> The crimes with which the Jews have been charged . . . were to justify the atrocities perpetrated against them. . . . They were supposed to have poisoned wells . . . murdered children for ritual purposes. They were falsely charged with a systematic attempt at the economic domination and exploitation of all mankind. Pseudo-scientific books were written to brand them an inferior, dangerous race. They were reputed to foment wars and revolutions for their own selfish purposes. They were presented at once as dangerous innovators and as enemies of true progress. They were charged with falsifying the culture of nations by penetrating the national life under the guise of becoming assimilated. In the same breath

they were accused of being so stubbornly inflexible that it was impossible for them to fit into any society.[17]

In addition, exploitation of the Jews, a minority group, served to reinforce the self-awareness of the majority group.[18]

Einstein's conclusion, on the basis of the overall form of his thinking on these matters, is so formulated that the particular phenomenon—in this instance, in Jewish history—reflects a universal phenomenon in human history as a whole. Thus, political reality is the product of a struggle between two trends, one optimistic and the other pessimistic. In the optimistic view, in the original sense of the German Enlightenment and British liberalism, society will attain the fullest development of its qualities if it is allowed to develop in complete freedom. The "free unfolding of the productive forces of individuals and groups" is the guarantee for the realization of progress, whereas the role of political forces is to safeguard the external conditions for this development in freedom. The pessimistic view, on the other hand, claims that freedom leads to the ruin of society and that only "blind obedience, and coercion" can ensure society's continued existence. Einstein concluded: "The adherents of this second trend are the enemies of the free groups and of education for independent thought. They are, moreover, the carriers of political anti-Semitism."[19]

B

Einstein's interpretation of the nature of anti-Semitism led to his concept of Zionism. Given the structure of Einstein's socioethical thinking, it followed that as Zionism was among other things a solution to anti-Semitism, it must necessarily maintain what anti-Semitism negates and build what it destroys—the universal values of "mutual toleration and respect," freedom for man and society, the equal status of men, and democracy. Furthermore, if the political reality is characterized, as we have seen, by a struggle between optimism and pessimism, and political anti-Semitism is an expression of pessimism regarding the nature of man, Zionism must necessarily develop according to the optimistic conception, that is, on "the free interplay of individuals and groups" without discrimination with regard to social class, ethnic group, religion, or nationality. This way of thinking led Einstein to conclude: "Palestine is not primarily a place of refuge for the Jews of Eastern Europe but the embodiment of the reawakening corporate spirit of the whole Jewish nation."[20] And that reawakening, as noted before, is structured in terms of a meeting between tradition and modern reality: "The community in Palestine must approach the social ideal of our forefathers as it is laid down in the Bible, and at the same time become a seat of modern intellectual life. . . ."[21]

The meeting between tradition and modernism, if expressed on the socioethical plane, requires political Zionism to establish a model society, a society whose sons will resemble their forefathers, "men who embodied the conscience of the western world, defenders of human dignity and justice."[22] For that reason, Zionism must keep in mind that its function is not "to create a political society"; rather "our aim is, in accordance with the old tradition of Jewry, a cultural one in the widest sense of the word."[23] These ethical criteria, Einstein added, are not applied at the expense of practical political criteria. On the contrary, the ethical approach, consideration for one's fellow man, elimination of the irrational psychological barrier between Jews and Arabs—these are the essential conditions if Zionism is to achieve its practical political aims. Consequently, "We—that is to say, the Arabs and ourselves—have got to agree on the main outlines of an advantageous partnership which shall satisfy the needs of both nations. A just solution of this problem . . . is an end no less important . . . than the promotion of the work of construction itself."[24]

The events in Hebron in August 1929 affected Einstein deeply and intensified his awareness of the urgent political necessity of finding a common language for Jews and Arabs in Eretz Israel and a solution for the tension between the two peoples through understanding and consent. The British mandatory policy earned Einstein's sharp criticism, but he felt the main responsibility for the future relations of Jews and Arabs belonged to the Jews themselves.[25] In a letter to Samuel Hugo Bergman, dated 27 September 1929, Einstein wrote that the riots in Eretz Israel proved once again how essential it was to create "a kind of true symbiosis" between Jews and Arabs. By symbiosis, Einstein meant the regular operation of joint organizations and institutions in the realm of administration, society, and economics. The separate existence of the two peoples, he added, would inevitably lead to dangerous tension.[26]

Einstein later criticized public statements by Zionist leaders such as Selig Brodetsky. In a letter of 25 November 1929 he complained to Chaim Weizmann that "the economic and psychological problems of the Judeo-Arabic symbiosis were completely bypassed, but handled as an episode of conflict . . . should we be unable to find a way to honest cooperation and honest pacts with the Arabs, then we have learned absolutely nothing during our 2,000 years of suffering, and deserve all that will come to us. . . ."[27] In a similar letter of the same date to Bergman, Einstein added that if the Zionist policy toward the Arab population did not change, a catastrophic development" was to be expected.[28] Bergman then wrote to Kurt Blumenfeld and expressed his fear that Einstein might become estranged from Zionism if he came to feel that Zionism could not be realized "without oppression of the Arabic people and in peaceful agreement with it. . . ."[29]

In a number of his letters to Bergman, Einstein consistently reiterated his view that there was no practical possibility or moral justification for carrying out Zionism in the absence of cooperation with the Arabs and that such cooperation must take place in many areas, including the community, economics, education. Moreover, the Arabs should be assisted in raising their standards of living, of health, and of education. For a better life and education are the sine qua non for the creation of a society based on reason. And such a society is essential for the growth of a free person—free from the point of view of his political status, free of political dependence or oppression, free of economic subjection, and free internally from the control of the irrational.[30]

The moral superiority of society to politics, which Einstein emphasized throughout his public life, not only required an attitude of understanding, cooperation, and equality with the Arabs but also led Einstein to approve the Zionist Labour and Kibbutz movement.[31] This favorable approach was based on his general conception of the nature of man and his relation to society. On the one hand, man is dependent on society, on the culture into which he is born and on the socioeconomic conditions obtaining in the society in which he lives. In all areas "his life is made possible through the labor and the accomplishments of the many millions past and present who are all hidden behind the small word 'society.' "[32]

Man's particular nature, however, is such that his dependence on society is neither passive nor automatic. Although heredity, biological condition, and cultural environment set the pattern of human lives, at the same time, thanks to man's ability to engage in "conscious thinking," "human beings are *not* condemned . . . to annihilate each other or to be at the mercy of a cruel, self-inflicted fate."[33] Far from saying that there is "free will" in empirical reality, Einstein believed that biological facts and geo-historical conditions have great impact on man. But through hard work, society is able to improve life, and that is what led Einstein to respect, and sometimes even to admire, the Labour-Zionist movement in Eretz Israel.

C

Einstein's socioethical conception of Jewish nationalism within the framework of Zionism reflected his general view of nationalism, and especially the need to restrain it. On the one hand, he himself in 1921 defined his position on Zionism as an expression of his admission that in political reality, nationalism is unavoidable: "I am a national Jew in the sense that I demand the preservation of the Jewish nationality as of every other. I look upon Jewish nationality as a fact, and I think that every Jew ought to come to definite conclusions on Jewish questions on the basis of this fact. . . . That was the main motive of my joining the Zionist movement."[34] On the other hand, Einstein constantly urged caution in regard to "narrow nationalism," "exaggerated nationalism," "excessive nationalism," and "the spell of nationalism," and he made it clear that "my Zionism does not exclude cosmopolitan views."[35]

The political and social events to which Einstein reacted in the course of four decades were varied and were even to him of diverse significance. The First World War was an expression of gratuitous hate between nations; Nazism and racism were a manifestation of the predominance of irrationalism; the use of atomic weapons from the end of World War II on was a major moral challenge for scientists and politicians; the McCarthy era was a test of the integrity of intellectuals in confrontation with political powers; and the State of Israel was a test of Jewish ethics and self-restraint. Despite all the variety in those events and processes and their relative historical significance, Einstein remained constant in his struggle in favor of curbing nationalism. As early as the spring of 1915 he agreed with his friend Romain Rolland that it was the duty of intellectuals "to restrain impassioned outbursts . . . opposing them with the voice of reason. . . ."[36] Also in that year he wrote in the same vein to Hendrik A. Lorentz, adding, "it would seem that men always need some . . . fiction in the name of which they can hate one another. Once it was religion. Now it is the State."[37] To the end of his days, Einstein rejected, in essence, the irrationalism in "exaggerated nationalism," for irrationalism, with its accompanying chauvinism, bitterness, and suspicion, was an obstacle to human progress, and thus also one of the causes of war.[38] Instead of "excessive nationalism," said Einstein, what was required was internationalism, that is, "a rational relationship between countries, a sane union and understanding between nations, mutual co-operation for mutual advancement without interference with the particular customs of any nation."[39]

D

The principle of retaining the individuality of each people within internationalism and as part of it was evident also in Einstein's view of the religious character of Judaism. Here, as in other matters, he dealt with the particular question of religion and Judaism as part of the universal problem of religion and reason, religion and science. Einstein's connection with Judaism and Zionism, as we have seen, was directed at Judaism not as an institutionalized religion but as a community and a historical tradition, which, together with classicism and Christianity, provided the foundation for the development of Western civilization. Einstein was of course aware that historically Judaism was based on religion, but here, as in all areas of his social thought, he formulated his stand from a critical rationalistic and universal point of departure.

Einstein viewed religion on two planes, the historico-institutional and the cognitive-symbolic. On the former, religion is defined as obsolete, unable to stand up to the rational criticism of modern man, and is in fact considered to be an obstacle to human progress. On the second, religion is defined as a manner of thinking and even feeling that complements science, enriching modern man's spiritual and aesthetic world. The two planes are explained as the product of a three-stage evolution, which Einstein termed the "stages of religious experience."

The first stage is that of the "religion of fear." Man is by nature beset with needs and fears—of hunger, of sickness, of death. Consequently, in order to enable man to overcome those fears "the human mind creates illusory beings more or less analogous to itself on whose wills and actions these fearful happenings depend. Thus one tries to secure the favor of these beings by carrying out actions and offering sacrifices which, according to the tradition handed down from generation to generation, propitiate them or make them well disposed toward a mortal."[40] The second stage is that of "moral religion," motivated by "the social impulses" of man. Here the "desire for guidance, love, and support prompts men to form the social or moral conception of God. This is the God of Providence, who protects, disposes, rewards, and punishes . . . the comforter in sorrow and unsatisfied longing: he who preserves the souls of the dead."[41] Although these two stages are represented in all institutional religions, in Judaism, and later in Christianity, the second stage is stronger. Thus, like the "religions of all civilized peoples, especially the peoples of the Orient," Judaism is to be viewed as a "moral religion."

In both stages, the conception of divinity is still anthropomorphic and as such suits the "religiosity of the naïve man. . . . a sublimation of a feeling similar to that of a child for its father, a being to whom one stands, so to speak, in a personal relation, however deeply it may be tinged with awe."[42] Religions of that kind, with all their social, legal, and denominational aspects, "I can only consider historically and psychologically; they have no other significance for me." However, Einstein continued, there is a third stage in the evolution of religions, latent or repressed in the first two stages, which has deep meaning for a person of scientific, rational critical thought: "I shall call it cosmic religious feeling."[43]

Although Einstein remained remote from and opposed to all sentimentalism, he considered aesthetic criteria relevant in this context, and he felt that the religiosity that could be meaningful to a rational man and to scientific thought was the "religious feeling [that] takes the form of a rapturous amazement at the harmony of natural law, which reveals an intelligence of such superiority that, compared with it, all the systematic thinking and acting of human beings is utterly insignificant reflection."[44] Moreover "The individual feels the futility of human desires and aims and the sublimity and marvelous order which reveal themselves both in nature and in the world of thought. Individual existence impresses him as a sort of prison and he wants to experience the universe as a single significant whole."[45] These statements are consistent in their content and fragmented in their literary form. Underlying them is the assumption that religion is a kind of "religiosity," a set of theoretical formulas and aesthetic impressions expressing rational man's awe at the pluralistic unity of the cosmos.

In the spirit of Spinoza's logic and pantheistic thought, Einstein defined religiosity as faith in the "rationality and intelligibility of the world," a faith based on the cognitive assumption that the world is rationally comprehensible. This religion was not conceived as a set of dogmas, rituals, or authoritative institutions, nor was the concept of God conceived as a "personal God who concerns himself with the fates and actions of human beings." Rather—and here too Einstein was close to Spinoza—God is a rational, logical concept, sometimes even metaphoric, as necessitated by the basic notion that the world is constructed according to the "orderly harmony of what exists." Thus, God is a rational, symbolic expression of the pluralistic unity of cosmic harmony. Consequently, there is no room for divinity in the traditional or institutional religious sense, or for a divinity that is a kind of "will or goal outside the human sphere." Einstein summarized his interpretations:

> My views are near those of Spinoza: admiration for the beauty of and belief in the logical simplicity of the order and harmony which we can grasp humbly and only imperfectly. I believe that we have to content ourselves with our imperfect knowledge and understanding and treat values and moral obligations as a purely human problem—the most important of all human problems. . . .[46]

In a similar vein, aesthetic feeling and a cosmic sense assume the form of rational harmony. Einstein wrote:

> The most beautiful experience we can have is the mysterious. It is the fundamental emotion which stands at the cradle of true art and true science. . . . A knowledge of the existence of something we cannot penetrate, our perceptions of the profoundest reason and the most radiant beauty, which only in their most primitive forms are accessible to our minds—it is this knowledge and this emotion that constitute true religiosity; in this sense, and in this alone, I am a deeply religious man. . . .[47]

In the course of many years, and quite consistently, Einstein made it clear that his spiritual proximity to Spinoza and his own conviction of the epistemological validity of causation and determinism do not necessarily lead to the relativization of man's ethical values. On the one hand, Einstein approved of Spinoza's view of both "the causal dependence of all phenomena" and the application of causality "not only to inanimate nature but also to human feelings and actions."[48] Moreover, Einstein also commended Spinoza's argument against man's notion of free will independent of causality. This notion "was an illusion resulting from our ignorance of the causes operative within us."[49] This, so it would seem at first sight, could not but lead to a deterministic interpretation not only of human knowledge but also of human ethics, thus practically undermining the notion of moral responsibility. Indeed, in a private letter referring to the horrors of World War II, Einstein says: "Objectively, there is, after all, no free will. . . . What need is there for a criterion of responsibility? I believe that the horrifying deterioration in the ethical conduct of people today stems primarily from the mechanization and dehumanization of our lives. . . ."[50]

On the other hand, however, the notion of ethical responsibility as one of the highest values in the human sphere was constantly and consistently an essential factor in the thought and life of Einstein. This notion was deeply rooted in the legacy of the Enlightenment: the individual, conceived in terms of a rational being, is endowed with the unique attribute of autonomy, that is, of the potential freedom and independence of reason. Seen from a logical point of view, man's rationality is a concept and a value that cannot be derived from empirical experience; hence, as Einstein argued against Bertrand Russell, it has to be grounded in reason itself.[51] This conclusion, which, as Einstein perhaps somewhat grudgingly admitted, is not altogether alien to Immanuel Kant, has far-reaching implications, for it makes autonomous ethics possible despite determinism. For if indeed the notion of man's inherent rational nature is not inductively gained from experience or sensation but rather from the a priori "instrumentality of thinking,"[52] there must be a realm of cognitive concepts and propositions separated from causation and determinism. This, then, is the realm of "the free creations of thought."[53] The freedom of thought, though, does not contradict scientific thought: rather, it complements it, as is indeed shown by the notions of religiosity and of ethics.

Religiosity and ethics are two different frames of reference and are independent of each other; neither can be determined by sense data. Religiosity, as we have seen, nourishes, enriches, or even inspires scientific thought, and Einstein was "of the opinion that all the finer speculations in the realm of science spring from a deep religious feeling and that without such feeling they would not be fruitful."[54] Thus, it is by the power of the intellect, of free creative thought, that man admires the magnificence of the structure bestowed upon nature and confronts it with a feeling of awe and humility.[55] Ethics, being "an exclusively human concern with no superhuman authority behind it,"[56] also enriches scientific activity, but in a different way: it sets the goals for science. The scientific method itself is neutral insofar as the purposes it may serve are concerned; it "would not have led anywhere, it would not even have been born . . . ,"[57] without the determining quality of man's ethical—or, unfortunately, unethical—decisions. Science itself cannot create ends, nor can it teach the merits of those ends. It does, however, supply the means by which those ends can be achieved once they have been determined by individuals or society. Moreover, the rules concerning ethical reasoning, its acceptability and logical validity, cannot be founded on a scientific empirical basis. Just as "the valuation of life and all its nobler expressions can only come out of the soul's yearning toward its own destiny," so should the "moral foundation for the personal conduct of life" be determined by man's ethical autonomy in the light of free thought.[58] What "Spinoza so often emphasized under the name of *amor intellectualis*"[59] exemplifies Einstein's own definition. Accordingly, one should preserve the "moral foundations of science," but one cannot limit man's inherent ethical autonomy by turning it around and speaking of the "scientific foundations of morality," for these foundations indeed are to be determined by man's reasoning faculty alone.[60]

With this statement, Einstein arrived at a crucial conclusion in the area of social ethics and religiosity, whereby "the cosmic religious feeling" and man's "deep conviction of the rationality of the universe"[61] do not contradict scientific thought, but, as we have seen, are likely to complement it. "For the scientific method can teach us nothing else beyond how facts are related to, and conditioned by, each other." However "knowledge of what *is* does not open the door directly to what *should be*." Religiousness and, independently from it, ethics are valid frameworks of evaluation, determining the aims to be achieved by the scientific faculties of mankind.[62]

SUMMARY

This analysis has focused on the form and content of Einstein's Jewish and universal social ethics as they are exemplified in his thought and life. Analysis of the form has shown that, according to Einstein's cognitive approach to social ethics, theoretical concepts are true, hence valid, only if they are applicable to actual social, empirical experience. Analysis of the content has shown that Einstein conceived of both Judaism and Zionism as particularistic frames of reference in which universalistic values are reflected and embodied. And, finally, we have seen that form and content together constitute a harmonious system, formulated in ethical terms, ordered according to rational rules, and expressed in the actual, empirical human condition.

This potential harmony of the human condition—peace—was the lifelong focus of Einstein's Jewish and universal social ethics. Peace is thus formulated as the harmony of the coordinative relationships between form and content, reason and faith, knowledge and understanding, theory and experience, truth and beauty, freedom and responsibility, autonomy and justice, Judaism and Mankind.

ACKNOWLEDGMENTS

I wish to express my gratitude to The Central Zionist Archives, Jerusalem, to The Jewish National and University Library, Jerusalem, and to The Weizmann Archives, Yad Chaim Weizmann, Rehovoth, for permission to use and to quote from their materials, and to their staffs for the help extended to me. I also wish to acknowledge the competent assistance of Mr. Giora Hon, Chelsea College, London.

The activities of Albert Einstein on behalf of institutions of higher learning, such as Hebrew University, Jerusalem, have not been dealt with in this article. Much source material on this matter is located in the Weizmann Archives, the Jewish National and University Library, and the Central Archives for the History of the Jewish People, Jerusalem. As to Einstein's connection with Brandeis University, see Israel Goldstein, *Brandeis University—Chapter of Its Founding*, New York, Bloch Publishing Co., 1951.

NOTES

1. Albert Einstein, *Ideas and Opinions* (based on *Mein Weltbild*, ed. Carl Seelig, Amsterdam, Querido Verlag, 1934), new trans. and rev. Sonja Bargmann, 5th printing, New York, Crown Publishers, 1960, p. 195. A similar definition of the meaning of Judaism was formulated by Einstein after a meeting with Dr. M. L. Perlzweig, chairman of the World Union of Jewish Students, of which Einstein was honorary president. A photocopy of the manuscript dated 3 October 1933, is preserved in the Central Zionist Archives, Jerusalem (hereafter cited as ZA), File K.11/224/3. The original is in the possession of Dr. Perlzweig. An English translation has been printed in Ronald W. Clark, *Einstein: The Life and Times*, New York and Cleveland, World Publishing Co., 1971, pp. 504-505; now also in *Albert Einstein, The Human Side: New Glimpses from his Archives*,

ed. Helen Dukas and Banesh Hoffmann, Princeton N.J., Princeton University Press, 1979, pp. 60-61.

2. *Ideas and Opinions*, p. 12.

3. Otto Nathan and Heinz Norden, eds., *Einstein on Peace*, New York, Schocken Books, 1968, pp. 150, 320. Cf. Virgil G. Hinshaw Jr., "Einstein's Social Philosophy," in Paul Arthur Schilpp, ed., *Albert Einstein: Philosopher-Scientist*, 2 vols., New York, Harper and Row, Harper Torchbooks, The Science Library, 1959, vol. 2, p. 653.

4. *Einstein on Peace*, p. 241.

5. Philipp Frank, *Einstein: His Life and Times*, trans. from a German manuscript by George Rosen, ed. and rev. Shuichi Kusaka, New York, Albert A. Knopf, 1947, p. 288; see also Einstein, *Out of My Later Years*, New York, Philosophical Library, 1950, p. 23; cf. Einstein's definition: "In the last analysis, everyone is a human being, irrespective of whether he is an American or a German, a Jew or a Gentile." *Einstein, The Human Side*, p. 61.

6. *Einstein on Peace*, p. 459. Cf. the plea for international understanding by Einstein, *New York Times*, 27 November 1947, p. 1, col. 7 (address to the Foreign Press Association of the U.N., 11 November 1947).

7. Einstein to Zvi Lurie, 4 January 1955, in reference to their conversation held on 29 December 1953; photocopy preserved in ZA, 12/50/K.

8. *Einstein on Peace*, p. 53; cf. Clark, *Einstein*, pp. 292-294. The paperback edition of Clark (New York, Avon Books, 1972) will henceforth be cited as Clark (Avon).

9. *Einstein on Peace*, p. 319.

10. *Ideas and Opinions*, p. 181.

11. Ibid., pp. 19-24, 46, 188, 197, 270, 276; Frank, *Einstein*, pp. 282-283. See also Philipp G. Frank, "Einstein, Mach, and Logical Positivism," in Schilpp, ed., *Albert Einstein*, vol. 1, pp. 269ff. Also, the letters by Einstein to Ernest B. Zeisler, 5 August 1936, 24 May 1951, 24 July 1951, and 10 December 1952, in the Schwadron Collection of the Jewish National and University Library, Jerusalem (hereafter cited as SC), "Einstein" Box, M/32/7.

12. *Ideas and Opinions*, p. 171; Einstein, *About Zionism: Speeches and Letters*, trans. and ed. Leon Simon, London, Soncino Press, 1930, pp. 27-28. See also *About Zionism*, p. 23: "Before we can effectively combat anti-Semitism, we must first of all educate ourselves . . . out of the slave-mentality. . . . We must have more dignity, more independence. . . ." On the impact of the Prague environment, see Clark, *Einstein*, pp. 137-141; also Clark (Avon), p. 179. In an illuminating letter to Menachem Ussischkin, dated 19 March 1929, Einstein once again stated that it was the "Berlin environment" that "enlightened" (*anfzuklären*) him to his "appurtenance to the Jewish people. . . ." The persons who supported him in this "turnabout" (*Wandlung*), Einstein added, were Chaim Weizmann and Kurt Blumenfeld; cf. SC, "Einstein" Box, M/32/5. Additional evi-

dence, especially on the period of Einstein's stay in Prague starting in the fall of 1910, is included in Samuel Hugo Bergman's *Personal Reminscences,* in the Bergman Archives of the Jewish National and University Library (hereafter cited as AB), 4°/1502/155/**a**.

13. *Ideas and Opinions*, p. 196; cf. pp. 178, 187-188. See also *About Zionism*, pp. 23ff. "Why Do They Hate the Jews?" appeared originally in *Collier's Weekly Magazine* **102** (26 November 1938), pp. 9-10, translated by Ruth Norden.

14. *Ideas and Opinions*, p. 197; see also p. 192.

15. Ibid.

16. Ibid., pp. 192.

17. Ibid., pp. 192-193.

18. Ibid., p. 193.

19. Ibid., pp. 197-198.

20. Ibid., p. 181. One of the major aims of Zionism, in Einstein's eyes, was the regeneration, or convalescence, of the Jew (*jüdische Gesundung*), and it seems to have been an essential and recurrent theme in his Zionism. He gave a typical summary of this approach in 1929: "The greatest enemies of Jewish national consciousness and Jewish dignity are fatty degeneration—by which I mean the loss of moral fibre which results from wealth and comfort—and a kind of spiritual dependence on the surrounding non-Jewish world. . . ." *About Zionism*, pp. 37-38; also p. 23. See also *Ideas and Opinions*, pp. 178, 187-188.

21. *Ideas and Opinions*, p. 180. Also see the message by Einstein to the Jewish Teachers Seminary: "the intellectual decline brought on by shallow materialism is a far greater menace to the survival of Jewry than the numerous external foes who threaten its existence with violence. . . ." *New York Times*, 8 June 1936, p. 22, col. 2.

22. *Ideas and Opinions*, p. 177.

23. Ibid., p. 178. Cf. "Jew and Arab," in *About Zionism*, pp. 51-64. In his 12 July 1938 letter to Haim Greenberg, then editor of the New York *Jewish Frontier*, Einstein reports on his meeting with Dr. Shatara, president of the Arab National League in New York, during which a possible agreement between Jews and Arabs to avoid the partition of Palestine was discussed; cf. ZA, A/259/16.

24. *Ideas and Opinions*, pp. 176-177; see also pp. 172-174.

25. "Palestine Troubles," *Manchester Guardian Weekly* 21 (1929), p. 314; reprinted in *About Zionism*, pp. 53-62. In his letter to Dr. Joseph Marcus, dated 8 March 1948, Einstein called the British policy in Palestine "perfidious" and the cause of the dangerous situation then prevailing in Palestine, even more than "our own failures and sins of omission"; in ZA, 12/50/K. On the moral responsibility of the Jewish people for the political situation in Palestine, see Einstein's letter of 1 June 1944 to Kurt Blumenfeld, president of the German Zionist Federation: "it would be nice if we could lay the entire blame for that on the British

. . ."; in ZA, A/222/111. Later, in his detailed letter of 29 January 1946 to Martin Buber, Einstein once again emphasized his criticism of British colonial policy in the Near East, which had been putting obstacles in the way of Jewish-Arab cooperation in Palestine since the 1920s. See *Martin Buber. Briefwechsel aus sieben Jahrzehnten*, edited and introduced by Grete Schaeder in consultation with Ernst Simon, and with cooperation of Rafael Buber, Margot Cohn and Gabriel Stern, Heidelberg, Verlag Lambert Schneider, 1975, vol. 3 (1938-1965), no. 79, pp. 98-100.

26. ZA, A/187/18/a.

27. In Clark, *Einstein*, pp. 402-403. Several months earlier, Einstein wrote that despite the August 1929 outbreak of violence in Hebron, which was a "tragic catastrophe," one must remember that "no irreconcilable differences stand in the way of peace between Jews and Arabs in Palestine. Let us therefore above all be on our guard against blind chauvinism of any kind. . . ." *About Zionism*, pp. 51-52. See also Einstein's public letter "To the Arabic World," published in *Falastin*, 28 January 1930 and reprinted in *Jüdische Rundschau* (Berlin), no. 9, 31 January 1930. The Hebrew press reacted skeptically, and even negatively, to Einstein's suggestion of joint endeavors. Although Einstein never received the harsh criticism leveled at members of the "Brith-Shalom" peace movement, his ideas were by and large rejected. See, for example, the official organ of the Zionist movement, *Haolam* (London), 1930, no. 15, p. 287, no. 18, p. 295.

28. ZA, A/187/18/a. Samuel Hugo Bergman, in a letter of 17 December 1929, agreed with Einstein that "a catastrophic development" was to be expected because Zionist leadership and public opinion were not aware that indeed "the Arab question is the key to our entire situation." Only "a generous cooperation" and "a forward looking program for Semitic cooperation" could help to settle the Jewish-Arab conflict. In his reply to Bergman of 19 June 1930 Einstein expressed his agreement with Bergman's views and with most of his suggestions and added that if the Jews did not realize the necessity of direct cooperation with the Arabic world, "the entire Jewish position . . . will become completely untenable . . ."; in AB, 4°/1502/(1928-1950). Cf. the Hebrew periodical, *Sheifotenu*, published by Brith-Shalom, Jerusalem, 1930, no. 5, pp. 15-18. In a letter to Martin Buber of 2 September 1929, Hans Kohn expressed his appreciation of Einstein's views on the Jewish-Arab issue and added that responsibility for the ethical-political situation in Palestine rested on the shoulders of the Jewish community; one should not shift this burden onto "fate and other people. . . ." Buber, *Briefwechsel*, vol. 2 (1918-1938), no. 307. p. 347.

29. Cf. Bergman's letter to Blumenfeld, 1 January 1930, ZA, A/187/18a. Also see the polemics of Einstein and Erich Kahler with Prof. Philip K. Hitti of Princeton University, in the *Princeton Herald*, 14 April 1944, pp. 1, 6, and 29 April 1944, pp. 1ff., reprinted in Erich Kahler, *The Jews*

Among the Nations, with an appendix, "The Jews and the Arabs in Palestine," New York, Frederick Ungar, 1967, pp. 130-149.

30. ZA, A/187/18a, and AB, 4°/1502.

31. Cf. Einstein's letter to Mrs. Lindheim, written in Pasadena, Calif., on 2 February 1933, SC, "Einstein" Box. Also see Samuel Hugo Bergman to Einstein, 8 October 1929, ZA, A/187/18a. In his letter to Zvi Lurie, quoted above (see note 7), Einstein stated that "the Kibbutz movement is an excellent example" of a just policy toward the Arab citizen in Israel; in ZA, 12/50/K, and *Einstein on Peace*, p. 638. Einstein's positive attitude to the communal way of life of the kibbutz was part of his much broader identification with the moral ideals of socialism, their roots in biblical prophecy as he saw it, and their contribution to the eradication of "the real source of the evil," namely, the "economic anarchy of capitalist society." "Why Socialism?" in *Out of My Later Years*, p. 128. Einstein's socialism is a topic of great significance and would require a special study.

32. *Out of My Later Years*, p. 126.

33. Ibid., p. 127.

34. *About Zionism*, p. 29.

35. Ibid., pp. 30, 52; *Ideas and Opinions*, pp. 188-189; *Einstein on Peace*, pp. 152, 240-241, 311ff., 406; Frank, *Einstein*, p. 149. Chaim Weizmann, reacting to the term "narrow nationalism," wrote to Einstein in detail on 28 April 1938 that despite the "inhuman provocation" by the Arabs and the international powers that supported them, the Jewish community in Palestine had succeeded in not sinking "to the level of our opponents"; in the Weizmann Archives, Yad Chaim Weizmann, Rehovoth.

36. Rolland to Einstein, 28 March 1915, reprinted in *Einstein on Peace*, p. 14.

37. Ibid., p. 11.

38. Ibid., p. 44; cf. p. 70. Also see Einstein's "Foreword" to Homer W. Smith, *Man and His Gods*, Boston, Little, Brown, 1952, pp. ix, x.

39. *Einstein on Peace*, p. 44. Also see the comment by Einstein on the need for "teaching, for instance, history without creating the obsession of the past . . . the spell of nationalism may be thus broken. . . ." *New York Times*, 11 August 1949, p. 8, col. 6.

40. *Ideas and Opinions*, p. 37. "Religion and Science" was written especially for the *New York Times Magazine*, and appeared there 9 November 1930, sec. 5, pp. 1-4. The German text was published in the *Berliner Tageblatt*, 11 November 1930, supplement, p. 1.

41. *Ideas and Opinions*, p. 37.

42. Ibid., p. 40.

43. Ibid., pp. 262, 38. See also "Is There a Jewish View of Life?" in *Opinion* 2 (September 1932), p. 7; reprinted in *Ideas and Opinions*, pp. 185-186.

44. *Ideas and Opinions*, p. 40.

45. Ibid., p. 38.

46. Banesh Hoffmann, with the collaboration of Helen Dukas, *Albert Einstein: Creator and Rebel*, New York, New American Library, 1973, p. 95. See also *New York Times*, 25 April 1929, p. 60, col. 4. Rabbi Herbert S. Goldstein pointed out, quite correctly, that Einstein used the term "God" in order to formulate the idea of cosmic unity as a rationally conceived equivalent for the term monotheism. Cf. Frank, *Einstein*, pp. 280-288; Clark, *Einstein*, pp. 413-414; *Ideas and Opinions*, pp. 38, 45, 52, 120, 174, 195, 262; Arnold Sommerfeld, "To Albert Einstein's Seventieth Birthday," trans. Paul Arthur Schilpp, in Schilpp, ed., *Albert Einstein: Philosopher-Scientist*, p. 102. Also see Banesh Hoffmann, "Albert Einstein," in *Leo Baeck Institute Yearbook XXI*, London, 1976, p. 284.

47. "What I Believe," *Forum and Century* 84 (1930), pp. 193-194, reprinted in *Living Philosophies*, New York, Simon and Schuster, 1931, pp. 3ff.; also reprinted under the title "Meeting Place of Science and Religion," in E. H. Cotton, ed., *Has Science Discovered God?*, New York, Crowell, 1931, pp. 91ff. See also *Ideas and Opinions*, p. 11. Similar ideas, especially on "truth and beauty," were expressed in the famous conversation of Einstein with Rabindranath Tagore; see "The Nature of Reality," *Modern Review* (Calcutta) 49 (1931), pp. 42-43. An authorized version of this conversation was published in "Tagore Talks with Einstein," *Asia* 31 (1931), pp. 138-142, and also, in abbreviated form, in *Asia* 37 (1937), pp. 151-152. Einstein himself, however, seems to have felt that his encounter with Tagore was rather disappointing; cf. *Einstein on Peace*, pp. 112, 655.

48. Einstein, "Introduction" to Rudolf Kayser, *Spinoza: Portrait of a Spiritual Hero*, New York, Philosophical Library, 1946, p. xi.

49. Ibid. Also Einstein's "Foreword" to Smith, *Man and His Gods*, pp. ix-x, and his argument that Spinoza "had fully recognized the senselessness of the question of an interaction of soul and body, as well as the problem which of both be the 'primary,'" in his "Foreword" to *Spinoza: Dictionary*, ed. Dagobert Runes, New York, Philosophical Library, 1951, p. vi. For a similar expression by Einstein of 5 February 1921, see *Einstein, The Human Side*, p. 40. On the question of determinism, see the letters by Einstein and Robert A. Millikan, published in the *New York Times*, 28 January 1931, p. 2, col. 2. Helen Dukas and Banesh Hoffmann pointed out the dissatisfaction of Einstein "with the quantum theory with its denial of determinism"; cf. *Einstein, The Human Side*, pp. 68, 69. At this point, questions emerge that are beyond the scope of our study. It would seem advisable to keep in mind the remark by Niels Bohr on "the importance of utmost caution" regarding a possible affinity of Einstein to Spinoza in matters other than those discussed here. Bohr, "Discussion with Einstein on Epistemological

Problems in Atomic Physics," in Schilpp, ed., *Albert Einstein: Philosopher-Scientist*, p. 237.

50. *Einstein, The Human Side*, pp. 81-82. Also see "My Credo" of 1932, reproduced in the impressive Catalogue for the "Exhibition Einstein 1879-1979," composed by Yehuda Elkana and Ady Ophir, Jewish National and University Library, Jerusalem, March 1979, pp. 48, 49, and Prof. Elkana's enlightening remarks on the rigidly classic epistemology of Einstein, according to which "human knowledge had to be deterministic" (p. 56).

51. Einstein, "Remarks on Bertrand Russell's Theory of Knowledge," trans. Paul Arthur Schilpp, in Schilpp, ed., *The Philosophy of Bertrand Russell*, 2d ed., Evanston, Ill., Library of Living Philosophers, 1946, p. 285.

52. Ibid.

53. Ibid., p. 287. See Russell's answer, ibid., p. 697.

54. Cf. "Science and God: A German Dialogue," *Forum* 83 (1930), p. 375.

55. *Einstein, The Human Side*, pp. 39, 43-44, 67-70. Also see "On the Moral Obligation of the Scientist—A Message from Albert Einstein," in *Impact*, Paris, UNESCO, 1950, vol. 1, p. 104. Here Einstein emphasized the notion of the ethical and social responsibility of the scientist, which is determined by man and his reasoning faculty; cf. p. 105. Similarly, on the "scientists and engineers" who "carry particular moral responsibility," see the communication by Einstein to the Fellows of the Society for Social Responsibility in Science, in *Science* 112 (22 December 1950), pp. 760, 761.

56. *Einstein, The Human Side*, p. 39.

57. From a broadcast recording for the Science Conference, London, 28 September 1941, reprinted in *Out of My Later Years*, p. 113.

58. "Science and God," p. 374.

59. Ibid., p. 375.

60. Ibid.

61. *Ideas and Opinions*, pp. 38-40. Also see the answer Einstein wrote to an enquiry by a Chicago rabbi on "The Religious Implications of the Theory of Relativity," 20 December 1939; in *Einstein, The Human Side*, pp. 69, 70.

62. *Ideas and Opinions*, pp. 41-42; cf. pp. 30, 46.

Fritz Stern

EINSTEIN'S GERMANY

THERE WAS NOTHING SIMPLE about Einstein, ever. His simplicity concealed an impenetrable complexity. Even the links to his native Germany were prematurely ambiguous. At a time when most Germans thought their country a hospitable home, a perfect training ground for their talents, Einstein was repelled; in 1894, as a fifteen-year-old, he left Germany and became a Swiss citizen. Twenty years later, a few weeks before the outbreak of the Great War, he returned to Germany and remained for eighteen years of troubled renown, years in which he appreciated what was congenial and opposed what was antipathetic in Germany. Long before Hitler, he felt unease. He could joke about his multiple, if uncertain, loyalties—the better perhaps to hide his feelings. In 1919, at the moment when fame first engulfed him, he explained in a letter to the London *Times*: "Here is yet another application of the principle of relativity for the delectation of the reader: today I am described in Germany as a 'German servant,' and in England as a 'Swiss Jew.' Should it ever be my fate to be represented as a *bête noire*, I should, on the contrary, become a 'Swiss Jew' for the Germans and a 'German savant' for the English."[1]

His fame, his capacity for homelessness, and the degradation of his country made Einstein a citizen of the world, seemingly detached from Germany. But I believe that his early encounters with Germany, his hostility to its official culture, shaped his public stance. My deliberately ambiguous title is meant to suggest that Einstein's Germany was both real and imaginary—that he had his own perception of reality. The German experience haunted Einstein to the very end, as it haunted so many of his generation later. It was the text of his political-moral education, the background against which he came to mold his unorthodox views and play his controversial public role.

In Einstein's time, Germany was the promise and later the nemesis of the world, the country that had a decisive bearing on world politics and where, for a moment that seemed a lifetime, the moral drama of our era was enacted. At certain critical moments, Einstein and even his closest colleagues described radically different responses. I believe this diversity will help to complicate our understanding of Germany, and this will be desirable, because Germany's past has often been treated with didactic simplicity. Einstein and Germany: they illuminate each other.

Before turning to my proper subject, however, I should like to express my unease—and not only unease at addressing so exacting an audience. The writing of this lecture posed special difficulties. Historians have a right to feel awe—a right perhaps rarely exercised these days—and I felt awe at my subject and at this occasion.

It is generally said of Einstein that he revolutionized modern physics and natural philosophy and that his genius had no equal save that of Newton's. But I shall not—and I could not—deal with what was central to him. I shall deal with the public figure, with the first scientist-hero to appear in the Western world. I shall concentrate on the thoughts that were important to this public figure who placed his scientific fame at the service of his moral indignation. The genius hovers in the background, and the occasional partisan in the foreground. In doing this, I was mindful of what Lionel Trilling has said:

> Physical science in our day lies beyond the intellectual grasp of most men. . . . This exclusion of most of us from the mode of thought which is habitually said to be the characteristic achievement of the modern age is bound to be experienced as a wound given to our intellectual self-esteem. About this humiliation we all agree to be silent; but can we doubt that it has its consequences, that it introduces into the life of mind a significant element of dubiety and alienation which must be taken into account in any estimate that is made of the present fortunes of mind?[2]

I felt this exclusion the more as I came to realize the intensity of the aesthetic joy that Einstein and his colleagues found in their discoveries, as their correspondence exemplifies. We are shut out from that knowledge and from that particular beauty. Lionel Trilling was abundantly right in calling this exclusion an unacknowledged wound.

Exclusion from substance was compounded by my more or less

accidental familiarity with some of the men around Einstein and, as a child, with the fringes of that milieu. From time to time I shall allude to some of these personal ties, which added puzzlement and poignancy to my efforts to understand even a part of Einstein's world. I read in Einstein's unpublished correspondence with the historian's habitual hope that the archives would yield some nuggets to shock or prod the mind; the letters were marvelously human, but Einstein remained elusive and enigmatic. The search has been fascinating and disheartening—and has fully borne out what a friend said at the very beginning: Einstein is the hardest person to say anything about. His own friends found him inscrutable, and not even their love of him offered a firm bridge of understanding.

In analyzing the scientific ideal that some historians cherished, Richard Hofstadter once said: "The historian is quickly driven to a kind of agnostic modesty about his own achievement. He may not disparage science, but he despairs of it."[3] Einstein would have agreed with this judgment, though drawn different conclusions from it. The one time I met him—in 1944, while an undergraduate at Columbia—he inquired after my plans, and I told him I was in a quandary, not knowing whether to continue my original purpose, which was to study medicine and thus follow in the footsteps of a father, two grandfathers, and four great-grandfathers, or to switch to history, an old interest turned into a new passion by the power of my teachers. To Einstein this was no quandary. Medicine, he said, was a science (which I doubt) and history was not—though it is significant that I cannot remember his rather harsh words about history. I chose not to follow his advice, but I will confirm his view that history is not a science, that it is an approximation of a time and space that we knew not.

∞

At the risk of risible compression, let me recall some of the characteristics of modern Germany, particularly those that would have impinged on Einstein's life and thought. He was born in the decade of Germany's unification, and he died a decade after its dissolution. The 1870s were a heady and extravagant time for a country whose historic experience had been defeat and division. For centuries, Germany had been a geographic expression, everybody's battlefield, Europe's anvil on which other nations forged their destinies. In the unbroken annals of defeat, Prussia had been the sole exception; and Prussia had evolved its own ethos of frugal duty, rectitude, and obedience. It was also, as Mirabeau had pointed out at the end of the eighteenth century, not as other countries, a state with an army, but an army with a state. In the Napoleonic era, even Prussia collapsed, but political impotence had its compensations: in the shadow of defeat, the Germans created a great literary and philosophical culture and a national identity based initially on intellectual-aesthetic, not political, achievement. That culture enshrined as a moral imperative the cultivation of the self and education—at least for the elite—as the prescribed path to self-formation. There were always two strains to this conception: the ideal of the harmonious being, the rational, aesthetically literate humanist on the one hand and the demonic, inexplicable, mysteriously creative genius on the other. Einstein fitted both categories. By piously and pedantically trying to inculcate the rational, German schools often encouraged the yearning for the irrational. By the mid-nineteenth century, it was widely believed that Germany had a special vocation for learning.

Unification under Prussian aegis, achieved in battle, directed by Bismarck, codified in a constitution that preserved the privileges of a governing elite—that kind of unification was a celebration of force and a denial of earlier hopes of freedom. As Nietzsche warned, this triumph could destroy the German spirit, drown it by the worship of practicality and power. The new Reich, rapidly industrializing, exuded power. But the country became still harder to govern; new social cleavages appeared next to old regional and religious divisions. Nationalism and militarism were a means of providing cohesion, of overcoming a sense of *unfulfilled* unity. Bismarck's Germany was an authoritarian state of uncertain viability, but it was also a government of laws, a haven of constitutionality as compared with primitive, autocratic Russia, a country without torture, callous and sentimental, rigid, efficient, hardworking, bent on achievement.

Whatever the shortcomings of the new Reich, Bismarck's generation had seen the fulfillment of the great national ideal. The next generation—epitomized by the young emperor, with his dreams and delusions that could never banish his anxieties—thirsted for its own glory, for its own imprint on history. Germany had become a giant in the center of Europe: it had the best army, the strongest economy, the most efficient industry. But what was its vocation, its purpose? In a celebrated phrase Max Weber warned that unification would be "little else than a piece of folly which was committed by the nation in her old days, and which, in view of its costly nature," should not have been embarked upon at all unless Germany would now take the next step and become a world power.[4]

What was it that Germany sought after 1890? It sought what every aspiring nation in Europe had sought before: recognized greatness, a measure of hegemony. Perhaps the Germans were more frantic in their search for greatness, but then their day had come late. They wanted grandeur, as others had wanted it before; they needed to exorcise centuries of dependency. Europe's competition for greatness, which involved more than political dominion, its ethos of heroic striving, was the very hallmark of its exacting and triumphant civilization. Europe had always been a crucible of genius. Should the Germans—the originators of the Faustian myth—restrain their will and not try for collective preeminence?

Let me cite one more example of this exhortation to greatness. Few scholars in Germany were as critical of the nation's development and of Bismarck's character as was the great historian Theodor Mommsen; few railed as much against the servility and political nonage of their fellow citizens, few combatted anti-Semitism as vigorously as he did. But he too felt the grip of greatness, and he too preached sermons of duty, as did the professoriate throughout Europe. I believe the call to greatness gnawed at Germans more deeply than at others. Witness Mommsen's rectorial speech:

> Of course we are proud of being Germans and we do not disguise it. Of all the boasts none is more empty and less true than the boast about German modesty. We are not at all modest and we do not want to be modest or appear to be. On the contrary, we want to continue to reach for the highest in art and science, in state and church, in all aspects of life and striving, and we want to reach for the highest in everything and all at once. There is no laurel wreath which would be too magnificent or too ordinary for us . . . we think it normal that our diplomats as our soldiers, our physiologists as our sailors stand everywhere in the front rank. . . . But even if we content ourselves in no way to be content, we are not therefore blind . . . in research and instruction . . . there is no standing still . . . if you don't go forward, you stay back and fall behind.[5]

Even in Mommsen, then, we find this call for greatness as the only alternative to decline.

The contradictions of imperial Germany have often been noted. Economic giant, master of the disciplined society, model of technical proficiency, nurturer of talent—and yet a nation that recklessly defied prudence in dealing with foreign nations, a governing class that suffered from the paranoid fear that Germany was threatened by subversion at home and encirclement abroad. This alternation between presumption and anxiety grated heavily on some Germans and on many foreigners.

In Germany as elsewhere, the generation before 1914 was prodigious in talent and achievement; it was then that Germany attained a preeminent place in the natural sciences. German universities, which had long thought scientific studies a secondary concern, suddenly discovered that their scientists had won world renown. Universities had not been inhospitable to talent—provided it came in politically respectable, male, preferably Protestant guise. Any deviation from the norm had to be paid for by a superabundance of talent and, on the whole, was admissible only in the newer and politically neutral fields, such as medicine and the natural sciences. In those fields, achievement was more easily measurable and more immediately useful. In medicine and physics, in particular, the barriers against Jews began to be breached early.

It is notoriously hard to account for creativity. Was the German flowering a result of Germany's having the highest rate of literacy and the highest per capita expenditure for public education? Was it that German industry decided early on to support scientific research with particular largesse? Was the system of the master and the apprentices, the professor and his school, productive not only of dependency and exploitation but also of a special bond? Did success have anything to do with the fact that German scientists had a particularly austere view of their profession, so much so that one of the early Nobel laureates, Wilhelm Ostwald, spoke of all the grief and loneliness that a true scientist must endure, "because every important discovery must be paid for by a human life. . . ."[6] No doubt there was much suffering, acknowledged and unacknowledged, but there was also an extraordinary measure of cameraderie and high spirits.

In our century, scientific achievement, whatever its causes, can be measured by the incidence of Nobel Prizes won. From the inception of that prize to the rise of Hitler, Germans garnered a larger share of prizes than any other nationality, about 30 percent. In some fields the share was higher still; of these German Nobel Prizes, German Jews won nearly 30 percent, in medicine, 50 percent. Germans and Jews collected a disproportionate share; and although Harriet Zuckerman has recently demonstrated how complicated the notion of disproportionateness is, it does seem clear that Germans and Jews shared a certain immodesty in talent.

The prominence of German Jews also says something about their place in culture, about the milieu in which they worked, which mixed, perhaps uniquely, hospitality and hostility—and perhaps both were needed for this extraordinary achievement. I have tried to suggest that Germans had a veneration for learning, a yearning for greatness, a lingering insecurity. German Jews shared these traits and found further sustenance for them in their own distinct past. Jews did not foster talent; they hovered over it, they hoarded it, they nearly smothered it. Elsewhere I have pointed out that the rise of German Jewry is one of the most spectacular leaps of a minority in the social history of Europe, but their new prominence was painfully precarious and recalled Disraeli's desperate boast to young Montefiore: "You and I belong to a race that can do everything but fail." It is impossible to talk about Einstein's Germany without talking of German-Jewish relations, and to this theme I shall return later.

∞

I have tried to suggest some of the contradictory aspects of German culture. At this point we may be more familiar with the darker sides, with possible portents of later disaster. Einstein seemed peculiarly attuned to these portents; his friends, as we shall see, relished the virtues of German life. Perhaps our own I. I. Rabi said it all when he remarked to me the other day that he had found German culture "brutal and brilliant," that he had come to post-1918 Hamburg "knowing the libretto but learning the tune." In the first third of this century German institutes of learning and research orchestrated many voices into one tune of discovery. The German contribution to our civilization was immense, and the greatness that eluded Germans in politics they realized in the realms of science and of art.

Let me now turn to Einstein and his experiences with Germany. We know little of his early life. He was no child prodigy; rather, his reticence in speaking for the first three years, his difficulty with learning foreign languages, and his mistakes in computation have been a source of endless comfort to the similarly afflicted or to their parents, though affinity in failure may not suffice for later success. He went through a brief but intense religious phase, the end of which, he said, left him suspicious of all authorities. His parents, secularized Jews, had little to do with his intellectual development; an uncle fed his mathematical curiosity. His father was an amiable failure, mildly inept at all the businesses he started. In 1894, his parents went to Italy to start yet another business, leaving the fifteen-year-old Albert behind in a well-known Munich gymnasium. The authoritarian atmosphere and the mindless teaching appalled him. There is more than a hint of arrogance about the young Einstein, and hence it does not strain credulity to believe that his teacher exclaimed: "Your mere presence spoils the respect of the class for me." He was a rebel from the start.

Encouraged by his teachers' hostility, he decided to quit school and leave Germany. His unsuccessful career facilitated his later fame in Germany: Erik Erikson has rightly referred to "the German habit of gilding school failure with the suspicion of hidden genius." It is often said that Einstein left school because he objected to its militarism. I find this unpersuasive: Bavarian militarism? I would suppose that there might have been stifling Catholicism, insolent, thoughtless authoritarianism, a repulsive tone—all of which would have sufficed to discourage a youth like Einstein. I suspect Einstein left so precipitously in order to escape serving in the German army; by obtaining Swiss citizenship in time, he could do so without incurring the charge of desertion. His first adult decision, then, was to escape the clutches of compulsion—and the image of Einstein as a recruit in a field-grey uniform does boggle the mind. He left Germany without regrets. His first encounters with that country had not been happy.

There followed the obscure and difficult years in Switzerland, the failures, the marginal existence, the Zurich Polytechnic, and, finally, the security of the patent office in Berne. From there in 1905 emerged the four papers destined to revolutionize modern physics and cosmology. They were published in the *Annalen der Physik*, and Max Planck was the first man to recognize the genius of the unknown author. The international scientific community took note as well, and Einstein finally received his first academic appointments. In 1914, while he was a professor at the Zurich Polytechnic, two German scientists appeared, Walter Nernst and Fritz Haber, in order to offer him an unprecedented position: salaried membership in the Prussian Academy of Sciences so that he would not have to teach, though he would have a chair at the university as well. When Nernst and Haber left, Einstein turned to his assistant, Otto Stern, and said: "The two of them were like men looking for a rare postage stamp." The remark was perhaps an early instance of that self-depreciatory humor, that modesty of genius.

As a native Swabian, Einstein found Prussian stiffness uncongenial; the gentler, less strident rhythm of southern Germany or Switzerland was more to his taste. He began his new German life in April 1914 with some trepidation at his "Berlinization," as he called it. Berlin was the world's preeminent center of the natural sciences, and Planck, Haber, and a dazzling array of talent rejoiced at having this young genius at the head of their circle. Three months later the war shattered the idyllic community. Einstein had returned to Germany in time to see the country seized by the exaltation of August 1914, when almost all Germans were gripped by an orgy of nationalism, by a joyful feeling that a common danger had at last united and ennobled the people.

The intoxication passed; the business of killing was too grim to sustain the unbridled enthusiasm of August 1914. The elite rallied to the nation—as it did elsewhere too. In the fall of 1914, ninety-three of Germany's best-known scientists and artists, including Planck, Haber, and Max Liebermann, signed a manifesto that was meant to repudiate Allied charges of German atrocities, but by tone and perhaps unconscious intent argued Germany's complete innocence and blamed all misfortunes and wrongdoing on Germany's enemies. The manifesto of the ninety-three has often been seen as a warrant for aggression, as a declaration of unrestrained chauvinism. I suspect it was as well the outcry of people to whom the outside world mattered and who intuitively sensed that the Allies would come to cast Germans as pariahs again. Some of the ninety-three probably hoped for continued respect across the trenches—and signed a document that had the opposite effect. It was not the last time that Germans confirmed the sentiments they set out to deny. With but few exceptions, intellectuals everywhere joined in this chorus of hatred and in the cry for blood. So did the guardians of morality and the servants of God, the priests who sanctified the killing as an act of mythical purification. In time, some of the ninety-three turned moderate—or perhaps they remained the patriots they had been—but others passed them on the right, in the nation's wild leap to pan-German madness.

Einstein was alone and disbelieving. The war that was to politicize everyone as the cause of universal grief politicized him as well. Before 1914 he had never concerned himself with politics; his very departure from Germany had been a youthful withdrawal from the claims of the state. Now, for the first time, he ventured forth from his study, convinced of the insanity of the war, shocked by the ease with which people had broken ties of international friendship and mutual respect. A pacifist asked him to sign a counter-manifesto addressed to Europeans, demanding an immediate, just peace, a peace without annexations. It was the very first appeal he ever signed. It was never published—for want of requisite signatures. Somewhat later he joined a tiny group of like-minded democrats and pacifists. In November 1915 the Berlin Goethebund asked for his opinion about the war, and he sent a message with this rather special ending: "But why many words when I can say everything in one sentence and moreover in a sentence which is particularly fitting for me as a Jew: Honor Your Master Jesus Christ not in words and hymns, but above all through your deeds."

His work remained his central passion. But intermittently he forsook his work in order to bear witness in an unpopular cause for what he took to be right. He had been a pacifist and a European of the first hour, never touched by the frenzy that ravaged nearly all. Convinced of Germany's special responsibility for the outbreak and the continuation of the war, he hoped for its defeat.

To understand Einstein's isolation, one must look at the responses of his friends and colleagues. Fritz Haber, for example, became the very antithesis of Einstein. Haber, Einstein's senior by nineteen years, was a chemist of genius, a born organizer, and in wartime an ardent patriot. Without Haber's process for fixing nitrogen from the air, discovered just before the war, Germany would have run out of explosives and fertilizers in the first six months of the war. During the war, he came to direct Germany's scientific effort; in 1915 he experimented with poison gas and supervised the introduction of the new weapon at the Western Front. In order to operate within a military machine that had no understanding of the need for a scientist, he received the assimilated rank of colonel. He relished his new role; the marshaling of all one's talents and energies in a cause one believes in and in the shadow of danger—that is a heady experience. Einstein, the lonely pacifist who had come to feel his solidarity with Jews, and Haber, the restless organizer of wartime science and a converted Jew—the contrast is obvious. For all their antithetical responses, Haber and Einstein remained exceptionally close and, on Haber's side, loving friends. Haber's life was a kind of foil to Einstein's, and it encompassed the triumphs and the tragedy of German

Jewry. I shall return to him because his relations with Einstein were so important—and because he happened to have been my godfather and paternal friend of my parents.

Einstein had been horrified at the beginning of the war, but I doubt that even he could have imagined the full measure of disaster: the senseless killing and maiming of millions, the starving of children, the mortgaging of Europe's future, the tearing of a civilization that appeared ever more fragile. For what? Why? Einstein blamed it on an epidemic of madness and of greed that had suddenly overwhelmed Europe—and Germany most especially. The old German dream of greatness had turned into a nightmare of blind and brutal greed. During the later phases of the war, Einstein was again totally absorbed in his work, but whiffs of hysteria would reach him—and always from the German side. I doubt that he knew of the excesses on the other side.

Einstein had been right about the war. At its end, many felt as he had at the beginning. The war was a great radicalizing experience, pushing most people to the left and some to a new, frantic right. If there had been no war, bolshevism and fascism would not have afflicted Europe. The war discredited the old order and the old rulers; antagonism to capitalism, imperialism, and militarism appeared everywhere. Lenin's Bolsheviks offered themselves as the receivers of a bankrupt system; bolshevism was a speculation in Europe's downfall. Liberal Europeans pinned their hopes on Woodrow Wilson, but that hope faded in the vengeful spirit of Versailles. The logic of events had brought many Europeans to share Einstein's radical-liberal, faintly socialist, thoroughly internationalist views.

For a short time Einstein had hopes for Germany. Defeat had brought the collapse of the old and the rise of a new, democratic regime, as he had expected. He supported the new republic, and in November 1918, at the height of the German Revolution, cautioned radical students who had just deposed the university rector: "All true democrats must stand guard lest the old class tyranny of the right be replaced by a new class tyranny of the left." He warned against force, which "breeds only bitterness, hatred and reaction," and he condemned the dictatorship of the proletariat in what was the first of his occasional bitter denunciations of the Soviet Union as the enemy of freedom.[7] At other times and in different contexts, he would sign appeals of what we have come to call "front organizations."

We now come to a fateful coincidence in the rise of the public

Einstein. In March 1919 a British expedition headed by Arthur Stanley Eddington had observed the solar eclipse. In November it was announced that the results confirmed the predictions of the general theory of relativity. It was in London that the President of the Royal Society and Nobel Laureate, J. J. Thomson, hailed Einstein's work, now confirmed, as "one of the greatest—perhaps *the* greatest of achievements in the history of human thought." Somehow it seemed as if Einstein's achievement would revive the old international community of science. The world listened. Almost overnight Einstein became a celebrated hero—the scientific genius, untainted by war, of dubious nationality, who had revolutionized man's conception of the universe, newly defined the fundamentals of time and space, and had done so in a fashion so recondite that only a handful of scientists could grasp the new mysterious truth.

The new hero appeared, as if by divine design, at the very moment when the old heroes had been buried in the rubble of the war. Soldiers, monarchs, statesmen, priests, captains of industry—all had failed. The old superior class had been found inferior; *Disenchantment* was the proper title for one of the finest books written about the war. "Before 1914," Noel Annan has asserted, "intellectuals counted for little";[8] after the war, and in a sense in the wake of Einstein, they counted for more. Einstein now became a force, or at least a celebrity, in the world.

After 1919, he appeared more and more often as a public figure. His views were continually solicited, and he obliged with his ideas about life, education, politics, and culture. He had a special kinship with other dissenters from the Great War; like Bertrand Russell, Romain Rolland, and John Dewey, he became what the French call *un homme de bonne volonté.* His views—rational, progressive, liberal, in favor of international cooperation, condemnatory of the evils of militarism, nationalism, tyranny, and exploitation—these views described as well a cast of mind characteristic of Weimar intelligentsia.

The intellectuals of Weimar—and this needs to be said at a time when Weimar is often portrayed as some sort of Paradise Lost—were a shallow lot in their moralizing politics. Their views often seemed utopian and simplistic, pious and fiercely polemical by turns. They were cynical, as Herbert Marcuse once put it to me about himself, because they knew how beautiful the world could be. They lived in a world peopled by George Grosz caricatures and three-penny indictments of bourgeois falsehood. It is perhaps too simple to say that they lived off the bankruptcy of the old order, but they did rather revel in the crudity of their opponents. It is not good for the mind to have dumb, discredited enemies. The real strength of Weimar were clusters of talent: Heidelberg around Max and later Alfred Weber; Göttingen in mathematics; the Bauhaus and the Berlin circles.

Einstein stood above these progressive intellectuals, in consonance with them, but usually more complicated, less predictable, and always more independent than they. But he too was a theorist without a touch of practical experience. Einstein offered his prescriptions the more readily because he had been so overwhelmingly right when the multitudes had been wrong. By 1919 he had not only overthrown the scientific canons of centuries; he had also defied conventional wisdom and mass hysteria in wartime. His views were often deceptively simple; they were not as naive as has often been alleged nor quite so profound as admirers thought. There was no reason to think that a scientific genius would have special insights into other realms. He had reflected on some issues and felt strongly on others; for the rest, it became clear that genius is divisible and can be compartmentalized.

Einstein's views and prescriptions were unassailably, conventionally well intended, but they often lacked a certain *gravitas*, a certain reality—in part, I think, because he approached the problems of the world distantly, unhistorically, not overly impressed by the nature or intractability of the obstacles to ideal solutions. He was not a political thinker; he was a philosopher, moralist, prophet, and the travails of the world would prompt him to propose or support social remedies. Sometimes these remedies would be blueprints of utopia addressed to people who had lost their footing in a swamp and were sinking fast.

At a much later time, in fact at a moment when Einstein had attacked the Nazi government, Max von Laue questioned whether the scientist should deal with political issues. Einstein rejected such considerations: "you see especially in the circumstances of Germany where such self-restraint leads. It means leaving leadership to the blind and the irresponsible, without resistance. Where would we be if Giordana Bruno, Spinoza, Voltaire, and Humboldt had thought and acted this way." Einstein's models were instructive, and Laue pointed out that they were not exact natural scientists and that physics was so remote as not to prepare its practitioners for politics in the same way that law or history did. On that letter, Einstein simply scribbled—don't answer.

Like so many thinkers of the 1920s, Einstein underestimated the force of the irrational, of what the Germans call the demonic, in public affairs. That is what so ill-prepared them for an understanding of fascism. In their innocence they thought that men were bribed to be fascists, that fascism was but frightened capitalism; in its essence, it was something much more sinister and elemental. In his social commentary, Einstein left out the very thing he once called "the most beautiful experience we can have: the mysterious."

What gave his views exceptional resonance was the magic of his person and his incomparable achievement. He was taken by many as a sage and a saint. In fact, as I have said before, he was an unfathomably complex person. In the complexity of nature he found simplicity; in the complexity of his own nature, the principle of simplicity ranked high. Indeed, it was his simplicity, his otherworldliness, that impressed people. His clothes were simple, his tastes were simple, his appearance was meticulously simple. His modesty was celebrated—and genuine, as was his unselfishness. How many scientists or academics ask for a lower salary than offered, in the absence (let it be quickly added) of private means? He was a lonely man, indifferent to honors, homeless by his own admission, solicitous of humanity, and diffident about his relations with those closest to him. At times he appeared like a latter-day St. Francis of Assisi, a solitary saint, innocently sailing, those melancholy eyes gazing distractedly into the distance. At other times he was playing with the press, finding himself in the company of the famous and the powerful despite himself.

In some ways, I believe, he came to invest in his own fame, perhaps unconsciously to groom himself for his new role. He lectured in distant lands, "a traveller in relativity." In 1921, after his first visit to the United States he said: "The cult of individuals is always, in my view, unjustified. . . . It strikes me as unfair, and even in bad taste, to select a few [individuals] for boundless admiration, attributing superhuman powers of mind and character to them. This has been my fate, and the contrast between the popular estimate of my powers and achievements and the reality is simply grotesque." This admiration would be unbearable except that "it is a welcome symptom in an age which is commonly denounced as materialistic, that it makes heroes of men whose goals lie wholly in the intellectual and moral sphere. . . . My experience teaches me that this idealistic outlook is particularly prevalent in America. . . ."[9] He knew that he had become a hero—

and was endlessly surprised by it. In 1929 he described himself as a saint of the Jews. He played many roles by turns, each, I think, completely genuinely; he was a simple man of complex roles.

In the simplicity and goodness that were his, I detect, perhaps wrongly, a distant echo of his encounters with German life. Could one imagine a greater contrast between his German surroundings and himself, between people so formal in their bearing, so attentive to appearance, so solicitous of titles, honors, externals, and himself? Did the insolence of office, the arrogance of the uniform push him into ever greater idiosyncratic informality? Was not his appearance a democratic rebuke to authority?

In the immediate postwar era, Einstein was friendly to the governments of Weimar and appalled by the vindictiveness of the Allies, who seemed to have caught what he had thought was a German disease. In all his public stands he had what Gerald Holton has called a "vulnerability to pity," and in the early 1920s he had a fleeting moment of pity for Germany. He refused to leave it at the time of trial. For years he was an uncertain member of the League of Nations' International Commission on Intellectual Co-operation, intermittently resigning when he thought the Commission too pro-French, too *Allied*. He hoped to restore an international community, Germans included. In the end he asked Fritz Haber to take his place. Successive German governments regarded him as a national asset, perhaps the sole asset in a morally and materially empty treasury. They saw in his travels and in his fame the promise of some reflected glory. But his own hopes gradually faded. He had warned Walter Rathenau against assuming the foreign ministry; Jews should not play so prominent a role. When rightwing assassins—widely hailed in Germany as true patriots—killed Rathenau, Einstein had reason to fear for his own life. The inborn servility of the Germans, he thought, had survived the successive shocks of 1918.

Immediately after the war and at the beginning of his popular fame, Einstein embraced several causes. Having embraced them, he would often embarrass and repudiate them as well. He was the antithesis of an organization man. Unstintingly he would help individuals and chosen causes, but I doubt that he would listen to them. He remained a detached theorist who saw the nature of the world wantonly violated. But at times his commandments contained visionary practicality. A pacifist during the war, he now became Germany's most prominent champion of organized pacifism. He hated militarism—blindly, as its defenders loved it—blindly. He condemned "the worst outgrowth of herd life, the military system. . . . I feel only contempt for those who take pleasure marching in rank and file to the strains of a band. . . . Heroism on command, senseless violence and all the loathsome nonsense that goes by the name of patriotism—how passionately I despise them!"[10] This, surely, is exemplary of the spirit of the 1920s, formed by the experience of the first war and soaked in the we-they antithesis that precluded understanding. It precluded the understanding that had led William James to plead for a moral equivalent of war, for something practical that would make peaceful use of the old martial virtues. Einstein insisted that "the advance of modern science has made the delivery of mankind from the menace of war . . . a matter of life and death for civilization as we know it." But Einstein did not grapple with the psychological issues, with people's desire for danger and comradeship. In his exchange with Freud about the nature of war he acknowledged that "the normal objective of my thought affords no insight into the dark places of human feeling and will."[11] For Einstein, war was a disease, a disorder planted by men of greed, to be abolished by men of good will through the creation of international sovereignty or through a revolutionary pacifism, that is, through the refusal of men to bear arms in peace or war. He called for resistance to war; but in 1933, almost immediately after Hitler's assumption of power, he renounced pacifism altogether—to the fury of his doctrinaire followers. In fact, he urged the Western powers to prepare themselves against another German onslaught.

His second great cause was Zionism, which he seems to have embraced during the war. By November 1919 the *Times* of London referred to him as an "ardent Zionist . . . keenly interested in the projected Hebrew University at Jerusalem and [he] has offered to collaborate." By the early 1920s he became a public advocate of Zionism—to the surprise and likely dismay of many of his colleagues. Assimilated Jews must have found this reminder of Jewish apartness painful; internationalists would have boggled at the implied argument for a new national community. But Einstein had come to feel a sense of solidarity with Jews, especially with Jewish victims of discrimination, and he seemed to believe in the existence of an ineradicable antagonism between gentiles and Jews, especially between Germans and Jews—with the fault by no means all on one side. Hence his view that Jews needed a spiritual home and a possible haven. He specifically cited the discrimination that talented Jews from Eastern Europe and from Germany suffered at German universities.

In 1921 Chaim Weizmann persuaded Einstein to join him on a trip to the United States to raise money for the projected Hebrew University in Jerusalem. For Weizmann, Einstein's support was critical; for Einstein, his visit to Jerusalem in 1923 was a deeply moving experience. Still, there were conflicts. Einstein railed against the mediocrity of the American head of the university; he saw him as a creature of the crass American-Jewish plutocrats for whom Einstein had contempt even as he helped to lighten their financial burden. He quarreled publicly with Weizmann over the policies of the Hebrew University and repeatedly threatened to withdraw his sponsorship. He urged a Palestinian presence that would promote, not injure, Arab interests. In 1929, at the time of major attacks on Jewish settlements, he again pleaded with Weizmann for Jewish-Arab cooperation and warned against a "nationalism à la prussienne," by which he meant a policy of toughness and a reliance on force:

> If we do not find the path to honest cooperation and honest negotiations with the Arabs, then we have learned nothing from our 2000 years of suffering and we deserve the fate that will befall us. Above all, we should be careful not to rely too heavily on the English. For if we don't get to a real cooperation with the leading Arabs, then the English will drop us, if not officially, then de facto. And they will lament our debacle with traditional, pious glances toward heaven, with assurances of their innocence, and without lifting a finger for us.[12]

Weizmann replied instantly, at the height of the Palestinian violence, with a four-page handwritten letter. He expounded his views, which were somewhere between Zionist extremists and the irenic Einstein—who, in the meantime, had criticized the Jewish stance publicly. Weizmann pointed to the recalcitrance of the Arab leaders, their fanaticism, their inability to understand anything but firmness. He pleaded with Einstein to cease his injurious attacks on the Zionists. Of course they would negotiate in time, Weizmann insisted, but "we do not want to negotiate with the murderers at the open grave of the Hebron and Safed victims." Einstein remained skeptical. Weizmann, desperate to retain his support, had written to Felix Warburg a year earlier: "There is really no length to which I would not go to bring back to our work the wonderful and lovable personality—perhaps the greatest genius the Jews have produced in recent centuries and withal so fine and noble a character."

At the time of the greatest need for a Jewish home in Palestine, immediately after Hitler's seizure of power, Einstein formally broke with the Hebrew University and with Weizmann. The correspondence between the two men suggests all the intractable issues about Jewish-Arab relations, all the differences between the safe outsider and the practical statesman. In April 1938, Einstein resigned his position and again warned against a "narrow nationalism." Once again Weizmann explained that at the moment when five million Jews faced, as he put it, "a war of extermination," they needed the support of the intellectual elite of Jewry, and not, by implication, public criticism. Einstein was not an easy ally. To some he must have appeared as a man of conscience and of unshakable principle; to others, as an uncompromising fanatic in purity, impervious to practical exigencies. As Robert Oppenheimer put it in his memorial lecture: "He was almost wholly without sophistication and wholly without worldliness. . . . There was always with him a wonderful purity at once childlike and profoundly stubborn."[13]

It would be hard to imagine three causes less pleasing to the bulk of the German professoriat than liberal internationalism, pacifism, and Zionism. Unlike many academics, Einstein took education with the utmost seriousness—and academics with magnificent irreverence. He had great faith in the possibilities of primary and secondary education; at one point he said that if the League of Nations could improve primary education, it would have fulfilled its mission. His ironic contemplation of universities found expression in private letters. He once complimented his close friend Max Wertheimer, the Gestalt psychologist: "I really believe there are very few who have been so little harmed by learning as yourself." In 1924 he wrote: "In truth, the University is generally a machine of poor efficacy and still irreplaceable and not in any essential way improvable. Here the community must take the point of view that the biblical God took towards Sodom and Gomorrah. For the sake of very few, the great effort must be made—and it is worth it!"[14]

Einstein's success—the enormous acclaim, especially abroad at a time when most German scientists were still banished from international meetings—caused much ill will at home. His opinions enraged the superpatriots. Some physicists condemned the fanfare surrounding the dubious theory of relativity; one fellow laureate attacked it as "a Jewish fraud." To anti-Semites, Einstein became a favorite and obvious target. The waves of hatred spilled from the streets into the lecture halls, and Einstein's occasional and sometimes ill-considered deprecations made things worse.

Germany frightened him again. His hopes for the Weimar Republic had dimmed. As early as 1922, his life was threatened. He traveled even more than before, but still he refused handsome offers from Leiden and Zurich, universities with which he had the closest ties. An American university also dispatched a letter of invitation, hitherto buried in the archives, dated 26 February 1923; the timing was delicately perfect because Germany found itself in the throes of foreign occupation and rampant inflation.

Dear Professor Einstein:

The duties and occupations of the professorship will be precisely what you wish to have them. Our aim will be to make it easy and convenient for you to pursue your personal studies and course of reflection without onerous academic or public burdens of any kind. The annual salary of the Professorship is $10,000 or about 40,000 gold marks. I wish to put this invitation before you on behalf of [Columbia] University with all possible urgency, and to beg you to accept it in the interest of science, of international comity, and, I trust, of the most successful prosecution of your own personal work.

Nicholas Murray Butler[15]

Alas for Columbia, Einstein declined, though pleased to have been asked "by your magnificent university." He stayed despite his misgivings about Germany; he stayed because Berlin in the 1920s was the golden center of physics; he stayed because proximity to Planck, Laue, Haber, and others was a unique professional gift, because, as he wrote Laue in 1928: "I see at every occasion how fortunate I can call myself for having you and Planck as my colleagues." In 1934 he wrote Laue that "the small circle of men who earlier was bound together harmoniously was really unique and in its human decency something I scarcely encountered again." In 1947 he wrote Planck's widow that his time with Planck "will remain among the happiest memories for the rest of my life."[16]

The unpublished correspondence among these men suggests even more than a professional tie. The letters bespeak a degree of humane collegiality, a shared pleasure in work, as well as a delicacy of sentiment, a candid avowal of affection, which in turn would allow for confessions of anguish and self-doubt, of melancholy as well as high spirits. They spoke of joys and torment, in close or distant friendship, in an enviable style. The letters also breathe a kind of innocence, as if science was their insulated realm, nature the great, enticing mystery, and one's labors of understanding exclusively an intellectual pursuit, remote from social consequences. Such clusters of collaboration and of friendship have always existed, I suppose, and they have made life better and infinitely richer. Germany may have had a special knack for breeding them.

Einstein's Germany included gentiles and Jews, working together in extraordinary harmony. And still it can be stated categorically that none of the Jewish scientists escaped the ambiguity, the intermittent hostility, that Jewishness produced in Imperial and Weimar Germany. Neither fame nor achievement, neither the Nobel Prize nor baptism offered immunity. Passions were fiercer in Weimar, in that cauldron of resentments. Most official barriers against Jews had been lowered, but new fears and hatreds came to supplement old prejudices. Three incidents may illustrate the uncertain temper of the time. In 1921, Haber begged Einstein not to go to America with Weizmann, on the ground that Germans would take amiss his travels in Allied countries with Allied nationals at the very time when the Allies were once again tightening the screws against Germany. To persuade Einstein, Haber warned that German anti-Semites would capitalize on his seeming desertion and that innocent Jewish students would be made to suffer. Anti-Semitism, rampant as it was, need not be goaded; Einstein's warning to Rathenau originated in a similar apprehension. Or take another incident. In 1920, a well-known physicist opposed the appointment of the later laureate Otto Stern: "I have high regard for Stern, but he has such a corrosive Jewish intellect."

Or consider this last example. In 1915 the king of Bavaria, confirming the Nobel laureate Richard Willstaetter's appointment to a professorship, admonished his minister: "This is the last time I will let you have a Jew." Ten years later, discussing with his colleagues a new academic appointment, Willstaetter proposed a candidate. A murmur arose: "another Jew." Willstaetter walked out, resigned his post, and never entered the University again, the unanimous pleas of his students notwithstanding. For the next fourteen years he had daily, hour-long telephone calls with his assistant so that she could conduct the experiments in a laboratory that he would no longer enter. A man of conscience and of courage, someone who did not blink at the reality of anti-

Semitism. But his stand in 1924 was his undoing a decade later. A devoted German, but now no longer a civil servant, he assumed that the Nazis would leave untouched a private scholar. He believed that some Jews had contributed to this new storm. He could not comprehend the radical newness of the phenomenon. In February 1938 he wrote my mother urging her that one ought not to leave Germany without the most careful reflection. He himself refused exile until the aftermath of Crystal Night forced him into it.

I cite Willstaetter's example among many precisely because of its contradictory nature: awareness of anti-Semitism could cloud one's perception of Nazism. If anti-Semitism had always existed, then perhaps Nazism was but an intensification of it. It is not uncommon these days to hear summary judgments about German Jewry, about their putative self-surrender, their cravenness, or their opportunism. These judgments often have a polemical edge; they are likely to do violence to the past and to the future: the myth of yesterday's self-surrender could feed the delusion of tomorrow's intransigence. If our aim is to understand a past culture, we must note that German-Jewish scientists thought Germany their only and their best home, despite the anti-Semitism that crawled all around them. They may have loved not wisely but too well, and yet their sentiments are perhaps not so much an indictment of themselves as a tribute to the appeals of Germany. We owe that past no less than what we owe any past: a sense of its integrity.

Let me hasten to the denouement. In 1932 Einstein left Germany provisionally, with the intention of returning to Berlin for one semester each year. Hitler's accession to power the next year changed all that. Einstein immediately denounced the new regime, and in response he was extruded from the Prussian Academy, his books were burned, his property seized. The first Nazi decrees for the purification of the universities would have allowed some Jews to maintain their positions. Einstein's non-Aryan friends spurned such sufferance and resigned. German physics was decimated, and a few remaining masters battled to defend some shreds of decency, some measure of autonomy. Laue once wrote Einstein that in teaching the theory of relativity he had sarcastically added that it had of course been translated from the Hebrew. Even such jokes—to say nothing of Laue's eulogies of Jewish colleagues—aroused Nazi wrath. The Nazis proscribed the very mention of Einstein, even in scientific discussions. They would have wished him to be an unperson.

For most, exile was hard; the habits of a lifetime are not easily shaken. For others, as the physicist Max Born put it, "a disaster turned out to be a blessing. For there is nothing more wholesome and refreshing for a man than to be uprooted and replanted in completely different surroundings."[17] Resiliency was a function of age and temperament. For Haber, exile was a crushing blow and led to a final irony in his relations with Einstein. By mid-1933 he wrote to Einstein that as soon as his health would allow it, he would go to Palestine, but in the meantime he begged Einstein to patch up his public quarrel with Weizmann. Einstein replied at length: "pleased . . . that your former love for the blond beast has cooled off a bit. Who would have thought that my dear Haber would appear before me as defender of the Jewish, yes even the Palestinian cause. The old fox [Weizmann] did not pick a bad defender." He then lashed out against Weizmann and concluded:

> I hope you won't return to Germany. It's no bargain to work for an intellectual group that consists of men who lie on their bellies in front of common criminals and even sympathize to a degree with these criminals. They could not disappoint me, for I never had any respect or sympathy for them—aside from a few fine personalities (Planck 60% noble, and Laue 100%). I want nothing so much for you as a truly humane atmosphere in which you could regain your happy spirits (France or England). For me the most beautiful thing is to be in contact with a few fine Jews—a few millennia of civilized past do mean something after all.[18]

The German patriot Haber died a few months later in Basel, en route to Palestine. And Einstein found a refuge at the Princeton Institute under conditions not dissimilar from what the Prussian Academy had offered him twenty years earlier. For as Erwin Panofsky has said of the Institute for Advanced Study, it "owes its reputation to the fact that its members do their research work openly and their teaching surreptitiously, whereas the opposite is true of so many other institutions of learning."[19]

Einstein's public life continued to be dominated by his fear of Germany. He warned the West against a new German onslaught. He abandoned the pacifism he had so fervently espoused and in 1939 signed the famous letter to President Franklin D. Roosevelt urging the administration to prepare the United States because Germany might develop nuclear fission for military purposes. In the winter of 1945, when Germany was desolate in defeat and when the Morgenthau spirit, if not the plan, had a considerable grip on American thinking, a fellow laureate and old friend, James Franck, asked Einstein to sign a manifesto of exiles that would appeal to the United States not to starve the German people. Einstein vowed that he would publicly attack such a plea. The German police—and he said this eleven months after the war—were still killing Jews in the streets of Germany; Germans had no remorse, they would start another war. Franck pleaded with him that to give up all hope for a moral position in politics would be tantamount to a Nazi victory after all. But Einstein, who had signed so many appeals that he himself once said he was not a hero in no-saying, scathingly rejected Franck's plea. For him, genocide was Germany at its most demonic; after Auschwitz he could muster no magnanimity. Even the righteous could not redeem the "country of mass murderers," as he called Germany. He rebuffed Laue's plea to help a young German physicist. He knew that Planck, who lost one son in the first war, had lost another whom the Nazis murdered because of his participation in the plot against Hitler. The serene Einstein, always the champion of the rights of the individual against the collectivity, now proclaimed the principle of collective guilt. At that moment, of course, the world shared Einstein's horror at German inhumanity. But in him the violence of sentiment, the total absence of that vulnerability to pity, puzzles, for it shows how desperately deep and all-consuming had been his antipathy to Germany.

Even his postwar laments about America, his horror at McCarthyism, were shaped by his image of Germany. America, he believed, was somehow following the path of Germany. The world of politics he saw through German eyes—always.

But let me end on a different note. Greatness in any guise is not in vogue today, not in my discipline and not in our culture. Historians feel that it is now the turn of the forgotten, both for reasons of retroactive justice and for heuristic purposes. The argument is compelling, the feeling comprehensible, though there is an implicit deprivation involved, for as Einstein in a rather German formulation put it: "the example of great and pure individuals is the only thing that can lead us to noble thoughts and deeds." We are uncomfortable even with the rhetoric of greatness, devalued as it so often has been. I would simply say that I find it inspiriting to look upon great peaks, as from an alpine village, and contemplate the distant mountains—cold, awesome, unattained and unattainable, mysterious.

It is often asserted that a culture must be judged by its treatment of minorities and deviants; a student of the German past would find this a cogent and, indeed, irrefutable argument. It is a necessary, but not a sufficient criterion. A culture must also recognize, recruit, and, in a sense, form talent; it must know how to coax talent into achievement. This too is a test of its virtue and of its instinct for survival. These are responsibilities that speak most directly to our universities, to every university.

NOTES

1. Quoted in Banesh Hoffmann, with the collaboration of Helen Dukas, *Albert Einstein: Creator and Rebel*, New York, 1972, p. 139.
2. Lionel Trilling, *Mind in the Modern World*, New York, 1973, pp. 13-14.
3. Richard Hofstadter, "History and the Social Sciences," in Fritz Stern, ed., *The Varieties of History*, 2d ed., 1973, p. 369.
4. Max Weber, *Gesammelte politische Schriften*, Munich, 1921, p. 29.
5. Theodor Mommsen, *Reden und Aufsätze*, Berlin, 1905, pp. 5-8.
6. Martin J. Klein, *Paul Ehrenfest*, Amsterdam, 1970, p. 77.
7. Otto Nathan and Heinz Norden, eds., *Einstein on Peace*, New York, 1960, p. 25.
8. Noel Annan, " 'Our Age': Reflections on Three Generations in England," *Daedalus*, Fall 1978, p. 83.
9. Albert Einstein, *Ideas and Opinions* (based on *Mein Weltbild*, ed. Carl Seelig), new trans. and rev. Sonja Bargmann, New York, 1954, p. 4.
10. *Einstein on Peace*, pp. 111-112.
11. Albert Einstein and Sigmund Freud, *Why War?*, Paris, 1933, p. 12.
12. Albert Einstein to Chaim Weizmann, 29 November 1929, The Weizmann Archives, Yad Chaim Weizmann, Rehovoth.
13. Robert Oppenheimer, "On Albert Einstein," *New York Review of Books*, 17 March 1966.
14. Albert Einstein to Max Wertheimer; Albert Einstein to Dr. Schwallie, 18 July 1924. Both in Einstein Archives, Institute for Advanced Study, Princeton, N.J.
15. Letter in the Archives of Columbia University, New York.
16. Letters in the Einstein Archives.
17. Max Born, *My Life and My Views*, New York, 1968, p. 38.
18. Letter in the Einstein Archives.
19. Erwin Panofsky, *Meaning in the Visual Arts*, New York, 1955, p. 322.

A NOTE ON SOURCES

It seems superfluous to list the standard works on and by Albert Einstein. For purposes of this essay I found *Ideas and Opinions* and *Einstein on Peace* particularly pertinent. I was also fortunate enough to be allowed to use the Albert Einstein Archive at the Institute for Advanced Study in Princeton, N.J., a treasure made still more valuable by the ever helpful advice and recollections of Helen Dukas, who is in charge of it. I also read the unpublished correspondence of James Franck and Albert Einstein, deposited at the University of Chicago Library. In addition to the books cited in the preceding notes, I found the following particularly useful: Allan D. Beyerchen, *Scientists under Hitler: Politics and the Physics Community in the Third Reich*, New Haven, 1977; Gerald Holton, *The Scientific Imagination: Case Studies*, Cambridge, 1978; Richard Willstaetter, *Aus Meinem Leben*, Weinhein, 1949; Harriet Zuckerman, *Scientific Elites: Nobel Laureates in the United States*, New York, 1977.

I benefited from conversations with Marshall Clagett, Felix Gilbert, Gerald Holton, Martin Klein, I. I. Rabi, and Malvin Ruderman. It was in long and frequent talks with Otto Stern that I first sensed how extraordinary those early days in Zurich must have been.

Einstein and the Nuclear Age

Paul Doty

EINSTEIN AND INTERNATIONAL SECURITY

NEXT TO PHYSICS, Einstein's most enduring passion was directed at abolishing war. This passion consumed increasing amounts of his time from approximately 1915 onward for four decades until his death. Retracing this part of his life can be rewarding at several levels. It reveals the mainsprings of his essentially humanitarian outlook. It shows the struggle and redirection that came on several occasions when pacifism conflicted with values he held even higher. And it shows the development of his abiding conviction that peace could become a normal part of the human condition only through the evolution of a world authority with sufficient military power to prevent war and sufficient prestige to require international arbitration of disputes. Despite his leadership and the adherence of many individuals to movements embracing this approach to international security, it is not the path that world politics has taken. Thus Einstein's centennial is an appropriate and timely occasion to review his own approaches to securing peace, to note the influence he had on scientists' involvement in peace issues, and to assess the nature and adequacy of the defenses of peace that have been fashioned in the Nuclear Age, which traces its origins back to his 1905 paper on special relativity, with its prophetic declaration of the equivalence of mass and energy.

∞

In the relative tranquility of pre–First World War Europe Einstein progressed from obscurity to renown as a scientist of the first rank—from his first publication in 1901, through a decade of brilliant work in Bern and Zurich, to professorships in Prague,

Zurich, and Berlin in 1911, 1912, and 1914, respectively. His preoccupation with science and the absence of international violence in this period meant that his potential for concern with broader issues and war itself were not yet challenged. Nothing in his writings of this period foretells the change that the First World War would trigger. Indeed, his detachment from the demands of everyday life and politics during the period became legendary.

The Great War began four months after Einstein moved to Berlin. His ready responses to this event suggest that his total detachment from the world about him was beginning to break down. He wrote to Paul Ehrenfest in December 1914, "The international catastrophe has imposed a heavy burden upon me as an internationalist."[1] Two months earlier he had signed a challenge to a manifesto circulated by German intellectuals that sought to justify the German invasion of Belgium and linked German culture to German militarism. The counter-manifesto was presumably the first political document that Einstein ever signed, and it foreshadowed many of the ideals to which Einstein was to give lifelong devotion. Thus we must conclude either that the first months of the war reoriented his concerns toward violence in world affairs or that such thoughts had been brooding, unannounced, for a longer time. That the counter-manifesto collected only four signatures reflects the extent to which the Berlin intellectuals were swept up with the war and how willing Einstein was to stand alone for his convictions even in a new and increasingly hostile community.

By early 1915 Einstein had become an active member of a newly formed antiwar group, the New Fatherland League, and began proposing actions at the international level to oppose the war. By 1917 his feelings against German nationalism had reached a very high level, and his pacifism had evolved toward concrete proposals for maintaining peace after the war. He wrote:

> the best method of resolving this dreary dilemma would be to form an enduring military arbitration pact among America, Britain, France, and Russia, with agreements on mutual aid and minimum and maximum limits of military preparedness. Such a treaty should include provisions for most-favored-nation treatment with respect to tariffs. Any nation should be allowed to join the treaty provided it has a democratically elected parliament in which the chief executive must command a majority. (20-21)

Despite Einstein's hatred of the war and his growing concern with approaches to preventing future wars, the war years were among his most productive scientifically, his most outstanding achievement being the completion of his general theory of relativity. Within a year of the war's end his general theory was verified by the determination of deflection of light by the sun in an eclipse studied by a British team. This example of international scientific collaboration so soon after the war and the impact of the grandeur and reach of Einstein's theory thrust him into a state of world acclaim. Thus, from 1919 onward he was the most famous living scientist, continuously sought after by the press and a great array of organizations wanting him to speak or comment.

It was probably inevitable that this sudden but sustained popularity would be put to the service of Einstein's strong views on pacificism, liberalism, and internationalism, even though such political involvement was contrary to European academic custom. His entry into public life was for the most part by this route. In accepting the political role thrust upon him, he immediately became the object of criticism and a lightning rod for anti-Semitic attacks. But his solid scientific reputation and his unworldly image ensured such adulation that his position in the public mind easily survived these mostly parochial attacks.

∞

By 1921 Einstein had traveled widely in Europe speaking on science, pacifism, and reconciliation and was induced to visit America to lecture on behalf of fund raising for Zionist organizations, a movement in which he had just enlisted. His lectures and interviews on relativity produced an enormous response and public receptions that often approached hysteria. His transition to a world figure was complete.

Returning to Einstein's involvement with matters relating to peace, it is natural to find him drawn to support the League of Nations. He early recognized the necessity of United States membership if the League were to succeed. But long before he abandoned hope in this, his disappointment was aroused by the French government's refusal in 1923 to submit the problem of Germany's war reparations payments to arbitration and its dispatch of military occupation forces into the German Ruhr.

It was this latter event that precipitated his resignation from the League's Committee on Intellectual Co-operation, which had been formed in 1922 of European intellectuals for the purpose of promoting closer communication among national cultures through supporting international cooperation among scientific and intellectual communities. He eventually rejoined, but frustrated by the lack of results, by political bickering over representation, by his own role in the German committee, and perhaps most of all by the irrelevance of the committee's work to the rising challenge of fascism, he finally withdrew.

In searching for a more activist role in combating fascism, he turned his efforts in 1928 to the support of organizations that encouraged individual war resistance. In that year he wrote: "people can most effectively fight the institution of war by establishing, in time of peace, an organization for absolute refusal of military service" (90). Depressed by the increasingly evident preparation for war, and unimpressed by the Kellogg-Briand Pact of 1928, Einstein committed himself more and more to the proposition that war could be prevented if even a small percentage of people in each country would declare their refusal to go to war under any circumstances. He was not dismayed by arguments such as that of his friend Jacques Hadamard, the French mathematician and pacifist, that many countries had found that not defending themselves had not prevented aggression. His fourth trip to America, in 1932, propelled him to the front ranks of world pacifists, and he found considerable support for his proposal that only 2 percent of male citizens pledging to refuse military service would make the pursuit of war preparations and war itself impossible. Lapel buttons reading "2%" spread quickly.

In this last year before night began to fall over Germany and then the rest of Europe, Einstein continued his efforts in support of war resistance but became increasingly concerned with the disarmament approach and in particular with critiques of disarmament efforts, which he considered quite inadequate. Nevertheless, he found the problem of how to proceed in any planned, programmatic way intractable. Hence, in a manner reminiscent of his quick, incisive thrusts at a physics problem, he concluded that disarmament could only be attained by a single stroke. Speaking at Pasadena in February 1932, he said:

> The greatest obstacle to disarmament has been the inability of most people to appreciate the enormity of the problem. Most objectives are accomplished in small steps. Think, for example, of the transition from absolute monarchy to democracy! But we are here concerned with an objective that cannot be attained slowly, step by step.
>
> So long as the possibility of war exists, nations will continue to insist on the fullest possible military preparedness to insure their victory in a future war and will find it unavoidable to educate the young people in warlike traditions and narrow nationalist vanities. . . . To arm means

simply to approve and prepare for war, not for peace. Hence, disarmament cannot come in small steps; it must come about at one stroke or not at all.

To accomplish so profound a change in the life of nations, a mighty moral effort and a deliberate rejection of deeply ingrained traditions are required. Anyone who is unwilling to let the fate of his country, in the event of conflict, depend without qualification upon the decisions of an international court of arbitration, and who is not prepared to see his country enter into treaties that provide for such a procedure without any reservation, is not really resolved to abolish war. This is a case of all or nothing. (163-164)

And, visiting the Geneva Disarmament Conference in May, he said:

if the implications were not so tragic, the methods used at the Disarmament Conference could only be called absurd. One does not make wars less likely to occur by formulating rules of warfare. One must start with the unqualified determination to settle international disputes by way of arbitration. What is involved are questions of morality and good will; it is not a problem for the so-called technical experts. (168)

The rise of Nazism in Germany and the continuing threat of war led Einstein to cast a wide net for help. He invited Freud to join with him in a public correspondence over how mankind could be spared the continuing menace of war. There resulted a remarkable exchange of letters. Einstein's letters restated in more elegant prose much of what has just been quoted. Freud's reply of September 1932 was longer and seemingly more profound. Two quotations are relevant:

Regarding the world of today the same conclusion holds good, and you, too, have reached it, though by a shorter path. There is but one sure way of ending war and that is the establishment, by common consent, of a central control which shall have the last word in every conflict of interests. For this, two things are needed: first, the creation of such a supreme court of judicature; secondly, its investment with adequate executive force. Unless this second requirement be fulfilled, the first is unavailing. Obviously the League of Nations, acting as a Supreme Court, fulfills the first condition; it does not fulfill the second. It has no force at its disposal and can only get it if the members of the new body, its constituent nations, furnish it. And, as things are, this is a forlorn hope. (195-196)

This argument may have been influential in securing Einstein's acceptance of the necessity for an international military force, for shortly thereafter he wrote that he agreed that the only military force that should be permitted to have truly effective weapons is a police force which would be subject to the authority of international organs and would be stationed throughout the world. The police formations [however] should not be composed of national troop units . . . [but] must be—both officers and men—international in composition. (205-206)

Another portion of Freud's long letter may have initiated another, slower change in Einstein's outlook:

The ideal conditions would obviously be found in a community where every man subordinated his instinctive life to the dictates of reason. . . . But surely such a hope is utterly utopian, as things are. The other indirect methods of preventing war are certainly more feasible, but entail no quick results. They conjure up the ugly picture of mills that grind so slowly that, before the flour is ready, men are dead of hunger. (200)

In any event, Einstein's reference to disarmament having to be accomplished in a single stroke was greatly muted thereafter.

With the Nazi rise to power in Germany early in 1933, Einstein's intellectual migration from a posture of war resistance, which had the virtue of being a prescription for action, to conceptions for a supranational authority backed by international military force, whose emergence seemed far away, left him in a vulnerable position. Yet he had the courage to put his values before the long-held convictions that he had developed in an earlier, simpler era. In July 1933 he wrote a letter that was published a month later:

What I shall tell you will greatly surprise you. Until quite recently we in Europe could assume that personal war resistance constituted an effective attack on militarism. Today we face an altogether different situation. In the heart of Europe lies a power, Germany, that is obviously pushing toward war with all available means. This has created such a serious danger to the Latin countries, especially Belgium and France, that they have come to depend completely on their armed forces. As for Belgium, surely so small a country cannot possibly misuse its armed forces; rather, it needs them desperately to protect its very existence. . . . Hence I must tell you candidly: Were I a Belgian, I should not, in the present circumstances, refuse military service; rather, I should enter such service cheerfully in the belief that I would thereby be helping to save European civilization. (229)

In replying to criticisms of this revision of his views, he argued that it was not his views but the situation in Europe that had changed and changed radically. Before the rise of Nazism, no serious aggression was foreseeable in Europe.

So long as Germany persists in rearming and systematically indoctrinating its citizens in preparation for a war of revenge, the nations of Western Europe depend, unfortunately, on military defense. Indeed, I will go so far as to assert that if they are prudent, they will not wait, unarmed, to be attacked. . . . They must be adequately prepared.

I take little pleasure in saying this, for in my heart I loathe violence and militarism as much as ever; but I cannot shut my eyes to realities. (230)

The criticisms continued, and Einstein tried to clarify his views for his disappointed pacifist friends:

I believe . . . that even now it is not too late to avert war by preventing German rearmament through diplomatic pressure. *But such pressure will require absolute military superiority on the part of Germany's neighbors.* To destroy such superiority or to prevent its achievement is tantamount to betraying the cause of European freedom. (236)

And, to complete the reversal, he had the honesty to write, after the annexation of Czechoslovakia in 1939, "It is no exaggeration to say that the British and, to some extent, the French pacifists are largely responsible for the desperate situation today because they prevented energetic measures from being taken at a time when it would have been relatively easy to adopt them." Clearly, Einstein's pacifism was tempered by the times. When fundamental values were not at risk, it was a strategy to prevent war. But peace was not above survival or saving European civilization from Fascism.

∞

The first half of 1939 saw a sudden, feverish activity in the physics laboratories of half a dozen countries, activity that would provide the finale for the war that was about to come. The triggering event was the discovery by Otto Hahn and Fritz Strassmann in Berlin that atomic nuclei could be split (fission), with a consequent release of large amounts of energy formed from the conversion of a small fraction of the original atomic mass into energy in accordance with Einstein's equation. Within weeks it was shown that more than one neutron was released per nucleus from the disintegrating uranium-235 atoms, with the result that a chain reaction that could produce very large amounts of energy suddenly became a clear possibility. Within six months, nearly twenty articles on fission had appeared in the journal *Nature*, and at least four governments had been alerted as to its possible military potential: Britain, Germany, the Netherlands, and the United States.

Einstein seemed unaware of the consequences of the chain reaction when visited by Leo Szilard and Eugene Wigner in July 1939. Although he remained skeptical of its development "in his lifetime," he consented to sign a letter to President Franklin D. Roosevelt recommending the assignment of responsibility within the government and possibly industry, as well as urging steps to ensure a supply of uranium. This was not the first alert to the government, and, despite its illustrious backing, it did not produce the acceleration the scientists thought to be needed. Consequently, Einstein signed another letter, delivered to Roosevelt in March 1940, which did succeed in producing an accelerated effort. But from here onward, Einstein had no further connection with the project, except possibly in some detailed matters compartmentalized from the main development that he undertook as a consultant to the Office of Scientific Research and Development in Washington in 1942-1943.[2] Whatever enthusiasm Einstein had for the project stemmed from his fear that the Germans would win the race for weapons development.

It was not until the summer of 1941 that the Maud Report from Britain, with its concrete plans, convinced the Americans that a chain reaction bomb could be made. Britain and the United States began the Manhattan Project on 6 December 1941, hours before Pearl Harbor. In retrospect, even without the acceleration of earlier plans that Einstein's letter had stimulated, the actual production of the first nuclear weapons would probably not have been delayed by more than a few months.

Although Einstein probably became acquainted with the vague outlines of the development of the bomb in the last year of the war,[3] its use came as a surprise. As the effects of the bombs dropped on Japan became known, Einstein joined those scientists who argued that the bombs should not have been used. This stand is consistent with his general view that he would consent to the use of force only "when confronted by an enemy who pursues the destruction of life as an end in itself" or "in the face of an enemy unconditionally bent on destroying me and my people." What attitude he would have adopted had the bomb been available during the European phase of the war has been debated, but there are no grounds for a determination. Surely it would have been consistent for him to argue that his role in promoting its development was based on the expectation that it would deter the Germans if they also developed it. Later, in 1953, he stated that the real danger at the time was that Germany under Hitler might become the sole possessor of this weapon.

In the postwar world, Einstein soon reached the conclusion that nuclear weapons were a profound risk to mankind and in time could bring an end to civilized society. His intuitive reaction was to devote his labors to the abolition of nuclear weapons and to resume his advocacy of a world government with power to judge disputes and power to prevent the outbreak of war. He proceeded on this route with a dedication and sense of responsibility that exceeded the intensity and political acuity of any other period.

Numerous scientists urged in 1946, after the United Nations had been founded, that all atomic weapons be assigned to the U.N. Security Council and that sufficient inspection be agreed upon to prevent circumvention. Einstein's broader concern showed in his objection to this direction, because it did not confront the issue of a sovereign state's ability to use war as an instrument of policy. He insisted, as before the war, that only a world government deriving its authority from directly elected representatives could bring an end to wars. And he clearly differentiated between the U.N. and a world government; the custodian of the world's nuclear weapons could only be the latter. Initially, Britain, the United States, and the Soviet Union should form the first stage of a world government to which all their military resources short of those needed for police action would be assigned. In his planning, a world government would also have to have the right and the power to intervene within a country in which a minority was oppressing a majority.

At no time did Einstein seem to reflect on the enormously complex problem of transferring military power from powerful nations to a world authority, nor did he seek out possible ways in which loyalty, effectiveness, and willingness to risk life could be maintained in a new organization. He was not ignorant of this crucial problem; rather, he feared tyranny by a world government much less than obliteration by nuclear war. And he believed that there must be some way of passing through the transitional period safely, if it were done quickly, and he often referred to the dispatch with which the United States was formed from the colonies, to the swiftness of India's change under Gandhi, and to the almost immediate vanishing of the institution of slavery once nations agreed to abolish it.

Early in public discussions of this approach, Sumner Welles argued that the Soviets would never participate, because such a world order would mark the end of the particular system they were developing. And a year later Andrei Gromyko indeed made this clear, although it was probably inevitable in any event, because by that time the Cold War was beginning to create a contentious environment. Scientists were scattered across this argument from one end to the other, but with the majority arrayed near the middle, favoring Einstein's approach but urging that the U.N. be used until the deep economic and political gulf that separated East and West could be narrowed.

Beyond this very basic argument, which Einstein was to carry on for years with diminishing support, he busied himself with more immediate matters. He tried unsuccessfully to persuade Soviet scientists to contribute to a public education effort focused on preparing a book, *One World or None,* for large circulation. He lent his prestige to efforts to declassify the enormous, war-induced backlog of classified scientific reports. And he examined, but could not accept, the Baruch Plan put before the U.N. by the United States in March 1946. Despite the radical nature of this proposal, it was still a composite of interim measures that characteristically were not enough for him. Although recognizing the virtue of world government and requiring the abandonment of the veto in the Security Council when dealing with nuclear matters, it was not a balanced, comprehensive program for abolishing war, building world government, and creating a powerful world judiciary. Later he agreed too that the staged process by which American commitments later in the process were contingent upon Soviet commitments being made early was unfair.

Here, as on other occasions, very large steps in the direction of his own thinking were not taken as elements of compromise but were instead politely rejected. Perhaps it was the mindset of a physicist that intruded or, more especially, the insistence that a complex problem must have a simple solution. In any event, courage and the longing for simplicity won out over compromise; the cost was a decline and narrowing of his support. Einstein never deluded himself on such matters. In the end, he doubted that any progress could be seen as the result of his efforts in the first decade of the Nuclear Age. But he would not embrace a more popular position if it weakened essential pillars in his conception of how the world should be.

By 1947 it became unmistakably clear that Einstein's conception of world government would not begin to take shape in his lifetime or become likely for a long time thereafter. He was not alone in this recognition. The shock of World War II, the climactic entrance of nuclear weapons into the world, and the widening sense that colonialism in which most of the world's peoples still lived was at an end—all of these created a willingness to look at new forms of world order as never before. Yet the divisiveness of the East-West split won out. In the mists of the Cold War, decisions were reached on both sides that the system of nation-states would continue and that nuclear weapons, despite the millionfold difference they make in explosive power, must be contained within the divided community of competitive sovereign states and not transferred to untried and untested world authority.

Efforts to expand the interest in world government continued nevertheless. Henry Usborne, a British Labour M.P., persuaded eighty other members of Parliament to support his version of a world government much like Einstein's. Indeed, it is a testimony to the visibility of world-government movements that the Soviets began to devote so much effort to criticizing it. The exchange of letters between Einstein and four Soviet scientists was a high mark in this debate. Although for several years it was argued that American policy was responsible for inducing Soviet insecurity and that Soviet thinking might change if this provocation were removed, the die was cast with respect to this issue, and Einstein's growing pessimism testified to this.

Nineteen hundred forty-eight appears as a climactic year. Henry Wallace's drive for the presidency strained or broke many peace-oriented alliances. There was even talk of preventive war. And later one learned that Stalin's turn toward greater oppression was beginning. For those committed to world government, the outlook was bleak, for no convincing strategy could be devised that would allow world-government initiatives to begin without involving the Soviet Union, and yet any initiative that did not involve the Soviets would be seen as a coalition against them. And the Soviets did not help by floating a proposal for general and complete disarmament without mutually guaranteed controls or, better in Einstein's view, supranational controls. It is not surprising that Einstein conceded that world government could not come overnight.

By his seventieth birthday in 1949, Einstein had withdrawn from many public activities. He had seen the the futility of the many international congresses on this subject, especially as they were almost routinely Communist-run. The Emergency Committee of Atomic Scientists, which he had chaired and which aimed at public education, had fallen on financial bad times. His own prescription now took the form: "To awaken people we must start with the conviction that security against military disaster can be realized only through patient negotiation."

Einstein retained in his last years an unshakable faith in the necessity that world government must in time come into being. His concern with individual liberties was heightened in this period: he urged noncooperation, rather than pleading protection by the Fifth Amendment, with congressional investigating committees. He believed that rationality was draining away from the American political scene and that the United World Federalists, the organization closest to his own views, was contracting to a small, ineffectual group. His fear of German rearmament under the NATO alliance came to a high pitch, clearly resonating with his early formed dislikes from the First World War and the Nazi period. Einstein's foreboding reflected a widespread mood in scientific circles. Despite the termination of the Korean War and the anticipation of an East-West thaw, the newly arrived hydrogen bomb and reports of its enormous power from tests in the Pacific cast a long shadow. With the failure to control fission bombs so evident, the challenge to face a new class of weapons up to a thousand times more powerful was met with increasing dismay.

It was in this situation that Bertrand Russell initiated an exchange of letters with Einstein early in 1955, searching for a way to reverse the trend. Their dialogue developed into a joint statement, the last document to which Einstein put his signature before his death. It was an eloquent statement of their perception of the erosion of the human condition and an appeal to begin the abolition of war by renouncing nuclear weapons as a part of East-West arms reduction. It urged that scientists assemble to appraise the new peril and plan its resolution along these lines.

∞

This revival in modern and urgent form of Einstein's long-held insistence that war must be abolished as an instrument of national policy is indeed his legacy to succeeding generations of scientists. A brief examination of how the scientific community responded tells a great deal about the interaction of scientists with the problem of nuclear war over the last quarter century.

Very briefly, the Russell-Einstein Manifesto consisted of statements defining the peril and outlining steps to be taken to counter it. The manifesto asserts that the peril transcends politics and puts in question the survival of the species. It warns of the long-term effects of fallout and the unappreciated scale of immediate and post-attack fatalities and injuries and warns that those scientists who are most knowledgeable are the most gloomy. The challenge is put: end war or end the human race. Distasteful limitations on national sovereignty would be required. Since any (presumably) East-West war would lead to the use of nuclear weapons, these must be renounced as a part of general arms reduction because this would greatly reduce tensions and the fear of a surprise attack. East-West issues cannot be decided by recourse to war.

The response to this appeal was the creation in 1957 of an organization known as Pugwash, or, more formally, as the Pugwash Conferences. The conference held in Mexico City in 1979 was the twenty-ninth in this series. These conferences and other developments under the Pugwash rubric represent the largest and most sustained attempt of the international community of science to respond to the threat of nuclear war; hence, a few observations on the nature and changing character of this response seem in order, because the conferences reflect the best documented attempts of an important, though ill-defined, sector of the scientific community to grapple with the central issue Russell and Einstein posed.

During the first decade (1957-1966) the conferences were dominated by the novelty of the experience they provided for the exploration of the arms control and disarmament terrain by Soviet and Western scientists coming to know each other for the first time. Leading scientists were well represented, particularly from Britain, the United States, and the Soviet Union. As a rough indication of the extent to which leading scientists were involved, two-thirds of those who participated in three or more of the sixteen conferences held in this first decade were members of their respective academies.[4] Governments were intensely interested in the proceedings and private discussions, and heads of state often sent messages to the opening session. The usefulness of the discussions were such that a bilateral group was formed by the American Academy of Arts and Sciences and the Soviet Academy of Sciences for more intensive discussions of specific arms control measures. This group met annually for a decade (1964-1975) in addition to and often reporting to the Pugwash Conferences.

An evaluation of such discussions can only be approximate at best. As examples of issues dealt with that may have had important consequences, the hot-line communication between Moscow and Washington, the limited nuclear test ban, and the banning of antiballistic missile defenses stand out. The discussion of banning antiballistic missiles began in earnest at the 1964 conference in Udaipur. It continued in the conferences and in bilateral meetings until the treaty was signed in 1972, but it was in 1967 that some Soviet scientists first came to support this approach and conveyed their estimate that it could be adopted as a national policy. The actual decision to negotiate such a ban probably depended on three factors: the instability that the deployment of a workable antiballistic missile system would create, the likelihood that the performance of real systems would be low and uncertain, and the estimate that the Soviets may have made that competing in an area of such high technology was not in their interests. One cannot know how much these unofficial discussions contributed to policy decisions, but it is certain that the attention that the problem received far exceeded that in any other forum at the time. To be balanced, it must be noted that numerous other ideas put forward have not led to any useful outcome. The most potentially useful were those on MIRVs (multiple independently targeted reentry vehicles), begun in 1967, but these did not lead to any significant attempts to control this very destabilizing development in nuclear weaponry.

The second decade of Pugwash saw numerous changes. Political scientists began to play a larger role: G. Arbatov, H. Bull, M. Howard, N. Inozemtsev, H. Kissinger, H. Morgenthau, and M. Shulman were frequent participants. The focus on arms control broadened to include East-West cooperation and regional security and then to study a fundamental cause of war: underdevelopment and overpopulation. The role of science and technology in development is now the second major theme of Pugwash. These changes meant that the role of scientists in devising means to prevent nuclear war—the original charge of the Russell-Einstein Manifesto—diminished. In part this was due to the increased attention this problem was receiving at governmental levels. But it was accelerated by the passing of many leading Soviet scientists within a short span: Artsimovitch, Blagonravov, Millionshchikov, Tamm, and Topchiev. Not only was the particular combination of talent and concern that these scientists represented not replaced, but in the West, too, one did not see successors to Cockcroft, Rabinovitch, and Szilard. With these developments came a drop in academy representation from two-thirds to one-half in the second decade and less in the most recent conferences. On the other hand, much has been gained from more vigorous and informed participation from Western Europe, East Germany, Egypt, Israel, Poland, and other nations. Twenty-eight nations now have national Pugwash groups.

Clearly, any such organization requires renewal and adaptation. For Pugwash this may come from its symposia and workshops, which began early in the second decade (1968) and now number nearly forty. These focus on specific topics and sometimes involve only regional representation. As such, they allow much more intensive examination of a problem and have occasionally resulted in important publications. The total effort in Pugwash seems to be shifting from the conferences to symposia and workshops, and this format may become the basis of a new style and influence.

Nevertheless, the future of Pugwash is clouded. The dilution of its original charge by the enlargement of its agenda, by its expansion to much wider national representation, and by the lowered concern and activism among the coming leadership generation in the global scientific enterprise suggests a much more pluralistic future and the likelihood that Pugwash will play a smaller role in arms control and disarmament. Indeed, this ambivalence mirrors the scientific community itself. Today the clarion call of the Russell-Einstein appeal is muted. The central problem is more complex, the investment of time needed to become proficient has grown enormously, and the irrational complications due to unforeseeable currents in domestic and international politics have diminished the intellectual attractions of involvement. Yet, if the present deterioration of East-West relations continues and if the challenge this presents reawakens in the new generation the profound concern and commitment that invested nearly all of Einstein's life, a new cycle will have begun.

∞

In reflecting on how the consciousness and special abilities of the future scientific community may be brought to bear on the permanent problem of preventing nuclear war, a quite subjective review may be in order. It emphasizes how much the nature of the problem has changed since Einstein's death and yet how central his main concern remains.

The assessment of our present position must begin with the acceptance that nuclear weaponry, like fire for the ancients, is here to stay. With tens of thousands of weapons now deployed and with an additional nation attaining nuclear capability on the average of every five to ten years, there is no possibility of turning back. The avoidance of the large-scale use of weapons has become a permanent requirement for the continuation of civilization.

By contrast, an observer in space looking over the whole of the earth and the whole of this century may tend to draw a different conclusion: that since the advent of nuclear weapons in 1945 the intensity and destructiveness of wars on earth have greatly diminished. The great convulsive world wars, in which nearly half of the people killed in all the recorded history of warfare died, have seemingly ended. Indeed, if the intensity of warfare continues in its present pattern to the end of this century, and taking into account the growing world population, one can conclude that a person living in the last half of this century has only about 8 percent of the chance of being killed in a war between nations that a person living in the first half had. If this correlation represents a causal relation, it is of course a benefit that is bought at the price of enduring the risk of an outbreak of wide-scale, highly destructive nuclear warfare. This is the nature of the bargain that an inquisitive mankind has struck.

In this first third of the century of the Nuclear Age there has come into view a firebreak between conventional war and nuclear war that appears to be increasingly widely accepted. This keeps conventional warfare from escalating to a large, quasi-global scale because the interests that become involved at a large scale are such as to threaten to jump the firebreak into nuclear warfare. A number of other collateral conditions help to stabilize this firebreak. For example, international trade has increased more than twentyfold (in constant dollars) since before World War II. Nations are vastly more interdependent: consequently, the interruption of trade that would accompany the onset of large-scale warfare stands as a persistent deterrent to such warfare.

With the harsher remnants of the Cold War dissipating by the end of the 1950s, more attention was given to the reinforcement of the cumulating restraints against the outbreak of nuclear war. The growing destructive potential of the nuclear arsenals of the two camps was itself a mighty deterrent. But beyond this, there developed the conscious strategy to negotiate limits and restraints on the nuclear weapons of both sides so that they would become less prone to use, less numerous, less rapidly improving, and even less costly. And if war did come, there was the hope that arrangements made beforehand would diminish the damage that would be done before such a nuclear war could be terminated.

This arms control approach, which has many aspects and many levels of operation, has come to replace the abrasive, dangerous strategy of confrontation of the Cold War years. It represents a new direction in East-West affairs, but the difference it actually makes is not clear and depends on how fast it is exploited—which is a mutual decision—and to what extent other events interfere with its execution.

The three principal levels of arms control efforts have been: to ban nuclear weapons from regions where they have not been deployed; to limit the testing, improvement, or production of certain weapons systems; and to freeze the numbers of nuclear weapons systems by category, then to negotiate maximum numbers allowed, and subsequently to reduce numbers. Since the first such treaty in 1959, which banned nuclear weapons from Antarctica, eighteen treaties and agreements have been completed that move in varying degrees in this general direction.

The techniques by which nuclear arms could be controlled were found to depend almost entirely upon the possibility of devising means of verifying the existence of such weapons on the other side without on-site inspection. This is a severe restraint, and strategic arms control has been possible only by developing surveillance satellites of extraordinary discrimination and by agreeing that control should be directed toward the weapons systems that launch the weapons rather than the weapons themselves, since the latter are too small to allow observation by surveillance. Thus, negotiation proceeded in terms of ballistic missiles and heavy bombers. This was the course taken in SALT I (banning antiballistic missiles), the Interim Agreement (freezing ballistic missile numbers) of 1972, and the SALT II Treaty now facing an uncertain future.

The weapons systems that need to be controlled are, of course, continuously changing in character and in number. Twenty years ago the expectation was for both sides eventually to amass roughly one thousand weapons deliverable at intercontinental range. Even with considerable attrition, this would have allowed either side to destroy the other's urban and industrial structure, and, since such a blow could be delivered after absorbing the other's attack, this second-strike capability served as an invulnerable deterrent to either side initiating such an attack.

This relatively stable situation began to erode about a decade ago with the emergence of the MIRV concept, whereby a number of warheads can be mounted in a single missile nose cone and delivered independently at various targets within a wide band. Parallel to this, another development was taking place: accuracy has improved to the point where the precision of delivery is of the same order as the size of the crater created by the explosion; that is, accuracy is reaching its practical limit. These developments have two consequences. First, the number of weapons commensurate with a thousand ballistic missiles has greatly increased, by roughly a factor of ten, meaning that many more targets than cities and industries can be covered. Hence, new strategies have been devised that target military units; the list of these is much longer than that of cities and industries. Second, the improved accuracy makes it possible to plan to attack the land-based ballistic missiles of the other side, those that had heretofore been of a secure, second-strike nature. The result is that the number of strategic warheads has leapt by an order of magnitude to 10,000, and land-based forces, the most inherently reliable part of the strategic force, have become in theory vulnerable.

These technological advances clearly complicate the arms control process. Indeed, it is becoming evident that arms control negotiations will have to move faster just to stay even with the technological advances. And those advances now coming into view present an equally difficult dilemma. The nuclear-armed, long-range cruise missile represents such a shrinking of the delivery system toward the size of the weapon itself that it may not be subject to verification by present means. And the emergence of mobile ballistic missiles may create an equally intractable verification problem.

Taken by itself, the SALT II Treaty accomplishes a great deal. It puts ceilings on the aggregate number of long-range delivery systems and sublimits on the number of those that can be MIRVed; it puts a ceiling on the number of MIRVs that various missiles can carry, thereby cutting that number off at about half of what it could otherwise reach; it limits new delivery systems to one for each side over six years, whereas they have been appearing at the rate of about one in every two years on the average; and it establishes a number of practices that make verification easier and more reliable. But judged against the gradient of constantly improving technological capability, the gains are not great, and in some important cases the treaty does not restore the situation at the end of SALT I in 1972.

Thus it follows that effective arms control will hang in the balance for some time even if the SALT II Treaty is ratified. The test will come when the decisions are made as to what to include in the SALT III negotiations or in a substantially revised SALT II. With the delays encountered in completing SALT II and with continuing technological progress, only partly sketched above, more and more vital issues have accumulated for attention in SALT III. If even the minimum are to be dealt with, the agenda will be very demanding indeed. To do less—and there are many advising a low-expectation approach on both sides—will demote the arms control approach to an exercise in consultative arms management, an exercise devoid of purpose and one that only promises a gradual raising of the present state of risk.

An important situation that favors arms control is the fact that it has become more and more integrated into the military planning of both sides, into the alliance arrangements in each bloc, and into Soviet-American relations. The defeat or demotion of arms control would affect all of these, for the most part adversely. On both sides, military expenditures are near the limit of what the publics are willing to support short of the stimulation that a severe East-West crisis or the introduction of new levels of coercion would provide. SALT II defines the military threat each side must position itself against, and meeting it will be easier and cheaper with SALT II in place. Despite some mild objections the allies of the United States are solidly in favor of the treaty; its demise would be a severe blow to the alliance relationship, and many doubt that such a breach could be repaired. Because of gradual erosion in our relations with the Soviet Union, the SALT II negotiation has come to represent the largest component of Soviet-American diplomatic business. Communication between the two sides in this channel, where with increasing frequency matters of considerable importance and sensitivity are discussed, is highly regarded as a means of resolving conflicts as well as improving the understanding of each other's positions. Collectively, these collateral features of the SALT process justify much of the actual negotiations and leave a positive residue when the agreements themselves fall short of objective need. Although one could wish for more, this achievement is a vast advance over earlier times, when neither the president in diplomatic channels nor Western scientists in contact with their Soviet counterparts, as envisioned in the Einstein-Russell Manifesto, could elicit any meaningful discussion of how to deal with the threat that nuclear war presented to both societies and the world.

∞

It is evident that the state of affairs with respect to the risk of nuclear war is, at least superficially, much different now than a quarter of a century ago, when in his last years Einstein became increasingly gloomy about the prospect. The military machines that project this threat are ever so much more awesome, but the dialogue over their control is considerably advanced, and there are some real accomplishments in the agreements that have been forged. However, at least four general problems fill the immediate horizon with heavy clouds. One is the failure thus far to curb the rate of technological advance to which we have referred and to begin a renovation of our arsenals in order to reduce the numbers of weapons and to increase the stability in crisis of those that remain. The second is the requirement to reach understandings between the two camps on the permissible level of intervention in the world's trouble spots; such areas are the tinderboxes of wars into which the two opposing superpowers can be drawn. The blocking of this route to wars that cross the firebreak is in the view of many as important as the technical control and reduction of nuclear weapons. The third problem is the preservation of the firebreak that separates conventional from limited nuclear war. The fourth is the need to keep the number of countries having nuclear arsenals to a minimum, since the danger in any crisis is compounded by the number of lethal actors.

Were Einstein alive today, he would probably be somewhat surprised that events have not developed in a worse fashion, but he would have little satisfaction with the present situation. The modus vivendi by which the risk of nuclear war is now restrained has none of the attributes that a physicist, or Einstein in particular, would find attractive. It is a sprawling, vague, uncertain effort interlaced with contradictory or unattended elements, difficult to judge and offering only modest hope for the long-term future. But by gradually establishing a tradition of non-use of nuclear weapons, one that in Einstein's time was not widely expected, it has succeeded in preventing large world wars, and the likelihood of nuclear war seems to have receded in the public mind, to judge from its disappearance from opinion polls listing current anxieties.

These aspects are tentative and may be illusory, but they have become the main feature of this era of the Nuclear Age. More important is the promise of more secure arrangements if the present process can be pursued with more determination and with a growing sense that Einstein's stricture remains the truly essential goal for mankind, that "above all nuclear war must be abolished from the instruments of nations." In terms of international treaties, this goal was not to be achieved in his lifetime, nor will it be in ours. But with the dedication to the common good that he exemplified and a commitment to diminishing both the physical potential and the political relevance of nuclear weapons, life on this planet may become increasingly secure. After all, it would be no less a high tribute to the human race that it is able to live in perpetuity with nuclear weapons but without their use, than simply without the weapons. And long hence, when most current political passions have subsided, nuclear weapons may find their natural abode within a carefully crafted world institution whose political structure has the stability, the sagacity, and the competence to take from humankind this enormous burden. Then Einstein may rest in peace.

NOTES

1. Otto Nathan and Heinz Norden, eds., *Einstein on Peace*, New York, Simon and Schuster, 1960, p. 2. Further quotations from this volume will be followed by parenthetical page references.

2. See Vannevar Bush to Frank Aydelotte, 30 December 1941, Technical Document no. 331, U.S. Atomic Energy Commission. See B. T. Feld, *Bulletin of the Atomic Scientists* **35**, no. 3 (1979), pp. 5-16.

3. Ronald W. Clark, *Einstein: The Life and Times*, New York and Cleveland, World Publishing Co., 1971, pp. 705-706, 709.

4. These data are derived from compilations by Joseph Rotblat, *Pugwash*, Prague, Czechoslovakia Academy of Sciences, and the *Pugwash Newsletter* of May 1977. See also J. Rotblat, *Scientists in the Quest for Peace*, Cambridge, Mass., MIT Press, 1972.

Bernard T. Feld

EINSTEIN AND THE POLITICS OF NUCLEAR WEAPONS

ALBERT EINSTEIN was the most highly appreciated and, at the same time, the least understood scientist of this era. For over a third of a century—from the end of the First World War until his death in 1955—he represented to the people of the world the epitome of the modern scientist. He was respected, even revered by all, from the Indian day laborer in the slums of Calcutta to the Queen Mother of the Belgians. His visage, from which honesty, kindliness, integrity, and wisdom glowed as from a beacon, was as well known, and for a longer period of time, as that of even the most charismatic movie idol.

Yet Einstein's modesty, frequently expressed with self-deprecating humor, was as genuine as it was legendary. In his response to a birthday greeting on his last living birthday from his very old friend and erstwhile fellow chamber-music player, the Belgian Queen Mother, he confessed, "the exaggerated esteem in which my lifework is held makes me feel very ill at ease. I feel compelled to think of myself as an involuntary swindler. If one attempts to do anything about this, one succeeds only in making matters worse."[1] Indeed, the ability to look at his notoriety with humor never left him, as, for example, in his acceptance of an award for constructive nonconformist thinking:

> It gives me great pleasure, indeed, to see the stubbornness of an incorrigible nonconformist warmly acclaimed. To be sure, we are here concerned with nonconformism in a remote field of endeavor [science]; it is a field in which no Senatorial committee has as yet felt compelled to tackle the important task of combating the dangers that threaten the inner security of the uncritical or intimidated citizen.

> As for the words of warm praise addressed to me, I shall carefully refrain from disputing them. For who still believes that there is such a thing as genuine modesty? I should run the risk of being taken for no more than an old hypocrite. You will surely understand that I do not find the courage to brave this danger. (591-592)

Also evident in the first part of this response is another of Einstein's incorrigible traits: his inability to think of himself as a serious contributor in any field other than theoretical physics. To the New York department store executives who were awarding him the prize, Einstein's nonconformism undoubtedly revolved about his lifelong opposition to militarism, his passionate pleas against the domination of the state over individuals and their consciences, his militant pacifism in post–World War I Germany, and his outspoken internationalism and espousal of world government in post–World War II McCarthyite America. Yet, to Einstein, his claim to noncomformist status seemed to rest solely on his youthful break with the scientific past (relativity), his middle-aged insistance on the sanctity of causality in a quantum-dominated period, and his stubborn pursuit of the holy grail of a unified field theory of gravitation and electromagnetism, which dominated the final decades of a life in which science was his abiding pleasure and consolation. In a world that constantly cried out to be saved, an overdeveloped conscience simply would not permit Einstein to devote himself exclusively to physics, his true and eternal love. And yet, the only activity in which he could ever take his contributions seriously was his science.

This intellectual dichotomy is poignantly illustrated by the "Autobiographical Notes," which Einstein prepared in 1946 for a volume dedicated to his seventieth birthday.[2] The notes start out in an autobiographical vein—"Here I sit in order to write, at the age of 67, something like my own obituary"—and follow his life until roughly the age of seventeen, when Einstein commenced his studies of physics in Zurich. From that point on, however, the notes evolve into the most lucid exposition of the history of physics in the twentieth century. Of course, since so much of this history bears the imprint of Einstein's genius, we obtain at the same time a fascinating picture of Einstein's intellectual development. But from the point of view of a biographer, what emerges clearly is that, as far as Einstein himself was concerned, once physics came in the door, the world of human affairs and events flew out the window.[3]

To the general public, of course, Einstein epitomized much more than pure science. Relativity represented to the popular mind the romance of the cosmos—and, to a considerable extent, it still does, though lately it has become rather vulgarized in the mass-media exploitation of visions of space travel. When in 1919 a British eclipse expedition led by Arthur Stanley Eddington observed the bending of light beams in the gravitational field of the sun—exactly as predicted by the general theory of relativity—the announcement created a scientific and popular sensation in a world exhausted by the ordeal of the First World War. Thus, in a letter from a British to a German scientist: "The talk here is of almost nothing but Einstein, and if he were to come here now I think he would be welcomed like a victorious general" (27, 29).

If relativity captured the public imagination in the period between the wars, the atomic bomb, which ended World War II, propelled Einstein even more directly into the limelight, but under circumstances that caused him acute moral, as well as personal, embarrassment. To the popular mind, Einstein was, and to a large extent remains, the "father of the atomic bomb." Is there any person so ignorant that he or she does not recognize his famous discovery of the equivalence between mass and energy, $E = mc^2$, which provides the theoretical basis for the release of nuclear energy? Add to this Einstein's audacious intervention with President Franklin D. Roosevelt to convince him of the military potential of the discovery of nuclear fission, and you have the two ingredients of that aspect of the Einstein legend that he worked most diligently, and unsuccessfully, to debunk.

Einstein's role in the successful American effort during World War II to develop a nuclear bomb was indeed peripheral. On this score, he tried hard to put the record straight. Thus, in response to a request from the editor of a Japanese journal demanding to know why Einstein, "as a great scientist who played an important role in producing the atomic bomb, . . . co-operate[d] in [its] production . . . although you were well aware of its tremendous destructive power?" he protested:

> My participation in the production of the atomic bomb consisted of one single act: I signed a letter to President Roosevelt, in which I emphasized the necessity of conducting large-scale experimentation with regard to the feasibility of producing an atom bomb.
>
> I was well aware of the dreadful danger which would threaten mankind were the experiments to prove successful. Yet I felt impelled to take the step because it seemed probable that the Germans might be working on

the same problem with every prospect of success. I saw no alternative but to act as I did, *although I have always been a convinced pacifist.* (583-584)

Yet the myth persisted that through his earlier discovery of the fundamental mass-energy relationship, Einstein somehow bore responsibility for the final outcome. With great patience, he explained:

Now you seem to believe that I, poor fellow that I am, by discovering and publishing the relationship between mass and energy, made an important contribution to the lamentable situation in which we find ourselves today. You suggest that I should then, in 1905, have foreseen the possible development of atomic bombs. But this was quite impossible since the accomplishment of a "chain reaction" was dependent on the existence of empirical data that could hardly have been anticipated in 1905. But even if such knowledge had been available, it would have been ridiculous to attempt to conceal the particular conclusion resulting from the Special Theory of Relativity. Once the theory existed, the conclusion also existed and could not have remained concealed for any length of time. As for the theory itself, it owes its existence to the efforts to discover the properties of the "luminiferous ether"! There was never even the slightest indication of any potential technological application. (623)

Even now, some forty years later, the importance of Einstein's role in the American development of nuclear weapons before the end of the war against Japan remains a source of legitimate controversy. Some feel it was seminal, that without his intervention the atomic bomb project would never have received the government attention and support that was indispensable to its success. Others believe that the physics was so compelling that once the early (and very inexpensive) demonstrations of the potentiality had taken place—and they would probably have happened even without government support—not even the most obtuse bureaucracy could have resisted the inevitable pressures of the cream of the American physics community.

The actual facts are now well known, and, irrespective of one's interpretation, it remains a fascinating story. The main actor in the drama was Leo Szilard, a student and colleague of Einstein's final decade in Berlin.[4] It was Szilard who approached Einstein for help in alerting the American government to the danger that the Nazis might turn the advantage of Otto Hahn's and Fritz Strassmann's discovery of the fission process into a weapon that could guarantee Hitler's domination of the world.

Actually, the original concern of Szilard and the other refugee scientists who were deeply concerned with the possibility of German progress toward a nuclear weapon[5] was that, with Czechoslovakia in German hands, the Nazis might be moving to obtain control over all available stocks of uranium. Since the major source was then from the Belgian Congo, Szilard and Wigner hoped that it might still be possible to arouse the Belgian government to the necessity of keeping its uranium stocks out of German hands. Szilard, recalling Einstein's longstanding friendship with Queen Elizabeth, thought that the best way to reach the Belgian government would be through a letter from Einstein to the queen. However, after some discussion, it was decided that it was more important to alert the American government and to seek its support in initiating research that could, on the one hand, settle the issue of the feasibility of a weapon based on nuclear fission and, on the other, if it did prove to be feasible, to ensure that the United States would be able to produce it ahead of Germany.

Szilard had found an intermediary, Dr. Alexander Sachs, an economist and unofficial advisor to President Roosevelt. On the basis of a shorter draft dictated in German by Einstein to Teller, Szilard prepared the following letter:

Sir:

Some recent work by E. Fermi and L. Szilard, which has been communicated to me in manuscript, leads me to expect that the element uranium may be turned into a new and important source of energy in the immediate future. Certain aspects of the situation seem to call for watchfulness and, if necessary, quick action on the part of the Administration. I believe, therefore, that it is my duty to bring to your attention the following facts and recommendations.

In the course of the last four months it has been made probable—through the work of Joliot in France as well as Fermi and Szilard in America—that it may become possible to set up nuclear chain reactions in a large mass of uranium, by which vast amounts of power and large quantities of new radium-like elements would be generated. Now it appears almost certain that this could be achieved in the immediate future.

This new phenomenon would also lead to the construction of bombs, and it is conceivable—though much less certain—that extremely powerful bombs of a new type may thus be constructed. A single bomb of this type, carried by boat or exploded in a port, might very well destroy the whole port together with some of the surrounding territory. However, such bombs might very well prove to be too heavy for transportation by air.

The United States has only very poor ores of uranium in moderate quantities. There is some good ore in Canada and the former Czechoslovakia, while the most important source of uranium is the Belgian Congo.

In view of this situation you may think it desirable to have some permanent contact maintained between the Administration and the group of physicists working on chain reactions in America. One possible way of achieving this might be for you to entrust with this task a person who has your confidence and who could perhaps serve in an unofficial capacity. His task might comprise the following:

a) To approach Government Departments, keep them informed of the further developments, and put forward recommendations for Government action, giving particular attention to the problem of securing a supply of uranium ore for the United States.

b) To speed up the experimental work which is at present being carried on within the limits of the budgets of University laboratories, by providing funds, if such funds be required, through his contacts with private persons who are willing to make contributions for this cause, and perhaps also by obtaining the cooperation of industrial laboratories which have the necessary equipment.

I understand that Germany has actually stopped the sale of uranium from Czechoslovakian mines which she has taken over. That she should have taken such early action might perhaps be understood on the ground that the son of the German Under-Secretary of State, von Weizsäcker, is attached to the Kaiser Wilhelm Institut in Berlin, where some of the American work on uranium is now being repeated.

Yours very truly,
(signed) A. Einstein

Although signed by Einstein on 2 August, the letter was not delivered until 11 October. Roosevelt responded promptly on 19 October with the appointment of a committee "to thoroughly investigate the possibilities of your suggestion regarding the element of uranium." But the government moved very slowly, and a worried Szilard prevailed upon Einstein to communicate with the president again in March 1940. As a result of this second letter, the president enlarged his advisory committee, which then proceeded to provide the funds and organization necessary for the experiments proposed by Fermi and Szilard. These culminated in Chicago on 2 December 1942 with the first man-made nuclear chain reaction.[6]

After his second intervention in early 1940, Einstein had no further connection with the uranium project,[7] which soon came under the control of the Manhattan District of the U.S. Army.

Considering the strict security and compartmentalization procedures instituted by General Leslie R. Groves, it is in fact highly unlikely that Einstein had any specific knowledge of the status of the project during the war. However, he must have obtained some inkling that there had been progress, and possibly even that there was some prospect of success, when Szilard again called on his help in March 1945. This time Szilard wanted a letter of introduction to President Roosevelt, which Einstein provided:

> The terms of secrecy under which Dr. Szilard is working at present do not permiit him to give me information about his work; however, I understand that he now is greatly concerned about the lack of adequate contact between scientists who are doing this work and those members of your Cabinet who are responsible for formulating policy. In the circumstances, I consider it my duty to give Dr. Szilard this introduction and I wish to express the hope that you will be able to give his presentation of the case your personal attention.[8]

The letter was received by the president, but he did not have the opportunity to act on it before his death. (It is tempting, if fruitless, to speculate on what might have been the course of history had Szilard succeeded in conveying to Roosevelt his concerns and those of his colleagues about the future effects on world politics of uninhibited American use of an atomic weapon in World War II.) Although Szilard finally succeeded in seeing James F. Byrnes, President Truman's choice for secretary of state, on 28 May 1945, the stage was already completely set for the testing and use of America's first nuclear weapons against Japanese cities. Furthermore, even if Byrnes had understood Szilard's message (a petition for a demonstration on an uninhabited target before using the atomic bomb against Japanese civilians), which apparently he did not, his subsequent account of the times indicates that he would not at all have been inclined to heed it.[9]

In any event, Einstein learned of the success of the Manhattan Project as did the rest of the world—by the public announcement on 8 August 1945 that a single bomb using the fission of an isotope of uranium had destroyed the Japanese city of Hiroshima. Einstein is alleged to have responded to this news with the time-honored lament, *"Oy Weh!"* (woe upon us).

There can be no question as to where Einstein would have stood with respect to the dropping of the bombs on Japan. In a letter of 19 March 1955 to Max von Laue, one of the few German physicists with whom he retained a lifelong friendship, he explained:

My action concerning the atomic bomb and Roosevelt consisted merely in the fact that, because of the danger that Hitler might be the first to have the bomb, I signed a letter to the President which had been drafted by Szilard. Had I known that that fear was not justified, I, no more than Szilard, would have participated in opening this Pandora's box. For my distrust of governments was not limited to Germany.

Unfortunately, I had no share in the warning made against using the bomb against Japan. Credit for this must go to James Franck. If they had only listened to him! (622)

The use of the bomb against Japanese civilians was especially painful to Einstein, because he had formed very strong friendships and a special fondness for the people during his visit to Japan in 1922. Not only was there not a single racist bone in his body, but Einstein never held the Japanese people responsible for the war in the sense that he held the German people responsible. Thus, in an uncharacteristically harsh response to an appeal for a "soft peace" from his old friend and fellow refugee James Franck in December 1945, he wrote:

I remember too well the campaign of tears staged by the Germans after the First World War to be fooled by its repetition. The Germans slaughtered several millions of civilians according to a well-thought-out plan. If they had murdered you too, some crocodile tears would undoubtedly have been shed. The few decent people who are among them do not change the picture as a whole. I have gauged from several letters received from over there, as well as from the information supplied by some reliable persons, that the Germans do not feel one iota of guilt or sorrow. . . . Dear Franck, do not involve yourself in this dirty mess! They will first misuse your kindness and then they will ridicule you for being a fool. But if I am unable to persuade you to refrain, I, for one, will certainly not get mixed up in this affair. Should the opportunity present itself, I shall publicly oppose the appeal. (367)

Nevertheless, he certainly would have opposed the bomb's use against Germany had it been ready in time,[10] as attested in a letter to a Japanese correspondent written on 23 June 1953:

I am a *dedicated* [*entschiedener*] but not an *absolute* pacifist; this means that I am opposed to the use of force under any circumstances, except when confronted by an enemy who pursues the destruction of life as an *end in itself.* I have always condemned the use of the atomic bomb against Japan. However, I was completely powerless to prevent the fateful decision for which I am as little responsible as you are for the deeds of the Japanese in Korea and China.

I have never said I would have approved the use of the atomic bomb against the Germans. I did believe that we had to avoid the contingency of Germany under Hitler being in *sole* possession of this weapon. This was the real danger at the time.

I am not only opposed to war against Russia but to all war—with the above reservation. (589)

Einstein's pacifism was, indeed, never of the doctrinaire variety. In the first instance, it seems to have developed from a great revulsion to the First World War and the chauvinistic attitudes that had brought it about. Thus, in March 1915 he wrote to Romain Rolland: "When posterity recounts the achievements of Europe, shall we let men say that three centuries of painstaking cultural effort carried us no farther than from religious fanaticism to the insanity of nationalism? In both camps today even scholars behave as though eight months ago they suddenly lost their heads" (13). But by the time the war was over, he was convinced that it was Germany that had to change:

In my opinion, Germany's salvation lies only in a swift and radical process of democratization, similar to the democratic institutions of the Western powers. It is only through the creation of a democratic constitution, whatever shortcomings it may possess, that one can effect a sufficient degree of decentralization of power to prevent a recurrence of the events of 1914. The present regime no longer enjoys confidence anywhere abroad.

This, briefly stated, represents my convictions. As one who was a pacifist even before the war, I feel I have the right to voice my beliefs now. (23)

As time passed and the League of Nations went from failure to failure in its attempts to establish a semblance of world order, Einstein's pacifism grew more confirmed and militant. Thus, in a visit to the United States in 1931, he stated in an interview:

I am not only a pacifist but a militant pacifist. I am willing to fight for peace. Nothing will end war unless the peoples themselves refuse to go to war.

Every great cause is first championed by an aggressive minority. Is it not better for a man to die for a cause in which he believes, such as peace, than to suffer for a cause in which he does not believe, such as war? Every war merely enlarges the chain of vicious circles which impedes the progress of mankind. A handful of conscientious objectors can dramatize the protest against war.

The masses are never militaristic until their minds are poisoned by propaganda. I agree with you that we must teach them to resist propaganda. We must begin to inoculate our children against militarism by educating them in the spirit of pacifism. The trouble with Europe is that her people have been educated on a wrong psychology. Our schoolbooks glorify war and conceal its horrors. They indoctrinate children with hatred. I would teach peace rather than war, love rather than hate.

The textbooks should be rewritten. Instead of perpetuating ancient rancors and prejudices, we should infuse a new spirit into our educational system. Education should begin in the cradle. Mothers throughout the world have the responsibility of sowing the seeds of peace into the souls of their children.

It may not be possible in one generation to eradicate the combative instinct. It is not even desirable to eradicate it entirely. Men should continue to fight, but they should fight for things worth while, not for imaginary geographical lines, racial prejudices and private greed draped in the colors of patriotism. Their arms should be weapons of the spirit, not shrapnel and tanks.[11]

But his pacifism, like his entire approach to politics, was pragmatic in nature. To Einstein, what was always important was the goal rather than the gesture. He held no illusions about the power of the individual swimming against the tide: it might be the only thing that conscience would permit a free spirit to do, but the individual must always bear firmly in mind the essential goal. This was especially clearly stated in a speech to the Los Angeles University of International Relations on 1 February 1932:

The greatest obstacle to disarmament has been the inability of most people to appreciate the enormity of the problem. Most objectives are accomplished in small steps. Think, for example, of the transition from absolute monarchy to democracy! But we are here concerned with an objective that cannot be attained slowly, step by step.

So long as the possibility of war exists, nations will continue to insist on the fullest possible military preparedness to insure their victory in a future war and will find it unavoidable to educate the young people in warlike traditions and narrow nationalist vanities. The glorification of the war spirit will proceed as long as there is reason to believe that situations may arise where that spirit will need to be invoked for the purpose of waging war. To arm means simply to approve and prepare for war, not for peace. Hence, disarmament cannot come in small steps; it must come about at one stroke or not at all.

To accomplish so profound a change in the life of nations, a mighty moral effort and a deliberate rejection of deeply ingrained traditions are required. Anyone who is unwilling to let the fate of his country, in the event of a conflict, depend without qualification upon the decisions of an international court of arbitration, and who is not prepared to see his country enter into treaties that provide for such a procedure without any reservations, is not really resolved to abolish war. This is a case of all or nothing.

It should be emphasized that previous attempts to insure peace have failed only because they were directed toward compromise solutions that were inadequate to cope with the task. Disarmament and security cannot be separated; they must come about simultaneously. Security will be achieved only when all nations commit themselves to abide by the decisions of an international authority. (163-164)

When Einstein became convinced of the enormous menace that Nazi Germany posed for the future of intellectual freedom and world peace, it was completely in character for him to turn to advocating the rearming of the democratic West, with a view to its collective opposition to Nazi hegemony:

The great task [of German intellectuals in exile] is to impress upon those countries that still adhere to the ideals of culture and freedom, the living conviction that it is terribly dangerous to witness events as passive spectators. Today it is still possible to crush those who have usurped power in Germany; it could still be done, without bloodshed, by economic means. Otherwise, within a few years, much sacrifice of human life will be unavoidable, and, even then, there will be no certainty of a favorable outcome. I have the impression that the enormity of the present danger is not properly appreciated abroad and that people are lulled by the hope that the specter will one day disappear of its own weight. . . . (222-223)

However, to most European pacifists, such a change amounted to gross heresy. To the barrage of outraged protest and criticism from his erstwhile pacifist comrades, Einstein replied with unfailing patience and understanding, as in his letter of 11 September 1933 to a Dutch friend:

I assure you that my present attitude toward military service was arrived at with the greatest reluctance and after a difficult inner struggle. The root of all evil lies in the fact that there is no powerful international police force, nor is there a really effective international court of arbitration whose judgments could be enforced. All the same, antimilitarists were justified in refusing military service as long as the majority of the nations of Europe were intent upon peace. This no longer holds true. I am convinced that developments in Germany tend toward belligerent acts similar to those in France after the Revolution. Should this trend meet with success, you may be sure that the last remnants of personal freedom on the continent of Europe will be destroyed.

While it is quite true that the deterioration of conditions in Germany is partially attributable to the policies of neighboring countries, there seems little purpose at this juncture in blaming them for these policies.

The plain fact is that the gospel of force and repression, currently prevailing in Germany, poses grave threats to the Continent of Europe and the independence of its inhabitants. This threat cannot successfully be combated by moral means; it can be met only by organized might. To prevent the greater evil, it is necessary that the lesser evil—the hated military—be accepted for the time being. Should German armed might prevail, life will not be worth living anywhere in Europe.

I believe, nonetheless, that even now it is not too late to avert war by preventing German rearmament through diplomatic pressure. *But such pressure will require absolute military superiority on the part of Germany's neighbors.* To destroy such superiority or to prevent its achievement is tantamount to betraying the cause of European freedom. (235-236)

It is far from surprising, therefore, that with World War II over and the intellectual battle lines forming—between those who would have taken advantage of the American nuclear monopoly to establish the dominance of American military influence over all national and international institutions and those who saw the atomic bomb as an opportunity finally to establish international institutions with the intention *and the power* to provide supranational control over future conflict and war—Einstein again plunged enthusiastically into the arena of public affairs. He firmly supported the fledgling atomic scientists' movement, organized into a Federation of Atomic Scientists, and encouraged its voice, *The Bulletin of the Atomic Scientists.*[12]

The battle in the immediate postwar days revolved about the atomic scientists' opposition to a bill before the American Congress. The May-Johnson bill, drafted by the War Department, would have perpetuated military control over all aspects of nuclear energy, both military and civilian. That battle was won with the establishment of a civilian Atomic Energy Commission, but the issues had by then shifted to questions of future international control of the atomic bomb and centered around a proposal originating in a Truman-appointed committee. The so-called Acheson-Lilienthal-Oppenheimer report was formalized for presentation to the United Nations by Bernard Baruch; hence this scheme is generally known as the Baruch Plan.

In support of the various national and international enterprises of the atomic scientists, Einstein agreed in May 1946 to serve as chairman of an Emergency Committee of Atomic Scientists.[13] Its program was based on an agreed set of principles concerning the dangers of atomic weapons:

These facts are accepted by all scientists:

1. Atomic bombs can now be made cheaply and in large number. They will become more destructive.
2. There is no military defense against the atomic bomb and none is to be expected.
3. Other nations can rediscover our secret processes by themselves.
4. Preparedness against atomic war is futile, and if attempted will ruin the structure of our social order.
5. If war breaks out, atomic bombs will be used and they will surely destroy our civilization.
6. There is no solution to this problem except international control of atomic energy and, ultimately, the elimination of war.

The program of the committee is to see that these truths become known to the public. The democratic determination of this nation's policy on atomic energy must ultimately rest on the understanding of its citizens. (395)

The immediate objective was to raise a fund of $1,000,000 in support of the various educational activities of the atomic scientists aimed at civilian national and international control of future nuclear energy activities. However, although the committee had a certain amount of success in raising money, as time passed it became increasingly clear to Einstein and his fellow committee members that the world was drifting in the wrong direction. In considering the root causes of their failure, Einstein and his colleagues were convinced that the seriousness of the problem demanded more drastic remedies. Their views moved more and more in the direction of "world government" as the only logical solution. Their analysis of the situation, prepared in June, 1947, was particularly cogent:

A year has passed since the founding of this committee and the publication of our first appeal to reason in the face of the overwhelming threat presented to civilization by the atomic bomb. During this year our hopes for international agreement on control of atomic energy have come to nothing. It is imperative that the American people understand this failure if any constructive solutions are to be arrived at in time.

Can thoughtful and well-informed men any longer expect fruitful agreements to come out of the discussions now going on in the United Nations Atomic Energy Commission? Are the American proposals for the international control of atomic energy fair? Are the Russians to blame for the fact that any agreement is further away than at the initiation of the discussions?

The analysis goes on to associate the need to control atomic energy with the need to build a secure, just, and stable world society. It calls on the American people to take the lead in this quest, concluding:

> But the American people should understand . . . that, in the long run, the creation of a supranational government, with powers adequate to the responsibility of maintaining the peace, is necessary.
>
> Is this realistic? We believe that nothing less is realistic. We know that the developments of science and technology have determined that the peoples of the world are no longer able to live under competing national sovereignties with war as the ultimate arbitrator. Men must understand that the times demand a higher realism which recognizes that "no man is an island," that our fate is joined with that of our fellow men throughout the world. We must contrive to live together in peace even at the cost of great material sacrifice; the alternative is the death of our society. As we approach what may be the last hour before midnight, the challenge is plainly before us. What will be our response?[14]

From then on, Einstein became ever more outspoken in his insistence that nothing short of an effective supranational authority able to exercise unquestioned control over international issues of war and peace was needed to save the world from nuclear disaster. He maintained that such an international authority could and should be established promptly through the United Nations, irrespective of whether all the major world powers were ready immediately to join:

> The time has come for the United Nations to strengthen its moral authority by bold decisions. First, the authority of the General Assembly must be increased so that the Security Council as well as all other bodies of the United Nations will be subordinated to it. As long as there is a conflict of authority between the Assembly and the Security Council, the effectiveness of the whole institution will remain necessarily impaired.
>
> Second, the method of representation at the United Nations should be considerably modified. The present method of selection by government appointment does not leave any real freedom to the appointee. Furthermore, selection by governments cannot give the peoples of the world the feeling of being fairly and proportionally represented. The moral authority of the United Nations would be considerably enhanced if the delegates were elected directly by the people. Were they responsible to an electorate, they would have much more freedom to follow their consciences. Thus we could hope for more statesmen and fewer diplomats.
>
> Third, the General Assembly should remain in session throughout the critical period of transition. By staying constantly on the job, the Assembly could fulfill two major tasks: first, it could take the initiative toward the establishment of a supranational order; second, it could take quick

and effective steps in all those danger areas (such as currently exist on the Greek border) where peace is threatened.

The Assembly, in view of these high tasks, should not delegate its powers to the Security Council, especially while that body is paralyzed by the shortcomings of the veto provisions. As the only body competent to take the initiative boldly and resolutely, the United Nations must act with utmost speed to create the necessary conditions for international security by laying the foundations for a real world government.

Of course there will be opposition. However, it is by no means certain that the USSR—which is often represented as the main antagonist to the idea of world government—would maintain its opposition if an equitable offer providing for real security were made. Even assuming that Russia is now opposed to the idea of world government, once she becomes convinced that world government is nonetheless in the making her whole attitude may change. She may then insist on only the necessary guarantees of equality before the law so as to avoid finding herself in perennial minority as in the present Security Council.

Nevertheless, we must assume that, despite all efforts, Russia and her allies may still find it advisable to stay out of such a world government. In that case—and only after all efforts have been made in utmost sincerity to obtain the co-operation of Russia and her allies—the other countries would have to proceed alone. It is of the utmost importance that this partial world government be very strong, comprising at least two-thirds of the major industrial and economic areas of the world. Such strength in itself would make it possible for the partial world government to abandon military secrecy and all the other practices born of insecurity.

Such a partial world government should make it clear from the beginning that its doors remain wide open to any nonmember—particularly Russia—for participation on the basis of complete equality. In my opinion, the partial world government should accept the presence of observers from nonmember governments at all its meetings and constitutional conventions.

In order to achieve the final aim—which is one world, and not two hostile worlds—such a partial world government must never act as an alliance against the rest of the world. The only real step toward world government is world government itself.

In a world government the ideological differences between the various component parts are of no grave consequence. I am convinced that the present difficulties between the United States and the USSR are not due primarily to ideological differences. Of course, these ideological differences are a contributing element to an already serious tension. But I am convinced that even if the United States and Russia were both capitalist countries—or Communist, or monarchist, for that matter—their rivalries, conflicting interests, and jealousies would result in strains similar to those existing between the two countries today.

The United Nations now, and world government eventually, must serve one single goal—the guarantee of the security, tranquillity and the welfare of all mankind.[15]

Although this view was openly and pointedly attacked by Soviet officials, as well as by some Soviet scientists, as playing into the hands of right-wing, anti-Soviet forces, Einstein stood his ground[16] and steadfastly maintained until the end of his life his adherence to the "world government" approach as the only means of saving humankind from the diastrous consequences of the uninhibited national exploitation of inevitable technological progress.

However, Einstein's unquenchable optimism notwithstanding, progress toward effective world order continued to be, if anything, retrograde. Efforts in the United Nations Atomic Energy Committee for agreement on a scheme of international control ground to a vitriolic halt; Gandhi's assassination removed from the international scene a moral force that was irreplaceable; the world-order movement in the United States, which had concentrated on the candidacy of Henry Wallace for the presidency, collapsed, a victim of the growing Cold War antagonism between the capitalist West and the communist East. Finally, at the end of 1948 the Emergency Committee, recognizing that "the education of mankind toward a clearer understanding of the implications of atomic energy, and full appreciation of the dangers and the hopes inherent in new discoveries, is a long-range task which cannot be solved on an emergency basis," decided to go out of business.[17]

Thus Einstein, entering his eighth decade, had every reason to retire from the battlefield. Instead, he chose to involve himself even more deeply in both the world-government movement and in the struggle to preserve individual rights and freedoms from the growing encroachments of the Cold War mentality and the McCarthy onslaughts on the intellectual integrity of the American academic community. This was a period in which an intellectual as fiercely independent as Einstein had to be both firm and discerning to avoid being used by either the pro-Soviet "left" or the red-baiting "right." Einstein managed always to steer a firm course between what would be referred to in today's vernacular as "a rock and a hard place." For example:

So long as security is sought through national armament, no country is likely to renounce any weapon that seems to promise it victory in the event of war. In my opinion security can be attained only by renouncing all national military defense.

Unfortunately, since both the United States and the Soviet Union persist in their mutually stubborn attitude, we cannot expect that either of them will make any constructive peace moves. Only a collective effort on the part of those nations which are militarily weak has any chance of success. I shall gladly use my humble energies in that direction, and I shall seek to enlist allies as far as my limited possibilities permit me to do.

If the nations which are militarily weak join forces, they can put pressure upon the two principal rivals to compose their differences peacefully by accepting the intercession of the United Nations and by pledging to submit to United Nations decisions without reservations. (516)

What was especially discouraging to Einstein was the failure of the United States to live up to its historic traditions in the field of civil rights. Referring in 1951 to the Nuremberg war crime trials, he wrote: "There is a curious inconsistency in a government which punishes aliens for *not* following their conscience in a given conflict, while penalizing its own citizens for *following* their conscience in the same kind of conflict. Apparently such a government holds the conscience of its own citizens in lower esteem than that of aliens" (542). The situation in America continued to worsen, and by 1954, even he could no longer contain his pessimism: "If I were a young man again and had to decide how to make a living, I would not try to become a scientist or scholar or teacher. I would rather choose to be a plumber or a peddlar, in the hope of finding that modest degree of independence still available under present circumstances" (613).

But eternal optimist that he was, Einstein could never give up hope. Thus, in the spring of 1955 he and Bertrand Russell, with typical youthful enthusiasm, launched a project aimed at starting a worldwide movement among scientists to reverse the Cold War trend toward nuclear war. The credo of this movement was the Russell-Einstein Manifesto signed by Einstein on 11 April 1955, exactly one week before his death.[18] The text of this manifesto serves well as a summing up of Einstein's view of the state of the world and of the responsibilities of scientists as world citizens:

In the tragic situation which confronts humanity, we feel that scientists should assemble in conference to appraise the perils that have arisen as a result of the development of weapons of mass destruction. . . .

We are speaking on this occasion, not as members of this or that nation, continent or creed, but as human beings, members of the species man, whose continued existence is in doubt. The world is full of conflicts;

and, overshadowing all minor conflicts, the titanic struggle between Communism and anti-Communism.

Almost everybody who is politically conscious has strong feelings about one or more of these issues; but we want you, if you can, to set aside such feelings and consider yourselves only as members of a biological species which has had a remarkable history, and whose disappearance none of us can desire.

We shall try to say no single word which should appeal to one group rather than to another. All, equally, are in peril, and, if the peril is understood, there is hope that they may collectively avert it. . . .

The general public, and even many men in positions of authority, have not realized what would be involved in a war with nuclear bombs. The general public still thinks in terms of the obliteration of cities. It is understood that the new bombs are more powerful than the old, and that, while one A-bomb could obliterate Hiroshima, one H-bomb could obliterate the largest cities, such as London, New York and Moscow.

No doubt in an H-bomb war great cities would be obliterated. But this is one of the minor disasters that would have to be faced. If everybody in London, New York and Moscow were exterminated, the world might, in the course of a few centuries, recover from the blow. But we now know, especially since the Bikini test, that nuclear bombs can gradually spread destruction over a very much wider area than had been supposed.

It is stated on very good authority that a bomb can now be manufactured which will be 2,500 times as powerful as that which destroyed Hiroshima.

Such a bomb, if exploded near the ground or under water, sends radioactive particles into the upper air. They sink gradually and reach the surface of the earth in the form of a deadly dust or rain. It was this dust which infected the Japanese fishermen and their catch of fish.

No one knows how widely such lethal radioactive particles might be diffused, but the best authorities are unanimous in saying that a war with H-bombs might quite possibly put an end to the human race. It is feared that if many H-bombs are used there will be universal death—sudden only for a minority, but for the majority a slow torture of disease and disintegration.

Many warnings have been uttered by eminent men of science and by authorities in military strategy. None of them will say that the worst results are certain. What they do say is that these results are possible, and no one can be sure that they will not be realized. We have not yet found that the views of experts depend in any degree upon their politics or prejudices. They depend only, so far as our researches have revealed, upon the extent of the particular expert's knowledge. We have found that the men who know most are the most gloomy.

Here, then, is the problem which we present to you, stark and dreadful and inescapable: Shall we put an end to the human race; or shall mankind renounce war? People will not face this alternative because it is so difficult to abolish war.

The abolition of war will demand distasteful limitations of national sovereignty. But what perhaps impedes understanding of the situation more than anything else is that the term *mankind* feels vague and abstract. People scarcely realize in imagination that the danger is to themselves and their children and their grandchildren, and not only to a dimly apprehended humanity. They can scarcely bring themselves to grasp that they, individually, and those whom they love are in imminent danger of perishing agonizingly. And so they hope that perhaps war may be allowed to continue provided modern weapons are prohibited.

This hope is illusory. Whatever agreements not to use the H-bombs had been reached in time of peace, they would no longer be considered binding in time of war, and both sides would set to work to manufacture H-bombs as soon as war broke out, for, if one side manufactured the bombs and the other did not, the side that manufactured them would inevitably be victorious.

Although an agreement to renounce nuclear weapons as part of a general reduction of armaments would not afford an ultimate solution, it would serve certain important purposes.

First: Any agreement between East and West is to the good in so far as it tends to diminish tension. Second: The abolition of thermonuclear weapons, if each side believed that the other had carried it out sincerely, would lessen the fear of a sudden attack in the style of Pearl Harbor, which at present keeps both sides in a state of nervous apprehension. We should, therefore, welcome such an agreement, though only as a first step.

Most of us are not neutral in feeling, but, as human beings, we have to remember that, if the issues between East and West are to be decided in any manner that can give any possible satisfaction to anybody, whether Communist or anti-Communist, whether Asian or European or American, whether white or black, then these issues must not be decided by war. We should wish this to be understood, both in the East and in the West.

There lies before us, if we choose, continual progress in happiness, knowledge and wisdom. Shall we, instead, choose death, because we cannot forget our quarrels? We appeal, as human beings, to human beings: Remember your humanity and forget the rest. If you can do so, the way lies open to a new paradise; if you cannot, there lies before you the risk of universal death.[19]

On 13 April Einstein was stricken with his last, fatal illness. At the time, he was working on the text of an address intended to be delivered on the anniversary of Israel's independence. Among his extrascientific concerns, none was closer to Einstein's heart or had greater call on his unbounded energies than the problems of the Jewish people and their attempts, finally, to establish a haven in the land of their forefathers. This last of Einstein's writings sums up his views on individual responsibility, almost as though it had been intended to serve as a last will and testament:

> I speak to you today not as an American citizen and not as a Jew, but as a human being who seeks with the greatest seriousness to look at things objectively. What I seek to acccomplish is simply to serve with my feeble capacity truth and justice at the risk of pleasing no one.
>
> At issue is the conflict between Israel and Egypt. You may consider this a small and insignificant problem and may feel that there are more serious things to worry about. But this is not true. In matters concerning truth and justice there can be no distinction between big problems and small; for the general principles which determine the conduct of men are indivisible. Whoever is careless with the truth in small matters cannot be trusted in important affairs.
>
> This indivisibility applies not only to moral but also to political problems; for little problems cannot be properly appreciated unless they are understood in their interdependence with big problems. And the big problem in our time is the division of mankind into two hostile camps: the Communist World and the so-called Free World. Since the significance of the terms *Free* and *Communist* is in this context hardly clear to me, I prefer to speak of a power conflict between East and West, although, the world being round, it is not even clear what precisely is meant by the terms *East* and *West*.
>
> In essence, the conflict that exists today is no more than an old-style struggle for power, once again presented to mankind in semireligious trappings. The difference is that, this time, the development of atomic power has imbued the struggle with a ghostly character; for both parties know and admit that, should the quarrel deteriorate into actual war, mankind is doomed. Despite this knowledge, statesmen in responsible positions on both sides continue to employ the well-known technique of seeking to intimidate and demoralize the opponent by marshaling superior military strength. They do so even though such a policy entails the risk of war and doom. Not one statesman in a position of responsibility has dared to pursue the only course that holds out any promise of peace, the course of supranational security, since for a statesman to follow such a course would be tantamount to political suicide. Political passions, once they have been fanned into flame, exact their victims. . . . (640-641)

On 18 April 1955, while still working on this document, Einstein died.

∞

If in some extra-universal dimension there should exist a version of the legendary paradise, Albert Einstein would certainly be present there in the first rank of the saints. In reviewing his life, what is overwhelmingly impressive is the degree to which he regarded himself as the servant of his fellow human beings. Nothing, of course, could interfere with his science: it was for that, in his view, that he had been put on this earth. Nonetheless, the time he devoted to societal obligations would have amounted to more than the full time of any lesser mortal. It was not just that he threw himself wholeheartedly into the causes in which he believed. There was no individual—whatever his or her status, educational or social background, race, or even language of expression—who could not reach Einstein and command his attention.

Einstein was a shy man and, fortunately, during most of his life, he was protected from the direct access of the many who constantly sought his advice and help or who just felt the desire to draw inspiration through contact with him. But he was always accessible through the mails. And with unfailing patience, kindness, and honesty—but never condescension—he responded to all, from semiliterate cranks to esteemed and learned colleagues (sometimes also cranks). No social issue, if he felt it was being honestly approached, was too trivial or too complex for him to try to illuminate it, but always with modesty and eschewing dogma.

Knowing that the human species is still capable of producing one of his ilk—and within the lifetime of many of us—cannot fail to bolster our faith in the ultimate triumph of good over evil in the ageless human struggle.

NOTES

1. *Einstein on Peace*, ed. Otto Nathan and Heinz Norden, New York, Simon and Schuster, 1960, p. 619. This volume is an invaluable source of documents from Einstein's voluminous correspondence and writings on social issues, beautifully translated from the (usually) German originals. Further quotations from this volume will be followed by parenthetical page references. Other sources of Einstein's extensive extrascientific writings include: *Mein Weltbild*, ed. Carl Seelig, Zurich, Europa Verlag, 1953; *The World as I See It*, New York, Philosophical Library, 1949; *Out of My Later Years*, New York, Philosophical Library, 1950;

Ideas and Opinions (based on *Mein Weltbild*), new trans. and rev. Sonja Bargmann, New York, Crown Publishers, 1954.

2. Paul Arthur Schilpp, ed., *Albert Einstein: Philosopher-Scientist*, Evanston, Ill., Library of Living Philosophers, 1949. The "Autobiographical Notes" are now available in a separate edition: *Albert Einstein: Autobiographical Notes*, Chicago, Open Court, 1979.

3. Thus, roughly two-thirds of Einstein's autobiography concerns the events of the first decade of the twentieth century: the generalization of classical statistical mechanics (Brownian motion), the revolutionary impact of the quantum hypothesis of Max Planck (photoelectric effect, specific heats of solids), and finally that grand synthesis of James Clerk Maxwell and Isaac Newton into the special relativity. This brings Einstein to 1908. For him the following seven years, which witnessed the complete breakdown of European order and the outbreak of the First World War, were consumed in the struggle to extend relativity into a world containing inertia (mass) and gravitation—the general relativity. For Einstein thereafter, two intellectual problems assumed overwhelming importance: the resolution of the contradiction between causal physics and the (to him) incomplete statistical description of nature inherent in the present formulation of quantum mechanics, and the extension of relativity to include electromagnetism (unified field theory). His autobiography ends, after a beautiful exposition of the problems of generalizing the general relativity, on a rather optimistic note concerning his new formulation of the equations of general relativity: "The theory here proposed, according to my view, has a fair probability of being found valid. . . ." "Autobiographical Notes," p. 93n.

4. Leo Szilard, an engineering student from Budapest, came to the University of Berlin in 1920. By then, Einstein was already well ensconced as one of the leading professors, soon to be awarded the Nobel Prize. Although Szilard's thesis advisor was Max von Laue, its subject matter, classical thermodynamics, was of great interest to Einstein, and the two became close colleagues and good friends. Both left soon after Hitler's rise to power—Einstein to the United States in 1932 to join the Institute for Advanced Study in Princeton, and Szilard to England in 1933 and then to the United States in 1938. Although they never jointly published any scientific papers, Einstein and Szilard together invented (*and patented*) a pump for the "Electrodynamic Movement of Fluid Metals particularly for Refrigerating Machines." For further details, see *The Collected Works of Leo Szilard: Scientific Papers*, ed. B. T. Feld and G. W. Szilard, Cambridge, Mass., MIT Press, 1972, and *Leo Szilard, His Version of the Facts: Selected Recollections and Correspondence*, ed. S. Weart and G. W. Szilard, Cambridge, Mass., MIT Press, 1978. Incidentally, this device, capable of moving conducting fluids through pipes by use of traveling electromagnetic waves, has been adapted for the circulation of liquid-metal coolants in the controversial plutonium-breed-

ing, liquid-metal, fast breeder reactors—an irony that would not have been lost on either Einstein or Szilard.

5. This group included Enrico Fermi, George Placzek, Edward Teller, Victor F. Weisskopf, and Eugene P. Wigner.

6. For details concerning the beginnings of the wartime atomic energy project in the United States, see *Szilard, His Version*, chaps. 3 and 4, and *Einstein on Peace*, chap. 9. The letters between Einstein and Roosevelt appear in the latter.

7. Of this fact there is no question; see *Einstein on Peace*, pp. 583-584. However, Einstein was by no means unwilling to help in the war effort. The physicist George Gamov recalled that from time to time during the war he consulted with Einstein for the Navy on some conventional ordnance problems. See A. P. French, ed., *Einstein: A Centenary Volume*, Cambridge, Mass., Harvard University Press, 1979, p. 79. In late 1941 Vannevar Bush, director of the Office of Scientific Research and Development, approached Einstein via the director of the Institute for Advanced Study, Dr. Frank Aydelotte, for help on a problem relating to the diffusion of gases through porous membranes. Einstein responded rather quickly with a simple solution, which apparently was too simple for what was needed. Dr. Bush pursued it no further, but not without making clear to Dr. Aydelotte his reasons for having "had to put up a very definite problem stated in academic terms." As he explained: "The reason that I am not going further is that . . . I do not feel that I ought to take him into confidence on the subject to the extent of showing just where this thing fits into the defense picture, and what the military aspects of the matter might be. . . . I wish very much that I could place the whole thing before him and take him fully into confidence, but this is utterly impossible *in view of the attitude of people here in Washington* who have studied into his whole history. It looks to me, therefore, that we might as well let the matter drop" (italics added). Bush to Aydelotte, 30 December 1941, declassified on 19 August 1970 by the U.S. Atomic Energy Commission as Technical Document no. 331; I am grateful to Arthur Steiner for calling this material to my attention. Interestingly enough, Aydelotte's problem actually related to the theory of the gaseous diffusion process, which turned out to be the most efficient one for separating the light uranium isotope but which did not work quickly enough to produce any bombs in World War II. Einstein was never given any indication of the point of Bush's question, and if he guessed it, he never let on publicly and certainly did not pursue the matter.

What this incident mainly illustrates is the paranoia of officialdom in wartime Washington, with respect not only to Einstein but also to all scientists of foreign birth. It is now abundantly clear that the refugee scientists on the Manhattan Project were essential to its success, at least as far as the plutonium aspects were concerned. (The problems of the uranium-235 bomb were essentially solved, at least in principle, by the

British group by the summer of 1941; see Margaret Gowing, *Britain and Atomic Energy, 1939-45*, London, Macmillan and Co., 1964. Of course, the British group also included Rudolph Peierls and other refugee scientists.) Moreover, the refugee scientists in America had an unblemished record of loyalty and scrupulous attention to "security." Yet, surprisingly, these aspects of the record are still occasionally confused by contemporary commentators; see, for example, Ronald W. Clark, *Einstein: The Life and Times*, New York and Cleveland, World Publishing Co., 1971, chap. 20.

8. *Szilard, His Version*, p. 205; also *Einstein on Peace*, p. 305.

9. James F. Byrnes, *Speaking Frankly*, New York, Harper and Brothers, 1947.

10. The war against Germany ended on 8 May 1945; the first test explosion in the New Mexico desert was on 16 July. It is interesting, if futile, to speculate whether, if the military bureaucracy had taken seriously the message contained in Einstein's original letter to Roosevelt and not dragged their feet for almost a year, the bomb might have been ready in time and then would have been used against Germany. As it was, in spite of the remarkably rapid achievement of an operating chain reaction and the subsequent production of weapons-grade plutonium in the Hanford reactors, the bomb that can be said to have ended the war with Japan—the one dropped on Hiroshima—was made of uranium 235, which had been painfully separated by entirely brute-force methods at Oak Ridge, Tenn., by processes devised and assembled by American-born physicists according to principles mainly developed in Britain and California. The plutonium bomb, the product of the Columbia-Princeton-Chicago groups, which followed the Fermi-Szilard approach, was (aside from the test explosion in New Mexico) the entirely superfluous and gratuitous bomb that was dropped on Nagasaki when the Japanese were already prepared to surrender and suing for peace. Thus, in some profoundly ironic sense, the popular legend of the birth of the atomic bomb has aspects that are more mythological than real.

11. Einstein, *The Fight Against War*, ed. Alfred Lief, New York, John Day Co., 1933; also *Einstein on Peace*, pp. 125-126.

12. For the definitive history of this period, see Alice K. Smith, *A Peril and a Hope: The Scientists' Movement in America, 1945-47* (1948), rpt. Cambridge, Mass., MIT Press, 1971.

13. His fellow members on this committee were: Harold C. Urey (vice-chairman), Hans A. Bethe, Thorfin R. Hogness, Philip M. Morse, Linus Pauling, Leo Szilard, and Victor F. Weisskopf. They were later joined by Harrison Brown.

14. *Bulletin of the Atomic Scientists*, August 1947, pp. 216-217; also *Einstein on Peace*, pp. 410-412.

15. *New York Times*, 23 September 1947; also *Out of My Later Years*, pp. 158-160.

16. See, for example, *Bulletin of the Atomic Scientists*, January and February 1948. See also *Einstein on Peace*, pp. 443-455.

17. *Bulletin of the Atomic Scientists*, December 1948, p. 354.

18. The other signers of the manifesto were: Max Born (Germany), Percy W. Bridgman (U.S.), Leopold Infeld (Poland), Frédéric Joliot-Curie (France), Hermann J. Muller (U.S.), Linus Pauling (U.S.), Cecil F. Powell (Britain), Joseph Rotblat (Britain), and Hideki Yukawa (Japan)—all but two, Nobel Prize winners in science. The manifesto led to the convening in July 1957 of the first Pugwash Conference on Science and World Affairs, an institution that continues to work for the achievement of the aims expressed in the manifesto.

19. *New York Times*, 10 July 1955; also *Einstein on Peace*, pp. 633-636. See also *Bulletin of the Atomic Scientists*, September 1955, pp. 236-237, and J. Rotblat, *Scientists in the Quest for Peace*, Cambridge, Mass., MIT Press, 1972.

Working with Einstein: Reminiscences by Associates and Friends

Peter G. Bergmann

REMINISCENCES

I FIND IT much easier to talk about physics than about physicists, but tonight I am to do the latter. Perhaps I should follow in the steps of Nathan Rosen, whom I also succeeded more than once, and tell you how I came to work with Einstein. I arrived in Princeton in 1936, not as a native-born American, but as a refugee from Hitler Germany. I had managed to complete my formal academic education in Prague, at the institute of Professor Philipp Frank. At the time, I did not know that he was a close friend of Professor Einstein.

I thought, when it was time to look for a place to go, that I had a great advantage over many young people: I did not need to worry about where I might offer my services for hire, because it was quite clear that for a person in my position there were no jobs at all. So I really could follow my personal preferences. I was not quite twenty-one years old. I wrote a letter to Einstein in Princeton, explaining that I was working on my dissertation, which was in the area of general relativity, and asking for permission to work with him. The reason I wanted so much to work with Einstein was not that I was taken with Einstein's fame, which, to be sure, was well established by that time. Rather, his research seemed to me very different from what most physics papers were like. It seemed to correspond to what I thought I would like to do myself.

At that time there was no air mail. When after a month I had received no response, I wrote a second letter, to the effect that I was really finishing my dissertation, and now what about it?

At that point I received a message inviting me, if I was coming to the States anyway, to get in touch with him and to come to the Institute for Advanced Study, which was a haven for people like myself. No money, but at least an opportunity to go, and if I should care to call on him—fine.

What I did not know at the time was that Einstein, on receiving my first message, had not simply ignored it. He wrote to Professor Frank, asking what kind of person was this Peter Bergmann. Apparently getting a fairly favorable response, he then wrote me to call on him. As it turned out, even though I came to Princeton with no expectation of receiving any salary, a salary and an assistantship were arranged a few months later, and I stayed on in Princeton for a total of five years.

Nathan Rosen has described something of Einstein's style of work, and I think that Ernst Straus undoubtedly went through the same routine: a couple of hours of intensive discussion in the morning, and equally intensive work at home, with paper and pencil, the rest of the day. I should add that very frequently in the afternoons my landlady's party line was busy with people talking about incomprehensible subjects in an incomprehensible language, subjects not at all suitable for juicy gossip.

On a day-to-day basis, I think that what impressed me most after I had gotten to know Einstein and his style a little better was his tremendous persistence. Once having perceived the really important problems, he would not let go. He might let go temporarily if a road seemed to be completely blocked, but he would pick up the problem again a few weeks later—in the meantime having worked on another, equally interesting problem.

A second thing that impressed me—and remember that I was very young and that Einstein was in his late fifties—was his tremendous creativity, even on the small day-to-day difficulties, his sheer inventiveness of new approaches, of new mathematical tricks.

A third thing—and I think it is something very rare today—was a tremendous capability of taking a step back in his scientific work and looking at it, as you might say, in the cold light of dawn. Whenever Einstein was working on a new model of field theory, he would be tremendously enthusiastic for weeks and months, and even for years, but inevitably the day of reckoning would come when Einstein would be the first to find a fatal flaw. With complete ruthlessness, he would discontinue this theoretical approach and come up with a completely new idea, usually within days.

Let me say a few words about whether Einstein, as he has been described, was a lonely man in the later part of his life. During my five years in Princeton, Banesh Hoffmann, Valya Bargmann,

Nathan Rosen, and Leopold Infeld were there. At any one time, there were one, two, or three people—no more—working with Einstein, and working with him very intensively.

Einstein did not give rise to what you might call a school; he was not the one to generate disciples. I remember distinctly a few times when I asked his advice on scientific strategy long after I had left Princeton. His response was generally: "In diesem Geschäft muss jeder in Gottes Namen seine eigenen Eier ausbrüten!" In this business, everyone, for God's sake, has to make his own decisions—has to hatch his own eggs, in more or less literal translation. Nevertheless, the people who were privileged to work with Einstein have formed ties of close friendship that have lasted over the years.

The style of work, the approaches that Einstein initiated, have not become rigid. They have been modified in the course of the development of physics, and an increasing number of people are devoting their interest to relativity. Nowadays, when we have one of our triannual conferences of the International Society for General Relativity and Gravitation, several hundred people come. The society is also sponsoring a journal for which there are now more manuscripts being submitted than the journal is capable of publishing.

So, I think that the fields that were essentially generated by Einstein are flourishing. There are young people who find them important and challenging, and I believe that Einstein's spirit continues to thrive among us.

Banesh Hoffmann

REMINISCENCES

LET ME START by telling about my first meeting with Einstein. It was around 1935 in Princeton. I had made some relativistic calculations, and a friend suggested that I go to see Einstein to ask his opinion of my work. The idea of my going to see Einstein seemed to me preposterous. I was far too scared. My friend almost had to push me to the door of Einstein's office. I knocked timidly, and Einstein called out the single word "come" with a friendly, rising inflection. I entered in fear and trembling, and there was Einstein sitting in a comfortable chair, sloppily dressed, his hair awry, a pipe in his mouth, and a sheaf of calculations on his lap. I managed to tell him why I had come. He must have seen my nervousness. He smiled and gently asked me to put my equations on the blackboard, and then came these words, which I shall always remember: "Please go slowly. I do not understand things quickly." This from Einstein! At once, as if by magic, all my fears left me.

It was my good fortune to collaborate with Einstein and Leopold Infeld. After the calculations had been made and the paper presenting the work had been drafted, we were faced with the problem of writing an introduction. The three of us discussed the matter at length, and Einstein proposed several items and arguments that should be included. Infeld and I took notes, and afterwards the two of us put Einstein's ideas in order and wrote out a draft of the introduction. The next day we read it to Einstein. He listened with rapt attention, looking at us with warm and growing admiration. At the end he beamed with pleasure as he congratulated us on what we had done. He was particularly impressed by the breadth and relevance of our background knowledge. He had quite forgotten that he had given us all the main ideas. He assumed that we had thought of them ourselves. It took us quite a while to convince him that he had supplied them, and when we did, he seemed a little saddened.

I now want to tell about a small dinner party that I was invited to at the house on Mercer Street. I have various reasons for telling of this, and I will reveal them later. While tea was being served—from a teapot and not via tea bags—Einstein pointed out that when one stirred one's tea, the tea leaves congregated at the middle of the bottom of the tea cup and not at the perimeter, where one might expect them to congregate. He asked me if I knew why.

I knew that Einstein had published an explanation, but, unfortunately, I could not recall the details. Perhaps, though, that was for the best, because Einstein found great pleasure in explaining it to me in detail, including its link to the meandering of rivers.

I wondered how to respond to this. Obviously, I would not be able to hold my own if I tried to trade science questions with Einstein. But all my life I had been attracted by beautiful geometrical facts. So I told him this one. Take any triangle *A, B, C*. Through *C* draw a line making 30° with *CB*, and through *B* draw a line making 30° with *BC*, and let them meet at *P*. On *CA* draw similar lines meeting at *Q*, and on *AB* similar lines meeting at *R*. Then, no matter what triangle *ABC* one starts with, *PQR* are the vertices of an equilateral triangle. I also showed him a lovely proof of this beautiful result,[1] and you never saw anyone so delighted, both with the theorem and the proof.

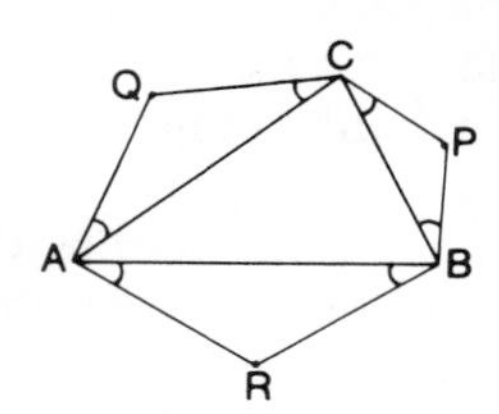

Einstein then asked me another science question. (I think I am telling these questions in their actual order.) Imagine a lighted candle in an enclosed elevator. The cable breaks and the elevator falls freely. What happens to the flame of the candle? Since the answer depended on Einstein's famous principle of equivalence, I was able to figure out what would happen: with zero gravity in the falling elevator, hot gases would not rise, and the candle flame would therefore be snuffed out.

Einstein was just as happy when I could answer one of his questions as he was when I could not.

It was now my turn again. I showed him this. Three equal circles are tangent to one another, and a larger circle is tangent to all three. From any point P on the large circle, tangent lines PT_1, PT_2, and PT_3 are drawn to the small circles. Then $PT_1 + PT_2 = PT_3$ (with appropriate changes as P moves so that PT_1 or PT_2 is the longest).[2]

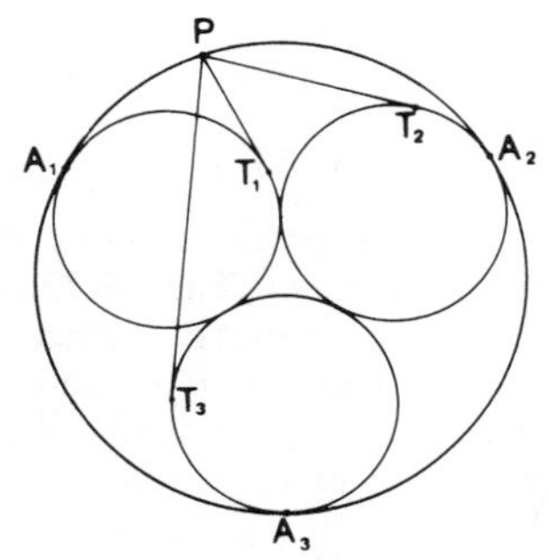

The last of the questions that Einstein asked me was this: Why is it that when you walk on dry sand your feet sink into it, but when you walk on damp sand it is firm? But if you walk on completely immersed sand it is not at all firm?

I had not the least idea. Here is the explanation that Einstein gave me. In damp sand there is a little water between neighboring grains, and the surface tension pulls the grains together, thus creating friction between them. With dry sand, as with completely immersed sand, there is no liquid surface between grains to provide surface tension.

I responded by stating Feuerbach's theorem,[3] and by then it was time to leave. I tell you all this partly because it was such fun and so spontaneous, and it shows Einstein so happy with beautiful geometrical results. But I have two other reasons for doing so, one of them a little mischievous.

First, I had assumed that the solution of the sand problem—and even the recognition that sand posed a problem—had been Einstein's, and it has occurred to me that perhaps I was wrong. So I now ask if any of you know whether Einstein was in fact the one who recognized and solved the problem of the sand, and if so, whether he published the solution.

My second reason is that I hope the story will have the same effect on you that it has had on me. When I see a lonely stretch of sand, whether it is firm or not, I always think of Einstein; and I hope that when you see a lonely stretch of sand, you, too, will think of Einstein. Unfortunately, you may also find yourselves thinking of me, and for that I apologize profoundly.

NOTES

1. The proof is this: Let K be the point such that AK, BK, and CK make equal angles of 120° with one another. Then P, Q, and R are the centers of the circles through BCK, CAK, and ABK. Since the line of centers of two intersecting circles is perpendicular to their common chord, QR, RP, and PQ are perpendicular to AK, BK, and CK, whence the result.

2. Here is an outline of a neat proof. Let A_1, A_2, and A_3 be the points of contact of the small circles with the large circle. Apply Ptolemy's theorem to the cyclic quadrilateral $PA_1A_2A_3$ to obtain $PA_1 + PA_2 = PA_3$. Now use the theorem that if three circles are coaxal and P is any point on one of them, the lengths of the tangents from P to the other circles have a constant ratio. Apply this to the three degenerate coaxal systems consisting of the large circle, a small circle, and their point of contact.

3. The nine-point circle of any triangle is tangent to the inscribed circle and to the three escribed circles.

Nathan Rosen

REMINISCENCES

IN MY CONTACTS with Albert Einstein, two occasions stand out particularly strongly in my memory: the first time I met him and the last time I saw him. So let me describe them both, with the addition of a little bit about what happened in between.

As for my first meeting with Einstein, I have to go back somewhat in time. After completing my undergraduate studies in engineering at MIT, I decided to spend a year doing graduate work there in theoretical physics. In addition to taking courses, I also had to do a master's thesis. At that time Einstein was working on unified field theories based on the concept of distant parallelism, and he published several papers on such theories. I decided to do my thesis research by looking for the static, spherically symmetric solutions of some of the field equations that he had proposed, for which I managed to get my master's degree. I then went on to work for the doctorate. At that time quantum mechanics was in a state of very active development. I was attracted by it and did my doctorate research on molecular structure, a field far removed from general relativity or unified field theories. Having obtained the doctorate, I spent the year 1932-1933 at the University of Michigan working on molecules with David M. Dennison. I then went to Princeton University to continue work on molecular structure with Edward U. Condon.

When I came to Princeton in the fall of 1933, I learned that Einstein had arrived shortly before. A few months later I decided to talk to him and to tell him something about what I had done in connection with his unified field theories. So one day I gathered up courage and went to see him. I knocked rather timidly on his office door, and he let me in. I found him working with his assistant, the mathematician Walter Mayer. Einstein greeted me in a very friendly manner and made me feel at ease. I told him about my work on the solutions of the equations of his unified field theories, and he seemed interested and asked many questions. After a while he began telling me about the work he was doing at that time and, before I realized it, the three of us were engaged in a lively discussion of his current problems. At the end of our conversation he invited me to come again the next day, which I did.

Before long I was coming every day. Walter Mayer, who was a pure mathematician, was not really interested in doing applied mathematics in connection with the unified field theories, and he gradually withdrew from the scene, so that after some time just the two of us, Einstein and I, were working together. We met almost every morning from ten o'clock to twelve, and we had what was for me a stimulating—and sometimes exhausting—session. Einstein would come full of ideas. He would propose new theories, build them up and investigate their consequences. What impressed me most, as I think back about him, was the clarity and simplicity of his thinking. He would start from some simple ideas as the foundation and then build up the theory step by step. When at a given stage there were several ways of continuing, he would choose what seemed to be the simplest. A word he used very often was "reasonable" (*vernünftig*). In building a theory, he would ask himself whether a certain assumption was reasonable before he would adopt it. Sometimes, when he was considering different possibilities, he would say, "Let me see, if I were God, which one of these would I choose?" And, as I said, he usually chose the simplest one.

At the end of our working session I would usually accompany him on his way back to his house, and we would talk about other matters, current events, politics, and so on. Then I would go home and would do homework on the things we had discussed. For example, if we had arrived at equations, I would try to solve them. If I found anything interesting, we would discuss it the next morning.

In this way we worked for almost three years, until the end of the summer of 1936, at which time I went off to a position abroad. After that, we maintained contact mainly by exchanging letters from time to time—not very often, because I knew how busy he was and how so many people were writing to him and taking up his time, but we kept in touch. If I happened to be in the vicinity of Princeton, I would drop in for a short visit.

The last time I saw him was in May 1953. I had accepted a position in Israel, and my family and I were on our way to settle there. Having come to New York, we went to visit him. Not having seen him for a number of years, I was struck by how much he had aged. He looked weak, I thought, but still he spoke vigorously and with the clarity of thought I had known in the past. We had a long conversation. He was greatly interested in Israel. He asked about our plans, and we talked about conditions in Israel and possible future developments. Then I asked about his work. He told me about the unified field theory he was working on, and I was impressed by his enthusiasm.

At the end, as we were parting, he said to me, "If I were a young man, I would also go to Israel like you." That is my last recollection of him. The visit was for me both memorable and inspiring.

Boris Schwarz

MUSICAL AND PERSONAL REMINISCENCES OF ALBERT EINSTEIN

I AM NOT a scientist. I am a musician who wants to share with you some of my musical and personal reminiscences of Albert Einstein.

These reminiscences reach back to the 1920s in Berlin. About 1923 I was introduced to Professor Einstein as a promising teenage violinist. Invited to come with my violin to the Einstein apartment in the Haberlandstrasse, I had to walk for some two hours from my home in the Grunewald because Berlin was in the grip of a transportation strike. With me was my father, Joseph Schwarz, a well-known concert pianist.

I still remember the apartment: one entered a long, rather narrow sitting room with a sofa facing large double doors that opened to the music room. As we arrived, Mrs. Elsa Einstein showed her motherly concern—perhaps I was too tired to play after the long walk?

After a short rest, I unpacked my violin and, with my father at the piano, I began to play: it was the first movement of the G-minor Concerto of Max Bruch. I stood in the open door near the grand piano while the Einsteins sat on the sofa in the sitting room. As I was rounding out the beautiful second theme—playing, as was my habit, with closed eyes—I suddenly became aware that Professor Einstein stood before me, and he said, "Da kann man sehen, dass er die Geige liebt [There one can see that he loves the violin]!" As soon as I had finished, the Professor unpacked his own violin, and we played the Double Concerto of Bach, followed by other pieces for two violins and piano. It was an unforgettable afternoon and the beginning of a relationship that I hesitate to call "friendship" because of the great difference in age. Yet I know that he enjoyed our music making as much as I did.

Our musical afternoons became a regular occurrence. Once or twice a month, my father and I visited the Haberlandstrasse, bringing stacks of chamber music, all for two violins and piano, and we would make music with the Professor, reading through the material we had brought. His musical taste was refined and selective: aside from Bach and Mozart, he loved to play the Baroque Trio Sonatas by Vivaldi, Handel, Corelli, Purcell, Telemann, and so many others. I remember that we alternated playing first and second violin. He was an experienced sightreader, with a steady rhythm, excellent intonation, and a clear and pure tone using a minimum of vibrato. There is an apocryphal story that Einstein, while playing with Artur Schnabel, lost his place in the music, whereupon Schnabel said mockingly, "Professor, can't you count?!" This, I suggest, is pure fiction: not once did the Professor miscount in the course of our music sessions.

We always played sitting down, and I can still see his characteristic stance, leaning forward, slightly hunched over the music stand, with a face of utter concentration and beatific joy. The drawing by Emil Orlik, which hangs in my music studio in New York, captures his playing position to perfection. On it, Einstein wrote a dedication to my parents: "Herrn und Frau Schwarz zur Erinnerung an frohe Stunden." I myself received as a present the Hermann Struck engraving of Einstein with the inscription "nach wundervollem Solo zugeeignet." I also have a photograph made in Einstein's music room, showing us while playing a Bach Trio Sonata, with Einstein's humorous rhymed inscription: "Dem Vater und dem Sohne—Das Spielen war nicht ohne [To the father and his lad—The playing was not bad]!"

During the 1920s, Professor and Mrs. Einstein became my patrons and helped me in every possible way. I remember that Professor Einstein lent me his own violin for a month because my instrument was so inadequate. Whenever I had learned a new piece of music that I thought the Professor would enjoy, I played it for the Einsteins. He would listen while sitting very still, in a most concentrated manner. His remarks on music were always perceptive but brief: he did not care to discuss music, he simply loved and enjoyed it, whether listening or playing. He preferred the style of the eighteenth century, with its purity, grace, and

Playing chamber music in the Professor's home in Berlin, ca. 1925. Left to right: Joseph Schwarz, Boris Schwarz, Albert Einstein. From the collection of Boris Schwarz.

Drawing by Hermann Struck, with Professor Einstein's dedication to Boris Schwarz, 1923. From the collection of Boris Schwarz.

balance; even in Beethoven he preferred the *young* Beethoven of the early period to the "Sturm und Drang" of later Beethoven. But Einstein was tolerant, even when I played him some virtuoso pieces needed in a concert program. When a piece of music gave him particular joy, he would laugh with that clear, childlike laughter that was so brief and stopped so abruptly. After a good, solid music session, we would retire to the dining room, sit around a big table, and enjoy Mrs. Einstein's "Kaffee und Kuchen."

Let me skip to the 1930s. The Einsteins left Berlin, I remained. Late in 1935 I received a letter from Princeton signed "Elsa Alberti." This was Mrs. Einstein's pseudonym when she corresponded with Germany. She inquired about our well-being and asked whether I would like to come to America in order to escape the Nazi terror; the Professor would send me an affidavit. By March 1936 I had the necessary affidavit from Albert Einstein, sent directly to the United States Consulate in Berlin, which entitled me to an American immigration visa. I still remember the American consul asking me suspiciously, "How well do you know the Professor?" But I proudly presented the photo of our music sessions, which was proof enough.

In July 1936 I arrived in New York, and the second phase of my friendship with the Einsteins began. That summer the Einsteins were at Saranac Lake, and I received a short handwritten welcome from the Professor:

Saranac Lake, 22.VII. [1936]

Lieber Boris Schwarz!

Willkommen in Amerika! Ich glaube und hoffe, es wird nun alles sich zum Guten wenden. . . .

Natürlich hätte ich Sie gerne bei Ihrer Ankunft gesehen. Aber die schwere Krankheit meiner Frau hat es nötig gemacht, dass wir uns bis zum Herbst hier im gebirgigen Norden des Staates N.York aufhalten.

Ich hoffe, Sie werden bald ein Zimmerchen in dem Empfangsheim angewiesen erhalten, das Frau Rabbi Wise ins Leben gerufen hat. Halten Sie dort das Zimmer schön in Ordnung wie eine gute Haustochter und nicht wie ein russischer Zigeuner, damit ich auch bei den Weibsleuten stolz auf Sie sein kann.

Guten Erfolg und herzlichen Gruss.

Ihr

A. Einstein.

[Welcome to America! I hope and trust that everything will now turn to the better. . . .

Of course I should have liked to see you right away after your arrival.

But the serious illness of my wife made it imperative for us to spend the summer in the hilly north of the State of New York.

I hope that you will soon receive a small room in the reception center organized by Mrs. Rabbi Wise. Keep that room nice and orderly like a good boy and not like a Russian gypsy so that I can be proud of you with the womenfolk.

Good success and cordial greeting.]

Another handwritten letter came from Princeton two months later:

Princeton, 20.IX. 36.

Lieber Herr Boris!

Deuten Sie mein Schweigen nicht als Mangel an Interesse und Sympathie. Mit der Korrespondenz gehts mir, wie einem Gaul, dem man zu schwer zum Ziehen gibt—er gibts auf und lässt sich lieber prügeln. Ich war recht enttäuscht, dass meine Bemühungen, eine Arbeitsstelle für Sie zu finden, nicht von Erfolg waren, zumal es ziemlich sicher in Aussicht gestellt war. So geht es aber oft hier. Im Ganzen ist es nicht gut, in N. York zu bleiben; dort ist an allem Überfluss, während im Innern viel mehr Möglichkeiten sind. Andererseits aber ermöglicht Ihnen N. York die Anknüpfung von Verbindungen, aus denen sich dann ein Arbeitsfeld für später ergeben kann. Es ist eben alles chaotisch und zufällig.

Wir haben viel mit der Krankheit meiner Frau durchgemacht. Gegenwärtig ist es eher etwas besser, aber sie lag den ganzen Sommer.

Ich freue mich, dass Sie schon Ihr "first paper" haben. Dies bietet schon einen gewissen Rückhalt nach der unsicheren Vergangenheit. Es wäre doch schön, wenn Sie allmählich dauernd Fuss fassen und auch Ihre Eltern aus dem deutschen Narrenhaus retten könnten. Ein Glück, dass wenigstens Sie einstweilen da sind.

Herzlich grüsst Sie Ihr

A. Einstein.

[Please don't interpret my silence as lack of interest and good will. With my correspondence, I seem like a nag which has too heavy a load to pull—it gives up and prefers to be whipped.

I was rather disappointed that my efforts to obtain a position for you were unsuccessful, particularly because it was almost promised. But that's the way it often goes here. Usually it is not advisable to remain in New York because of an abundance of everything, while there are more possibilities in the interior of the country. On the other hand New York offers you the opportunity of making contacts which can open up potential fields of activity. It's all rather chaotic and accidental.

The illness of my wife caused us many worries. Now she is somewhat better, but she was in bed throughout the summer.

I am glad that you already have your "first paper." It gives you some support after the uncertain past. It would be so good if with time you could establish yourself here so that you could rescue your parents from the German madhouse. How fortunate that you at least are here.]

This letter, full of everyday advice, shows the practical side of Professor Einstein's mind, which was aware of reality and its problems.

Finally, on 12 November 1936, I received a telegram from Princeton:

EXPECTING YOU TODAY BETWEEN FOUR AND FIVE OCLOCK
GREETINGS ALBERT ELSA EINSTEIN.

It was a joyous and, as far as Mrs. Einstein was concerned, a tearful reunion. She was very ill, and this was the last time I saw her alive. I remember that I brought my violin and that I played, accompanied at the piano by Valery Bargmann, then a young assistant to Professor Einstein and, as it happened, a former piano student of my father.

I needed assistance to find a job, to establish myself sufficiently to rescue my parents from Berlin. Professor Einstein was never too busy to help, to advise, to write letters of recommendation. He was proud that I was able to establish myself quickly so that my parents arrived in New York within seven months on the strength of my own affidavit. He extended the same warm welcome to my parents that he did to me, lending support and encouragement whenever needed. I remember that he promised to attend our first concert in Philadelphia. It was arranged that I should call for him by car on my way from New York. When I arrived at Mercer Street, fully dressed for the concert, I found the Professor in his most comfortable house outfit—pullover without shirt, moccasins without socks. As I looked at him questioningly, he said with a beatific smile, "Ich bin fertig [I am ready]!" Miss Dukas raised her hands in mock despair, "Aber Herr Professor, Sie müssen sich doch etwas umziehen [But Professor, you ought to change a bit]!" Patiently, he went upstairs and changed into a dressy attire with a wing collar. After the concert, a photo was taken in front of the fireplace of the Settlement Music School in Philadelphia and given to me with the following rhymed inscription:

Hier sitz ich und streck meine Füsse raus,
Und doch sieht man deutlich, es war nicht zu Haus.

A. Einstein 1942

[Although I sit here and stretch my feet out
It wasn't at home; of that there's no doubt.]

Of course, his facetious comment referred to his dressy outfit.

In 1941 I applied for an appointment at Queens College in New York, and Professor Einstein willingly wrote two letters on my behalf, one to the president of the college, the other to the chairman of the music department, and he sent me the carbon copies. In both letters Einstein remembered our happy music sessions of years past. I might add that I received the appointment and that I held the post for thirty-five years, happily justifying the Professor's confidence.

During the 1940s and 1950s I visited Einstein in Princeton about once a year, and I always brought my violin. In his later years, however, he had to give up his own violin playing for reasons of health. I remember sending him my phonograph records of the Six Bach Sonatas for violin and harpsichord, which he acknowledged with pleasure. We corresponded occasionally: now it was I who received his requests for advice for new immigrants whom Einstein wanted to help.

Two letters from Einstein are particularly dear to me: his handwritten letters of condolence when my parents died, my mother in 1943, my father in 1945. He expressed sorrow and consolation, choosing his words with such artless simplicity and yet such deepfelt sensitivity that the meaning transcended the personal grief and touched on his philosophy of life and death. He wrote as he spoke, with utmost directness and clarity, in graceful and polished sentences, every comma, every dot in place, yet without any pedantry. These two letters are too personal to be read in public, except for one brief paragraph referring to sudden death (my father had died in an accident). Here are Einstein's words:

Aber es bleibt Ihr Trost, dass ein plötzlicher Abschied von dieser besten aller Welten dasjenige ist, was man einer geliebten Person in erster Linie wünschen muss, dass es nicht geht wie in Haydns Abschiedssymphonie, wo ein Instrument des Orchesters nach dem andern sich verflüchtigt. So wird das Bild, das wir alle von diesem liebenswerten Manne haben, ungetrübt durch die Erinnerung an einen langsamen Verfall, als eine lichte Erinnerung weiterleben. Er war ein ächter Künstler, auf den

Sie mit Recht stolz sein dürfen und es bleibt ihnen auch der frohe Gedanke, dass Sie ihm alles waren, was ein Sohn seinem Vater nur sein kann. . . .

[It is your comfort that a sudden farewell to this best of all worlds is something that one must wish a beloved person above all, so that things do not happen like in Haydn's Farewell Symphony, when one instrument of the orchestra vanishes after the other. Thus the picture that we retain of this lovable man will remain unblemished by the memory of a slow decline, it will continue to live as a most cherished reminiscence. He was a true artist of whom you can rightly be proud, and you can cherish the thought that you were everything to him that a son could ever be to his father.]

This is how I remember Albert Einstein, whom I saw some six months before his final illness and death—"unblemished by the memory of a slow decline"—on the sunny porch in Princeton, posing for a photo with my wife and my little son. And as always when I left him, I felt ennobled by his presence.

Ernst G. Straus

REMINISCENCES

My mother told me that the first time that Albert Einstein came to our house was when I was only three years old, on the occasion of the founding session of the Council of the Hebrew University. It was October 1925, and our house was just around the corner from the Oktoberwiese, the place for the big annual fair in Munich. According to my mother, Einstein said that he would much rather be there than at a session. He did indeed leave early, and my mother had her suspicions as to where he went, but there is no proof.

I had not planned to talk about Albert Einstein the public man, but I was somewhat goaded into it, because I got the feeling that this is the time that everybody wants to convert Pegasus to a draft horse and hitch him to his own cart. I know that this is utterly wrong from Einstein's point of view, so I shall mention a few quotations, to add to that pile of cards that, as Professor Erikson said, everybody seems to shuffle.

In talking about Albert Einstein as a dedicated member of any political, national, or other grouping, we should recall this general attitude that he expressed again and again: "In the long run, I have never been able to tolerate belonging to any group, any nation, any party, even any committee."

There is one story that impresses me as particularly typical. An FBI man came to his office and said: "Mr. X applied for a position as a physicist at the Aberdeen Proving Grounds, and he gave you as a reference." Einstein said: "Yes, he is a very talented young physicist, the government is very lucky to get him." The FBI man said: "You don't understand. I am not interested whether he is a good physicist, I want to know if he is a good American." Einstein was taken aback and said: "Oh no, he is much too intelligent to be a good anything." I must say in all fairness to the FBI man that he caught himself fairly shortly after that and said: "I think we still don't understand each other—I want to know if he is going to commit espionage." Einstein said: "Of course not." And everybody was happy.

I want to mention one more incident. During the period I was with him, 1944-1948, there were very exciting things happening in the political world, and we usually talked about this during the twenty-minute walk from Einstein's house to the Institute, unless we had a research topic that could be discussed without a blackboard. One frequent topic concerned the emerging United Nations Organization. Einstein had much greater hopes than I did that it would develop into a kind of world state, or, as he called it, a supranational authority. Once he mentioned some suggestion that had been sent to him, which would convert the United Nations from an organization of governments more into an organization of people. I, mainly to be polite, said: "Oh, that sounds like a good idea." He suddenly stopped in his tracks, turned to me, and said: "There are absolutely no good ideas in politics. The ideas are all obvious, the only problem is to get people to act on them."

When I think of the talks we have heard that relate to Einstein's political ideas and his scientific ones, it seems to me that this is the main point of difference. He would turn every scientific idea in all directions, never ceasing to look at it from a new angle, to criticize it again, to pick it up again, and to examine it. In his political ideas, on the other hand, he felt that the idea is clear, that the only problem was to state it. I think that the constant preoccupation that marked his scientific work simply did not happen in his political and social thought.

Most revealing in this respect is the autobiography—he used to call it his obituary—that he wrote at the time I was working with him, around 1947. If you read this really remarkable document, which, he felt, was to contain the core of his life story, you note that he mentions only one relative, his father, and that only in connection with the compass that he showed to the five-year-old Albert. It mentions no pacifism, no socialism, no Zionism, no atomic bombs and thereby, I think, goes to the essence of what was most important in Einstein's life. As he puts it himself: "What is essential for a man of my kind is how he thinks and what he thinks, not what he does or what is done to him."

There is one aspect of Einstein's political life that I knew; but

I only learned at this symposium that it was a part of his public stand as well. I mean his uncompromising hostility to Germany and all things German, which outlasted the destruction of Nazism. It seems inconsistent with his deep sense of justice and humanity. Perhaps it was an attempt to limit the incomprehensible horror by attributing it to a particular national trait. I only bring it up because I want to balance it with an event that puts the harsh words in a somewhat different perspective.

After the end of the war a distinguished mathematician came to visit Princeton from Germany. Several professors at the university and at the Institute for Advanced Study objected, since he, not a German by birth, had chosen to become a German citizen after German armies had overrun his homeland. A copy of this protest was in Einstein's mailbox for his signature. He asked me to find out what it was about, and when I told him, he was genuinely shocked. "You mean they want me to sign something just to harm another human being?" With that he dropped the letter in the wastepaper basket.

Now, I would like to say a few words that may add one little card to Professor Erikson's stack, a story about Einstein's childhood that I think everybody quotes from Frank's book. I still have not read Frank's book, but I do have a version of the story from Einstein himslf.

He said that when he was between two and three years old, he formed the ambition to talk in whole sentences. If somebody asked him a question and he had to answer, he would form a sentence in his mind and then try it out on himself, thinking that he was whispering it to himself. But, as you know, a child is not very good at whispering, so he said it softly. Then, if it sounded all right, he would say it again to the person who had questioned him. Therefore, it sounded, at least to his nursemaid, as if he said everything twice, once softly and once loudly, and she called him "der Depperte," which is Bavarian for "the dopey one." The nickname stuck, and that, at least in Einstein's mind, was the cause of all the stories about his slow development.

As to the story about his dyslexia, I had not heard it, and he never commented on it, as he presumably had not heard it either.

Another comment you frequently hear concerns his solitude. From an objective point of view, I think that this is quite an inaccurate description of his life. At the time that I was with him, he had in his home a sister with whom his relations were excellent, a stepdaughter who had grown up with him and knew him as really her only father, because her own father had died when she was quite young, and a devoted secretary who had been with him for decades. He had many friends, including some extremely close friends, at least in Wolfgang Pauli and, in particular, in Kurt Goedel. I think his relations with me were also very good. When he talked about solitude, he really meant what he used to emphasize whenever he felt like giving fatherly advice, saying that if you want to be a happy man, you should tie your life to a goal, not to other people and not to things. I think his connection of his life to his goal was a passion so intense that by comparison any relation to people certainly pales. But I think that this would be the case even if we compare it with the most passionate relation between any other people. I have never seen a relation among human beings that would approach the passion of Einstein's relation to his goal. So, I think the solitude is one of comparison, not of objective reality.

Even more interesting, and I think slightly a self-deception, is Einstein's view that thinking, the quest for truth, is an entirely solitary occupation. Those of us who have related our reminiscences of Einstein have offered, to a certain extent, counterexamples. All his life, he had friends or assistants with whom he worked. Thus his prescription that a great scientist should live where he can be completely alone and still have enough to eat really did not fit the way he worked and lived. But I know from at least one case that this was a very strongly held conviction.

Customarily, we worked together in the morning. In the afternoons, we worked separately, unless there was something exciting, in which case he would telephone—in the beginning I could not afford a telephone, so he used to send telegrams. On one occasion, we came back the next morning, and each of us had made some advance on our problem without having solved it. We were batting it around, and after about half an hour, we had solved the problem. Einstein's reaction was strange. He said: "This has never happened to me—we thought together. Two people being able to think! I never thought that this was possible. I was always convinced that thinking was a singular occupation."

Now, he was very quick to praise, but never in this particular form. Still, I thought that it was a new way of saying something nice. As was our custom, we picked up Goedel to go home for lunch, and Einstein told the same story to Goedel, saying that this exceptional and unbelievable thing had happened to us. Goedel, who was really totally solitary and would never talk with anybody while working, also thought that this was a miracle. So they strengthened each other in this particular conviction.

Perhaps in this context I should comment on Einstein's ideas of the proper occupation for a scientist. The prescription that he should be a lighthouse keeper is directly related to the conviction that thinking is a solitary process, best done in the absence of human company. During the McCarthy period he said that if he had to choose an occupation now, he would become a plumber. People have interpreted this to mean that he had become disillusioned with physics and all the evil that it had brought on the world. The correct explanation is entirely different. He always had the feeling that there is something wrong in making a living by research. In his own case, he felt that the work in the Swiss patent office was ideal and that his other occupations, which involved preparing classes and teaching, were relatively worse ways of doing research. Especially during the McCarthy period, it seemed to him that teaching not only interfered with research, because it made you think about something related, which nevertheless was not the same thing, but it also subjected you to evils like the Un-American Activities Committee, which in his opinion could not touch plumbers, who were too independent. He was convinced that science at its best should be a hobby and that one should make a living at something else. I do not think that he ever meant that while he was making his living as a plumber he would not go home to take out his pad and start writing.

Since Banesh Hoffmann told of a little mathematical puzzle that he shared with Einstein, I will tell you mine, which may even shed some light on Einstein's approach to mathematics. The first day I came to work with him he asked me to tell him a theorem that I had proved recently. I picked one of which I was rather proud. Given any closed curve C in the plane and any triangle, there are three points, P, Q, and R, on C that form the vertices of a similar triangle.

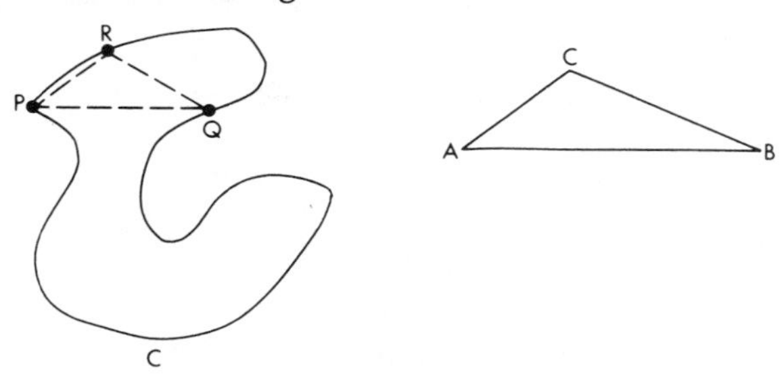

Einstein was aghast. He said: "But this is so unnatural! Similar triangles are metric concepts and closed curves are topological [he said analysis situs] concepts." For those of you who are not specialists, he meant that similarity of triangles is the kind of property you preserve if you make good maps, while closed curves are the kind of objects preserved if you draw your maps on a sheet of rubber and allow people to push and pull. He felt that since these two groups of transformations are not the same, you should not mix the concepts. From the point of view of the physicist who made the demand of invariance under certain groups of transformations the core of his search for the laws of nature, this is presumably the correct approach. On the other hand, most mathematicians would probably agree with me that an interplay of invariants under different groups is of special interest.

This difference between the approach of the physicist and that of the mathematician impressed me frequently, especially since Einstein would ask me to explain not only my own mathematics but also such questions as why, for example, John von Neumann was considered a great mathematician. He was not impressed by ideas that struck him merely as "very clever," but he would appreciate beauty and depth when he perceived it.

The one man who was, during the last years, certainly by far Einstein's best friend, and in some ways strangely resembled him most, was Kurt Goedel, the great logician. They were very different in almost every personal way—Einstein gregarious, happy, full of laughter and common sense, and Goedel extremely solemn, very serious, quite solitary, and distrustful of common sense as a means of arriving at the truth. But they shared a fundamental quality: both went directly and wholeheartedly to the questions at the very center of things.

Einstein often told me that when he was a young student, he had hesitated between mathematics and physics, and then it had occurred to him that in mathematics there are many beautiful questions, all of them seeming to be of equal importance, and that he would never be able to decide which were central and which were peripheral, whereas in physics, he could see what the central questions were. I think he probably misjudged himself. He had an incredible grasp of the physical side, far beyond his powers on the mathematical side, even in the areas that he was working on, and that was the only reason I could be of any use to him. But he said: "Since I have met Goedel, it has become clear to me that it is possible also in mathematics to recognize what the central problems are."

Einstein's quest for the central problem for the ultimate correct field theory is generally considered to have failed. I think that this did not really surprise Einstein, because he often entertained the idea that vastly new mathematical models would be needed, that possibly the field-theoretical approach through the kind of mathematics that he knew and in which he could do research would not, could not, lead to the ultimate answer, that the ultimate answer would require a kind of mathematics that probably does not yet exist and may not exist for a long time. However, he did not have the slightest doubt that an ultimate theory does exist and can be discovered.

Index of Names

Abraham, Max, 8, 9, 12, 13, 16, 19, 20, 23, 83
Adler, Fritz, 287
Ahad Ha'am (Asher Hirsh Ginsberg), 285
Aichelburg, P. C., 247
Annan, Noel, 330
Arbatov, G., 360
Arendt, Hannah, 261
Aristotle, 179, 290
Artsimovitch, L. A., 360
Avogadro, Amedeo, 43, 47, 55
Aydelotte, Frank, 367, 391

Bach, Johann Sebastian, 154, 209, 409, 410, 415
Bacharach, A., 167, 173
Bacon, Sir Francis, 211, 217, 223, 229, 242, 254, 273, 288
Bacon, Roger, 276
Bargmann, Valery, 35, 398, 414
Barthes, Roland, 275
Baruch, Bernard, 380
Bedini, Silvio A., 274
Beethoven, Ludwig van, 209, 291, 410
Bell, John Stewart, 72
Bellarmine, Cardinal Robert, 264
Ben-Gurion, David, 293-295
Benjamin, Walter, 208
Bergia, Silvio, 250
Bergman, Shmuel Hugo, 204, 304, 305, 315
Bergmann, Peter G., 398
Bergson, Henri, xiii, 190-191, 264, 265
Berkeley, Bishop George B., 74, 222
Besso, Michele, 41, 73, 76, 145, 157
Bethe, Hans A., 392
Bismarck, Otto von, 322, 323
Blagonravov, A. A., 360
Blake, William, xvi, 155
Bloch, Marc, 271
Bloomfield, Leonard, 261
Blumenfeld, Kurt, 284, 305, 313, 314
Bogdanov, A. A., 128
Bohm, David, 72, 73
Bohr, Niels, xxvii, 65, 66, 68, 69, 70, 72, 74, 75, 76, 84, 85, 86, 107-108, 127, 148, 180, 212, 291, 317
Boltzmann, Ludwig, 14, 40, 41, 44, 46, 49, 50, 52, 53, 54, 55, 247
Bolzano, Bernardo, 181
Born, Hedwig, 231
Born, Max, xviii, 64, 67, 68, 69, 73, 112, 203, 213, 222, 243, 249, 286, 289, 340, 393
Brace, D. B., 9, 24
Brans, Charles, 36
Braque, Georges, 146
Brecht, Berthold, 208, 244
Bridgman, P. W., xiii, 267, 393
Broad, C. D., 82
Broch, Hermann, xvi, 171
Brodetsky, Selig, 304
Broglie, Louis de, 64, 65, 66, 222
Broglie, Maurice de, 65
Brown, Harrison, 392
Buber, Martin, 315
Bucherer, A. H., 20, 21, 26
Bull, H., 360
Burtt, E. A., 68
Bush, Vannevar, 367, 391
Butler, Nicholas Murray, 337
Byrnes, James F., 375, 392

Çapek, Milic, 270
Cartan, Elie, 34

Cassirer, Ernst, 146, 187-188, 190
Chase, Stuart, 262, 266, 267
Clauser, J. F., 72
Cockcroft, J. D., 361
Cohen, I. B., 245
Cohen, Robert S., 247
Cole, Frank Nelson, 243
Compton, Arthur Holly, 83
Conant, James B., 271, 272
Condon, Edward U., 405
Copernicus, Nicolaus, xxi, 161, 164

Darwin, Charles, 161, 165, 282, 291
Davidson, Randall (Archbishop of Canterbury), xii
Davy, Humphry, 256
Democritus, 212
Dennison, David M., 405
Detienne, Marcel, 220
Dewey, John, 259, 263, 266, 271, 330
Dicke, R. H., 36
Dirac, P.A.M., 244
Dorling, J., 63
Drude, Paul, 22
Dugas, R., 248
Dukas, Helen, xxiii, 414
Durrell, Lawrence, xvi, xvii, xxx

Eban, Abba, 294, 295
Eddington, Arthur Stanely, 80, 81, 82, 86, 103, 108-121, 127, 128, 130, 207, 223, 247, 330, 370
Ehrenfest, Paul, 62, 71, 104, 110, 348
Einstein, Elsa, 230, 409, 412
Einstein, Maja, *see* Winteler, Maja
Eisenhower, Dwight D., 289
Eitingon, Max, 167
Eliot, T. S., xvi, xx
Epicurus, 205
Erikson, Erik H., 326
Euclid, 154

Faraday, Michael, xxvii, 10, 11, 12, 13, 16, 19, 24, 28, 247, 256
Faulkner, William, xvii, xviii, xx, xxx
Ferenczi, Sandor, 167
Fermi, Enrico, 373, 374, 391
Feyerabend, Paul, 136, 218, 246
Fine, Arthur, 66
Fitzgerald, George Francis, 81
Flandern, T. C. van, 89
Fok (Fock), V. A., 109, 110, 120-129
Föppl, August, 12, 13, 16, 25, 26
Francis of Assisi, Saint, 160, 332
Franck, James, 239, 286, 341, 343, 376
Freud, Sigmund, 160, 161, 162, 165, 166, 168, 169, 170, 282, 334, 351, 352
Fridman (Friedmann), A. A., 86, 109, 125, 132

Galileo Galilei, xxi, 30, 249, 253, 264, 268, 276, 281, 291
Gandhi, Mahatma, 355, 384
Gardner, Michael R., 72
Geertz, Clifford, 210, 217, 218
Gellner, E., 243
Gessen, B. G., 128
Gibbs, Josiah W., 40, 44, 54
Gillispie, Charles C., 120
Ginzburg, Vitalii L., 126
Goedel, Kurt, 420, 422
Goethe, Johann Wolfgang von, 156, 160
Goitein, David Yeheskel, 294, 295
Goldstein, Rabbi H. S., 317
Goltsman, A., 128
Goodman, Nelson, 245, 249
Grassmann, H. G., 36, 37
Greenberg, Haim, 314
Gromyko, Andrej, 356
Grossmann, Marcel, 54, 102, 105
Groves, General L. R., 375
Grünbaum, Adolf, 194, 195

Haber, Fritz, 284, 326, 327, 328, 333, 337, 338, 340
Habicht, Conrad, 48, 55, 56, 60, 61
Hadamard, Jacques, 140, 141, 142, 350
Hahn, Otto, 353, 372
Haldane, Lord Gerald, xii, 203, 283
Hamilton, P. N., 41
Hawking, S., 35
Hegel, G.W.F., 179, 246
Heisenberg, Werner, xxviii, 66, 68, 70, 75, 76, 84, 180
Helmholtz, Hermann von, 5, 14, 40
Hertz, Heinrich, 5, 13, 14, 40, 41
Herzl, Theodor, 284
Hess, Moses, 285
Hitler, Adolf, 287, 288, 319, 324, 336, 339, 341, 355, 372, 376, 377, 390

Hoffmann, Banesh, 31, 60, 187, 317, 398, 421
Hofstadter, Richard, 321
Hogness, Thorfin R., 392
Holenstein, Elmar, 144
Holt, R. A., 72
Holton, Gerald, 139, 149, 155, 164, 190, 212, 213, 246, 249, 268, 269, 333
Horne, M. A., 72
Horton, Robin, 243
Howard, M., 360
Hume, David, 18, 222, 282, 287, 290, 301
Huxley, T. H., 282
Huygens, Christiaan, 253

Infeld, Leopold, xxv, 31, 127, 393, 399, 401
Inozemtsev, N., 360

James, William, xxvii, 334
Jeans, James, xvii, 108, 109, 121
Joliot-Curie, F., 373, 393
Jordan, Pascual, 128

Kahler, Erich, 315
Kaluza, Th., 35, 70
Kant, Immanuel, 4, 191, 193, 198, 200, 202, 221, 243, 247, 282, 290, 310
Kapitsa, Peter L., 109, 129
Kaufman, Bruria, 32
Kaufmann, Walter, 8, 9, 20, 21, 83
Kepler, Johannes, xxi, 164, 212
Keswami, G., 59
Kissinger, Henry, 360
Klein, Martin J., 60
Kobzarev, I. Iu., 127
Kochen, S., 72
Kohn, Hans, 315
Korzybsky, A., 275
Kramers, Hendrik Anthony, 69
Kuppermann, Aron, 72

Landau, Lev, 109, 126
Laue, Max von, ix, 75, 331, 337, 339, 340, 341, 375, 390
Leibniz, Gottfried Wilhelm, 214, 215
Lemaitre, Georges, 86
Lenin, Vladimir Ilyich, 121, 127, 129, 329
Leontovich, M. A., 129
Lévi-Strauss, Claude, xiv
Liebermann, Max, 327
Liouville, Joseph, 40, 41
Lobachevsky, N. I., 125
Locke, John, 258, 259, 272
Lodge, Oliver, 164
Lorentz, Hendrik A., ix, 5, 6, 7, 8, 9, 13, 14, 15, 16, 17, 18, 19, 20, 21, 23, 28, 83, 228, 247, 306
Loschmidt, Joseph, 46, 55
Luria, Alexander, 219
Lurie, Zvi, 313, 316

Mach, Ernst, xii, xxi, 4, 5, 18, 49, 50, 57, 212, 213, 282, 287, 288
Majakowskij, Vladimir, 149
Maksimov, A. A., 128
Mann, Thomas, xvi, xx
Marcus, Joseph, 314
Marcuse, Herbert, 330
Marx, Karl, 127, 129, 183
Maxwell, James Clerk, 5, 19, 28, 46, 62, 63, 64, 224, 247, 390
Mayer, Walter, 70, 405, 406
Mendeleev, Dmitri Ivanovich, xxvii
Merleau-Ponty, Maurice, 264, 265
Michelson, Albert Abraham, 8, 81
Mill, John Stuart, 178, 288
Millikan, Robert A., x, 63, 317
Millionshchikov, M. D., 360
Minkowski, Hermann, xx, xxvii, 29
Mommsen, Theodor, 323
Mondrian, Piet, 146
Morgenthau, Henry, Jr., 341, 360
Morley, Edward Williams, 8, 81
Morse, Philip M., 392
Mozart, Wolfgang Amadeus, 154, 209, 410
Muller, Hermann J., 393

Nernst, Walther, 46, 55, 65, 66, 284, 326
Neuman, Johann von, 422
Newton, Sir Isaac, viii, xii, xv, xx, xxi, xxiii, xxiv, xxvii, 4, 5, 6, 7, 8, 16, 17, 18, 19, 28, 29, 30, 80, 100, 103, 104, 120, 122, 156, 187, 189,

Newton, Sir Isaac (*cont.*)
206, 207, 212, 213, 242, 271, 281, 283, 286, 291, 320, 390
Nicholas of Cusa, 176
Nietzsche, Friedrich, 183, 197, 200, 322

Ogden, C. K., 266, 275
Ong, Walter J., 274
Oppenheimer, J. Robert, 271, 272, 336
Ortega y Gasset, José, xxviii
Ostwald, Wilhelm, 49, 57, 324

Panofsky, Erwin, 340
Pauli, Wolfgang, xvii, 64, 225, 247-248, 420
Pauling, Linus, 392, 393
Penrose, R., 35, 36
Perlzweig, M. L., 312
Perrin, Jean, 51, 56, 57
Pestalozzi, Johann Heinrich, 156
Piaget, Jean, xxii-xxiii
Picasso, Pablo, xiv
Placzek, G., 391
Planck, Max, xii, xv, xxi, 3, 13, 14, 46, 47, 55, 56, 61, 62, 64, 65, 68, 122, 134, 145, 214, 246, 247, 282, 284, 291, 326, 327, 337, 340, 341, 390
Plato, xxvii, xxviii, 208, 220
Podolsky, B., 71
Poincaré, Henri, 5, 8, 9, 10, 14, 15, 18, 19, 20, 21, 23, 213
Popper, Sir Karl, xiii
Powell, Cecil F., 393
Premack, David, 243
Protagoras, 188
Purcell, Edward Mills, xi

Quine, Willard van Orman, 243, 250

Rabi, I. I., 325
Rabinovitch, J., 361
Rathenau, Walter, 300, 333, 338
Rayleigh, Lord J.W.S., 9, 62
Reichenbach, Hans, 188
Renninger, Mauritius, 72
Richards, I. A., 266, 275
Riemann, G.F.B., 87
Rolland, Romain, 306, 330, 377
Roosevelt, Franklin D., xxii, 340, 354, 371, 373, 374, 375, 376
Rosen, Nathan, 399
Rosenfeld, Leon, 71
Rosenthal-Schneider, Ilse, 207, 247
Rotblat, J., 393
Routh, H. V., xvii
Russell, Bertrand, xxii, 222, 241, 282, 310, 330, 358, 359, 385
Rutherford, Lord Ernest, xvii, 65, 291
Ryle, Gilbert, 218, 219

Sachs, Alexander, 373
Samuel, Lord Herbert, 283
Sapir, Edward, 147
Sartre, Jean-Paul, xvi
Schapiro, Meyer, xxx
Schlick, Moritz, 191
Schnabel, Artur, 410
Scholz, Heinrich, 181, 182, 183, 187
Schopenhauer, Arthur, 241
Schrödinger, Erwin, 66, 67, 73, 75
Semkovskii, S. Iu., 128
Seneca, 209
Shapiro, I. I., 89
Shimony, Abner, 72
Shmidt, O. Iu., 109
Shteinman, R. Ia., 128
Shulman, M., 360
Siedentopf, H., 56
Sitter, Willem de, 80, 81, 86
Smoluchowski, Marian von, 48, 53
Soddy, Frederick, xvii
Solovine, Maurice, 222, 223, 239, 242
Solvay, Ernest, 65
Sommerfeld, Arnold, 66, 228
Specker, E. P., 72
Spees, A. A., 26
Spinoza, Baruch, 282, 289, 308, 309, 317, 331
Sprat, Bishop Thomas, 254, 256, 263
Stalin, Joseph, 120, 129, 357
Stefan, Josef, 45
Stern, Otto, 66, 326
Strassmann, F., 353, 372
Straus, Ernst G., 136, 238, 249, 398
Svedberg, Th., 57
Szilard, Leo, 361, 372, 373, 374, 375, 376, 390, 391, 392

Tagore, Rabindranath, 317
Talmey, Max, 170

Tamm, Igor, 109, 126, 360
Tarski, Alfred, 196
Teller, Edward, 373, 391
Thomson, Sir Joseph John, xii, 40, 109, 330
Trilling, Lionel, xxv, 39, 320
Truman, Harry S., 375, 380

Urey, Harold C., 392
Usborne, Henry, 357
Ussishkin, Menachem, 313

Vavilov, S. I., 109
Vernant, Jean-Pierre, 220
Veselov, M. G., 129
Visciglini, Mario, 105
Voltaire (François-Marie Arouet), 283, 290, 331
Vygotsky, L. S., 219, 220

Waals, Johannes Diderik van der, 46, 55
Wald, František, 50
Wallace, Henry A., 357, 384
Warburg, Felix, 335
Weber, Alfred, 331
Weber, Max, 217, 322, 331
Weinberg, Steven, 238
Weisskopf, Victor F., 391, 392
Weizmann, Chaim, xxii, 237, 283, 287, 294, 299, 304, 313, 316, 335, 336, 338, 340
Weizsäcker, Carl Friedrich von, 374
Welles, Sumner, 356
Wertheimer, Max, 140-141, 336
Weyl, Hermann, 33, 34, 38, 66, 70, 87, 88
Wheeler, John A., 135
Whitehead, Alfred North, xiii, 206, 207
Whittaker, E. T., 59
Whorf, Benjamin Lee, 147
Wien, Wilhelm, 8, 22
Wigner, Eugene P., 354, 373, 391
Williams, William Carlos, xvi
Willstaetter, Richard, 338-339
Winteler, Jost, 143, 144, 145
Winteler, Jost Fridolin, 144
Winteler, Maja (Einstein's sister), 139, 143, 152, 154

Yukawa, Hideki, 393

Zahn, C. T., 26
Zeisler, Ernest B., 301
Zuckerman, Harriet, 324

Subject Index

Aarau, 143, 144, 145
Aberdeen Proving Grounds, 417
absolute, the, xv, 122, 131, 175, 176, 177, 224
absoluteness, 189, 205
absolute values, 87, 119
absolutism, 80, 131
acceleration, 8, 28, 100, 123, 124; gravitational, 29, 30, 31
Acheson-Lilienthal-Oppenheimer Report, 380
alchemy, xxi
algebra, 36, 93; Grassmann, 36, 37; hypercomplex, 36
American Academy of Arts and Sciences, 359
anthropology, 108, 216; cultural, xiv
antimatter, 85
anti-Semitism, 284, 285, 286, 301, 302, 303, 323, 337, 338, 339, 349
antisymmetric particles, 37
Arabs, 236, 237, 283, 286, 289, 295, 299, 304, 305, 335
arbitration, 347, 348, 349, 351, 378, 379
arms control, 359, 360, 361, 363, 364, 365
art, artists, viii, ix, xiv, xv, 115, 145, 207, 210, 240, 290, 291, 300, 309, 323, 325, 327
astronomy, x, xx, 28, 30, 80, 89, 108, 111
asymmetric theory, 32, 34
asymmetry, 17, 18, 149, 177, 224, 225, 227, 233, 243
atheism, 117, 130, 251
atom, xxvii, 14, 43, 50, 69, 97, 98
atomic bomb, *see* bomb, atomic
atomic clock, 89
Atomic Energy Commission, 380, 391
atomic phenomena, 74
atomic problems, 88
atomism, 50, 212, 231
authoritarianism, 326, 333

Baruch Plan, 356, 380
behavior, 199, 200, 201
behaviorism, 209, 243
Belgian Congo, 373, 374
Belgium, 348, 352, 373
Berlin, 103, 284, 327, 331, 337, 339, 348, 353, 372, 390, 409, 412, 414; intellectuals, 348
Berlin (Prussian) Academy of Sciences, 65, 104, 250, 326, 339, 340
Bern, 347; Patent Office, 40, 283, 326
big bang theory, 87, 88
bolshevism, 329
Boltzmann's constant, 42, 46, 55. *See also* constant, universal
Boltzmann's relation, 61, 63
bomb, 354, 373, 376, 386, 391, 392; atomic, 290, 299, 371, 372, 375, 376, 380, 381, 386, 418; chain reaction, 354; fission, 358; hydrogen, 358, 386, 387; nuclear, 371, 386; plutonium, 392; uranium, 391. *See also* weapons
boundary conditions, 86, 124, 125
breeder reactors, 391
Brownian (molecular) motion, *see* motion, Brownian

Canada, 374
causality, xiii, xviii, 4, 46, 69, 70, 73, 101, 102, 131, 176, 185, 203, 271,

causality (*cont.*)
291, 309, 370, 390
chain reaction, 353, 372, 373, 374, 392
child, childhood, 151-161 passim, 165, 291, 419
China, 376
Christianity, 297, 307, 308
circuit: closed, 12; electric, 4, 11; motion of, 4
civil discourse, 258, 259, 263, 266, 267
clocks, xi, xviii, 4, 29, 70, 81, 99, 112, 254, 260; atomic, 89
Cold War, 356, 357, 363, 384, 385
collision processes, 67
colloidal particles, 47
common sense, 85, 112, 129, 210, 229, 234, 237, 246, 254, 256, 257, 259, 261, 262, 263, 265, 266, 270, 287, 422
communality, invariance of, 163
communism, 258, 266, 383, 384, 385, 386, 387, 388
compass, 154, 418
complexity, 36, 211, 213-216, 219, 221, 227-231, 240, 242, 243, 319, 332, 333
Compton effect, 63, 84
conductor, 16, 17, 19
conservation: of energy, 101, 212, 213; laws, 102, 212
constant: atomic, 46, 88; Boltzmann's, 42, 46, 55; Faraday, 47; Planck's, 46, 88; Stefan-Boltzmann, 45; universal, 44, 45, 46, 49, 51. *See also* universal constant *k*
continuity, 33, 69; invariance of, 163
continuum, 68, 70, 75, 164, 211, 213, 219, 227, 228, 230-233, 243, 247, 248
contraction, 8, 9, 10, 15, 19, 29, 81, 87; Lorentz-Fitzgerald, 81; of measuring rods, 81, 260
coordinates, 101, 186; Cartesian system of, 186; complex, 36; harmonic, 126, 129, 131; local time, 7, 13, 15, 16, 17; spatial, 17; system of, xiv, 101, 125, 126, 170; transformations of, 100, 101
corpuscular structure, 259
correlation, 176, 177, 179, 181, 182, 183, 184, 196, 197
corresponding states, (Lorentz) theorem of, 7, 8, 10, 16, 17
cosmos, xxvii, 145, 193, 214, 216, 217, 219, 220, 240, 277, 308, 371
cosmology, viii, x, xx, 36, 86, 88, 326
Coulomb's law, xi
covariance, principle of general, 101, 102, 103, 104
credo, 232, 240, 250, 318
criteria, absolute, 218
cubism, xiv, 146
culture, cultural, viii, xiii, xxvi, xxvii, 184, 185, 207, 209, 210, 216, 217, 234, 254, 261, 263, 265, 266, 268, 270-272, 286, 289, 305, 325, 330, 339, 341, 342, 379; analysis, 216, 217, 218, 219; attitudes, 269; conditions, 197, 202; crisis, 260; development, 285; effects, 273; effort, 377; mutation, 282; national, 349; nationalism, 285; political, 107-132 passim; popular, 271; process, 210; public, 262-267, 273; relativism, xiii; role of science, 272, 273; of science, 261, 271; twentieth-century, 139, 282; Western, 206, 211
cunning reason, 220, 246. *See also* *metis*
curiosity, 159, 160, 326
current, electric, xi, 3, 10, 11, 12, 13, 19
Czechoslovakia, 353, 373, 374

Darwinism, 282
democracy, 254, 256, 257, 258, 261, 262, 266, 268, 269, 273, 287, 288, 297-299, 303, 327, 329, 333, 348, 350, 377-381; democratization, 256, 271, 377
density: of charge, 6; of matter, 86, 239
deontic, the, 182, 192, 193, 199, 201
dependence, 176, 177, 182, 183, 184, 199
derivation, 177, 178
description, 218; thick, 210, 212, 218, 219, 226, 246; thin, 219
determinism, 68, 84, 85, 102, 119, 183, 243, 245, 291, 309, 310

diameters, atomic, 46
diffraction, 62, 110
diffusion, xi, 391
dimensions, atomic, 49
disarmament, 350, 351, 352, 357, 361, 378, 379. *See also* arms control
displacement, 47, 51
distance, 28, 29, 30, 33, 86, 87, 88, 89, 90; action at a, 28, 29, 227, 233, 277; invariant, 34; relative, 34
distribution, 41, 61, 63; law of, 40, 43, 46, 47; of matter and energy, 121
double star, 31
Dulong-Petit theorem, 64, 65
dynamics, 4; laws, 32; theory, 11
dynamic synchrony, 148
dynamo, 4, 10, 11

eclipse expedition, xii, 81, 82, 83, 103, 206, 207, 330, 349, 371
education, 266, 305, 324, 330, 336, 356, 358, 378, 381, 384
Egypt, 295, 360, 388
Eidgenössische Technische Hochschule, 15, 40, 143, 156, 284, 326
einstein (unit of light energy), x
Einstein-Bose statistics, x, 37
Einstein–de Sitter model, 86, 87
Einstein-Podolsky-Rosen paradox, 71, 72, 73
Einstein tensor, x
Einstein viscosity equation, x
electric charge, 30, 47, 68
electricity, 20, 70, 225, 267
electrodynamics, ix, xiii, 3, 5, 18, 19, 29, 32, 61, 99, 224, 247
electromagnetism, 4-14, 16, 17, 19, 28, 34, 87, 97, 99, 110, 225, 226, 370, 390. *See also* induction; waves, electromagnetic
electron, 4, 5, 6, 8, 9, 10, 13, 14, 16, 18, 20, 21, 83, 85, 88, 97, 98, 225
electrostatics, xi, 9
emancipation, 285, 301
Emergency Committee of Atomic Scientists, 358, 380, 384
energy, 13, 29, 40-46, 60, 62, 64, 67, 69, 70, 71, 97, 98, 121, 227, 353, 373; atomic, 299, 381, 384, 388; conservation of, 53, 69, 247; distribution of, 63; international control of atomic, 380, 381, 384; kinetic, 227; mass-energy relationship, 372; nuclear, 371, 380, 381; photoelectronic, 63; radiation, 61
Enlightenment, xxiv, 165, 170, 254, 266, 271, 273, 281, 303, 310
ensembles, 41, 44, 73
entangled systems, 71, 74
entropy, 40, 41, 46, 50, 52, 61, 62, 247
epic theater, 205, 208, 209, 213, 244
episteme, 220
epistemology, xxi, 113, 131, 207, 211, 214, 215, 217, 220, 224, 227, 229, 231, 232, 233, 238, 241, 243, 245, 246, 247, 261, 268, 309
equations, 85, 88, 114, 406; field, 29, 34, 38, 86, 104, 125, 226, 228, 405; Klein-Fock, 110; linear-differential, 41, 68, 69, 124, 125; Lorentz (force), 5, 6; Maxwell, 94, 95; Maxwell-Lorentz, 5, 6, 7, 16, 23; of motion, 41; ponderomotive, 38; Schrödinger, 73; transformation, 17, 18
equipartition (of energy): law of statistical mechanics, 62; theorem, 40, 52
equivalence, 19, 100, 101, 147, 402; of acceleration and gravitation, 124; of gravitational and inertial mass, 123, 124; of inertial reference frames, 125; of mass and energy, 347, 371; of reference systems, 18
esthetics, 30, 31, 100, 102, 103, 165, 179, 234, 288-300, 307, 308, 309, 320, 322
ether, xxi, xxiv, 5, 6, 7, 8, 10, 12, 15, 18, 66, 81, 95, 99, 112, 164, 226, 228, 374
ether-drift, *see* experiments
ethics, viii, 165, 176, 179, 194, 195, 196, 199, 200, 201, 273; social, 297-312 passim. *See also* Jews, ethics
evolution, 282
experiments: electromagnetic, 7; ether-drift, 7, 15, 18, 19, 20; mechanical, 16, 100; Michelson-Morley, 8, 9, 81, 269; optical, 7, 9, 20, 99; Rayleigh-Brace, 9. *See also* thought-experiment

falsification, xiii, 20

Faraday constant, 47
fascism, 266, 290, 299, 329, 332, 350, 353
Federation of Atomic Scientists, 380
Fermi-Dirac statistics, 37
Feuerbach's theorem, 403
fields, 28, 92, 31, 33, 247; affine, 248; electric, 6, 8, 13, 16, 29; electromagnetic, 5, 8, 28, 31, 33, 34, 35, 64, 87, 88, 89, 94, 95, 110, 225, 226; guage vector, 70; gravitational, 28, 29, 30, 31, 32, 33, 35, 87, 88, 89, 100, 101, 123, 124, 125, 225, 226, 371; laws, 31, 32, 34; magnetic, 6, 8, 12; pervasive, 28; physical, 34; theory, 30, 34, 37, 68, 69, 70, 75, 398, 423. *See also* equations, field; unitary field theories
fission, 354, 372, 375; nuclear, 340, 353, 371, 373
fluctuations (in energy), 43, 44, 45, 46, 47, 49, 50, 52, 53, 60, 63; theory of, 52, 53
force: fields of, 64; gravitational, 87; intermolecular, 42; lines of, xxvii, 10, 11, 12, 13; long-range, 87; resistive, 51
frame of reference, free-falling, 31, 34
frames, inertial, 6, 7, 16, 19, 28, 30, 125
France, 281, 288, 348, 349, 352, 353, 373, 379
Freudian slip, xviii, 167
fusion: of fields, 35; of mathematical structures, 33, 34

games, 142, 154
gases, 14, 40, 43, 46, 47, 52, 97; degeneracy of, 66; diffusion of, 391; poison, 328; theory of, 44, 49, 50
generator, xi, 12
geodesic, 31
geometry, 10, 30, 32, 33, 84, 87, 88, 91, 92, 93, 154; Cartan, 34; Euclidean, 4, 185, 229; geometrization, 33; Lobachevskiian, 125; non-Euclidean, 291; Weyl's, 34, 88, 89
German citizenship, 156, 288
Germany, xxii, 103, 153, 284, 287, 301, 302, 319-343 passim, 349, 350, 351, 352, 353, 354, 355, 371, 373, 375, 376, 377, 379, 380, 392, 414, 418; citizenship, 156, 288; culture, 319, 322, 325, 348; education, 155, 284; ghettos, 285, intellectuals, 348, 371, 379; Jews in, 284, 286, 324, 325, 328, 339; nationalism, 288, 348; people, 302, 376; rearmament, 353, 358
Gesellschaft für positivistische Philosophie, xii
God, xviii, xx, xxvii, 115, 120, 129, 131, 151, 153, 158, 163, 176, 191, 205, 212, 228, 230, 239, 240, 249, 251, 260, 289, 300, 308, 309, 327, 336, 406
good, goodness, 176, 181, 182, 333, 389; absolute, 176
Göttingen, 331
Grassmann algebras, 36, 37
gravity, gravitation, x, 29, 31, 32, 34, 70, 100, 101, 103, 110, 124, 194, 213, 225, 228, 370, 390, 402; red shift, 103; supergravity, 37; theory of, 37, 86, 90, 122, 127, 242. *See also* acceleration, gravitational; equivalence; fields, gravitational; force, gravitational; mass, gravitational; waves, gravitational
Great Britain, 103, 108, 118, 120, 237, 286, 304, 335, 348, 353, 354, 355, 359, 371, 390, 391, 392. *See also* eclipse expedition
Greek drama, 205, 206, 207, 208, 209, 216, 220, 244

Hanford reactors, 392
harmony, 205, 214, 215, 217, 222, 224, 225, 233, 287, 289, 291, 308, 309, 337, 338; Leibnizian, 215, 220
heat, 225; molecular theory of kinetic, 39, 42, 48, 64, 65; specific, 64, 65, 390
heat bath, 42, 43
Hebrew University of Jerusalem, 283, 289, 299, 334, 335, 336, 417
Hebron, 236, 304, 335
Heidelberg, 331
hidden-variable theories, 72
Hiroshima, 375, 386, 392
Holocaust, 299
hydrogen, 47

hypercomplex numbers, 36

idealism, 129, 130, 209. *See also* Jews, ideals
independence, 176, 288
indeterminacy, principle of, xx, 70
indeterminism, 69
India, 355, 369
induction, electromagnetic, 4, 10, 11, 12, 13, 16-19, 22
inevitability, 205, 206, 207, 209, 216
infinity, 31, 85, 125, 126, 185, 228, 229, 239
innocence, 288, 332, 335, 337
Institute for Advanced Study, 340, 343, 390, 397
integration, 191, 192
intention, 200
interaction: of light and matter, 98; of physical bodies, 28
interactionist realism, 217-219 passim, 221
interference, 62, 98
internationalism, 285, 286, 307, 329, 334, 336, 348, 370
International Labor Organization, 299
International Society for General Relativity and Gravitation, 339
intuition, 4, 12, 21, 33, 97, 98, 100, 103, 141, 144, 145, 214, 215, 219, 220, 221, 223, 224, 227, 230, 232, 233, 235, 239, 254, 263, 265, 290
intuitional leap, 221-224 passim, 229, 230, 238, 240, 241
invariance, xv, 141, 144, 146, 148, 170, 422; relativistic, 148. *See also* communality, invariance of; continuity, invariance of; distance, invariant; natural laws, invariance of; structure, invariant
Invariantentheorie, xv
ion, 47
irreversibility, 41
isolation, 157
Israel, State of, 281-292 passim, 294, 295, 299, 306, 387, 388, 407; presidency of, 293-296 passim

Japan, 250, 354, 371, 372, 375, 376, 386, 392
Jerusalem, 288, 335
Jews, 119, 156, 236, 237, 283, 284, 285, 286, 288, 297, 299, 302, 304, 306, 324, 325, 328, 333-336, 338-341, 387; affiliations, 156; community, 284, 302; ethics, 298, 299, 300, 306, 311; ideals, 286; identity, 156, 301; nation, 285, 303; nationality, 306; philosophy, 300; relations with Arabs, 336; values, 301; world, viii, 281
Jewish Agency, 299
Judaism, 153, 297, 298, 300, 301, 307, 308, 311, 312
Judeo-Christian tradition, 298

Kantianism, 4, 12, 221, 223; neo-, 4, 14, 18
kinematics: Newtonian, 95; of relativity, xi
kinetic-molecular theory, 39, 42, 48, 64, 65
kinetic theory of gases, 42, 49, 50
Klein-Fock wave equation, 110
knowledge, absolute, 197
Korea, 376
Korean War, 358

language, linguistics, viii, 139-150 passim, 154, 159, 188, 221, 258-262, 266, 267, 272
laser, xi
Latitudinarians, 256
League for Human Rights, 250
League of Nations, 167, 289, 336, 349, 351, 377; Committee on International Cooperation, 299, 333, 349
Leibnizian harmony, 215, 220
light, 8, 14, 60, 62, 67, 94, 96, 224, 225, 259, 270; bending/deflection of, 81, 103, 349, 371; emission, xxi, 13, 62, 224; interaction with matter, 98; Maxwellian theory of, 98; particles of, 14, 64; signals, 81; sluggish/frozen, 96, 97; speed/velocity of, 6, 7, 10, 14, 15, 16, 17, 18, 19, 28, 96, 112, 122, 188; transformation of, 60, 62; waves, 15, 64, 93, 95, 100, 156, 224; wave theory of, 6, 18, 62, 98. *See also* quantum, quanta, light
lightning, 257, 264

Liouville's theorem, 40, 41
Lorentz-Einstein theory, 20, 21, 83
Lorentz-Fitzgerald contraction, 81
Lysenkoism, 128

McCarthyism, McCarthy period, 306, 341, 370, 384, 421
magic, 104, 242, 244, 262, 263, 332
magnetism, xi, 3, 4, 7, 10, 11, 12, 13, 16, 17, 19, 225
magnitudes, atomic, 43, 49
Manhattan Project, 354, 374, 375, 391
manifesto of the ninety-three, 327, 348
Mars, 89
Marxism, 109, 110, 121, 126, 128-130, 288
mass: atomic, 353; gravitational, 100, 123; inertial, 100, 123, 390
materialism, xiii, 116, 119, 121, 123, 128-131, 282
mathematics, mathematicians, xxiv, 3, 9, 16, 29, 33, 82, 83, 111, 115, 120-122, 127, 140, 144, 145, 153, 159, 160, 184, 187, 213, 214, 215, 222, 225, 227, 228, 254, 255, 262, 266, 290, 331, 405, 406, 421, 422, 423
matter: dielectric, 3, 7, 13; magnetic, 3
Maud Report, 354
Maxwellian theory of electric processes, 97
Maxwell-Lorentz electrodynamics, 68
May-Johnson bill, 380
mechanics, xiv, xxi, 4, 5, 6, 7, 8, 12, 14, 16, 17, 19, 31, 32, 40, 41, 49, 69, 73, 100, 225; causal theory of quantum, 73; Galilean, 229; matrix, 66; molecular, 65; Newtonian, 99, 100, 277; quantum, 64, 66-73, 75, 84, 85, 102, 108, 110, 120, 121, 238, 290, 405; relativistic, 29; statistical, xi, 14, 32, 40-42, 44, 49, 50, 52, 60; wave, 66
mechanistic world view, 119
Mercury, 81, 103, 104
metaphors, xxvi, xxvii, xxviii, 111
metaphysics, xiii, 4, 68, 201, 234, 241, 243, 247, 250, 260, 262, 265, 266, 267, 272, 283
metis, 220, 221. *See also* cunning reason
miracle, 242, 251, 421
molecules, 44, 46, 47, 50, 52; collisions, 48; structure, 405
momentum, 69, 71, 73, 74, 101; developmental, 170
moral, the, morality, 182, 197, 198, 199, 200, 208-211, 229, 234, 235, 286, 289, 292, 297-300, 305, 308-311, 332, 334, 341, 351, 371, 380, 388
moral authority, 382
moral code, 235-238 passim
moral effort, 351, 378
moral imperative, 322
moral indignation, 320
moral instability, 284
moral values, 282, 283, 288, 297, 298
motion, movement, xxiii, 5, 6, 11-19, 28, 41, 50, 51, 69, 73, 94, 96, 99, 100, 103, 104, 125, 187, 225, 226; absolute, 8, 28; accelerated, 125, 246; Brownian, xi, 14, 47, 48, 49, 51, 52, 53, 60, 390; inertial, 19, 125; thermal, 47, 48
motivation, 100, 199, 200
Munich, 152, 155, 284, 291, 326, 417
music, xvii, 153, 154, 207, 209, 210, 409-416 passim
mysterious, the, xxviii, 160, 164, 222, 240, 241, 242, 287, 288, 309, 322, 330, 332, 338, 341
mystical, the, 156, 158
mysticism, xxiii, 110, 111, 114, 116, 118, 130, 260
mystique, 263
myth, 149, 205-251 passim, 272, 339, 372; of simplicity, 241-244 passim
mythology, 167, 168, 263, 327, 392

Nagasaki, 392
national home, 285, 336
nationalism, 234, 282-285, 287, 288, 306, 322, 327, 335, 336, 350, 377, 378
natural law, invariance of, 260
Nazism, xxx, 287-300, 302, 306, 331, 339, 341, 352, 353, 358, 372, 373, 379, 412
Netherlands, the, 103, 354

neutron, 353
New Fatherland League (antiwar group), 348
noematic, the, 182, 183
North Atlantic Treaty Organization (NATO), 358
Nuclear Age, vii, viii, 347, 356, 362, 367
nuclear war, 355, 358-363, 366, 367, 382, 385
nuclear weapons, *see* weapons, nuclear

ontogeny, 162, 163, 164
ontology, 131, 149
optical phenomena, xx, 6, 7, 62, 96, 98
optics, 5, 15, 19, 20, 164, 276. *See also* experiments, optical
oscillation, 148
oscillator, 52
osmotic pressure, 51, 52

pacifism, 287, 288, 300, 328, 334, 336, 340, 347, 348, 349, 350, 353, 370, 372, 376, 378, 379, 418
Palestine, Palestinians, 236, 283, 285, 286, 289, 303, 304, 335, 336, 340
particles, xxiv, 8, 14, 28, 30, 31, 32, 34-37, 47, 49, 52, 61, 71, 73, 85, 87, 110, 226, 233; displacement of, 47; of light, 97-100 passim; Newtonian, 99; radioactive, 386; suspended in liquid, 47, 48, 50, 51, 52; suspended in gas, 44
peace, xxii, 234, 258, 260, 272, 295, 296, 299, 311, 328, 347, 349, 353, 377, 378, 379, 382, 383, 385, 387, 388
perihelion, 81, 103, 104
phenomenalism, 282
philosophy, philosopher, vii, viii, xi, xii, xiii, xv, xxi, xxiv, xxvi, xxviii, 3, 4, 10, 12, 59, 67, 68, 75, 80, 82, 107, 109, 110, 121, 122, 128, 129, 131, 153, 179, 186, 187, 189, 197, 214, 216, 217, 221, 234, 243, 254, 259, 271, 281, 282, 287, 288, 293, 320, 331, 415
phonological opposition, 146
phonological units, 146
photochemistry, x, xi
photoelectric effect, x, 60, 61, 390
photon, 61, 67, 70, 71
physics: atomic, 32, 37; causal, 390; elementary particle, 37; mechanical, 50; Newtonian, 61, 122, 271; statistical, xi, 39, 40, 145
Planck's constant, 46, 88
Planck's radiation law, 61
Platonic ideas, 233
Platonic theory of vision, 164
play, playfulness, 108-143 passim
plutonium, 390, 391, 392
positivism, xxi, xxvii, 123, 207, 209, 218, 243, 246, 251
Prague, 146, 150, 284, 347, 397
probability, probabilism, 40, 41, 42, 52, 61, 63, 67, 69, 84, 86, 102, 224, 228
psychoanalysis, 151-173 passim, 249
psychology, vii, viii, 97, 102, 155, 185, 219, 249, 308, 344, 378
Ptolemy's theorem, 404
Pugwash conferences, 359, 360, 361, 367, 393
Pythagorean theorem, 91, 92, 93

Quakerism, 103, 117, 118, 131
quantum, quanta, xxv, 36, 53, 60, 64, 65, 69, 72, 370; of action, 62, 68; elementary, 148; field theory, x, 61, 69; laws, 32, 35; light, 14, 61, 62, 63, 84, 97, 288; of radiation, 69; theory, 32, 35, 37, 53, 64-68, 70, 73, 74, 115, 228, 243, 247, 291, 390. *See also* mechanics, quantum; physics, quantum

radar waves, 89
radiation, 3, 4, 14, 45, 46, 53, 60-66, 68, 69, 248; blackbody, 44, 45, 47, 52, 53; cavity, 14; field of, 62; monochromatic, 61; Rayleigh-Jeans formula for, 62, 65; spectrum, 52; thermal, 44
radium, 373
reality, realism, 73, 74, 75, 127, 198, 206-209, 211, 233, 235, 238-241, 254, 256-261, 263, 265, 266, 268, 270-273, 283, 286, 287, 290, 291, 301, 319, 331, 332; absolute, 211;

reality, realism (*cont.*)
interactionist, 217-221 passim; naive, 222; physical, 68, 71, 73, 74, 75, 121, 122, 126, 128
reductionism, 202, 209, 243
reference systems, inertial, *see* frames, inertial
relatedness, 175-204 passim
relativism, xiv, 122, 130, 166, 175-204 passim, 218, 282; relativization, 309
relativity, ix, xxiii, xxv, xxviii, 79-83, 85, 96, 97, 175-204 passim; absolute, xv; ethical, xiv; Galilean, xiv, xxi; general theory of, xii, 30, 31, 32, 34, 67, 68, 70, 81, 83, 100-104, 120, 122, 125, 127, 225, 226, 247, 283, 330, 349, 371, 390, 397, 405; of human experience, 169; situational, 144; special theory of, ix, xi, 15, 27, 30, 31, 39, 61, 66, 82, 85, 94, 96, 99-102, 120, 122, 125, 224, 247, 269, 347, 372, 390
religion, religiousness, ix, xiii, 110, 111, 115-121, 129-131, 153, 154, 184, 194, 207, 210, 211, 218, 231, 240, 260, 268, 285, 289, 291, 298, 306-310, 325; cosmic religious feeling, 308; fanatic, 377; of fear, 307; moral, 307, 308
rest, absolute, 28, 187
Riemannian geometry, 30, 32, 34
rotation, 11, 12, 29, 34
Royal Society, 103, 206, 254, 255, 256, 274, 330
Russell-Einstein Manifesto, 358, 359, 360, 361, 366, 385

Saturn, 253
signs, 140, 141, 142, 155. *See also* symbols
simplicity, simplification, xii, xv, xx, xxv, 3, 12, 28, 31, 87, 100, 103, 104, 205-251 passim, 284, 309, 331-333, 356, 406, 415
simultaneity, xiv, 15, 18, 61, 187, 188, 264, 265; absolute, 28, 29, 112
socialism, 282, 285, 287, 329, 418
solar system, 28, 87, 89
solitude, 143, 268, 289, 332, 419, 420, 421, 422
Solvay conferences, 53, 65, 70, 84
Soviet Academy of Sciences, 109, 359
Soviet Union (USSR), 108-110, 120-131 passim, 322, 329, 348, 355-357, 359, 360, 377, 381, 384, 385; relations with U.S., 365; scientists, 356, 357, 359, 360, 384
space, xviii, 4, 14, 27, 29, 33, 62, 86, 87, 88, 97, 110, 186, 265; absolute, 4, 5, 187; exploration of, 89; Friedmann-Lobachevsky, 125; Galilean, 125; Minkowsky's, 61, 82; travel in, 371
space-time, xiii, xvi, xx, xxvii, 15, 29, 30, 33-36, 61, 70, 100, 101, 122-126, 148, 155, 162-164, 185, 188, 189, 227, 288, 321, 330
speech development, 151, 152, 157
spin, 37, 85, 110
stars, 31, 81, 82, 86
statistics, 32, 227; averages, 221; Einstein-Bose, x, 37; Fermi-Dirac, 37; interpretation, xxvi; laws, 239; quantum theory, 72; theory, 43, 47, 228. *See also* mechanics, statistical; physics, statistical; thermodynamics, statistical
Stefan-Boltzmann constant, 45
Stefan-Boltzmann law, 45
Stoicism, 208, 231, 249
Stoke's law, 51, 61
Strategic Arms Limitation Talks (SALT), 363, 364, 365. *See also* arms control
structuralism, 191
structure, invariant, 36
supergravity, 37
supersymmetry, 37
supranational order, *see* world authority, government
Switzerland, 326, 327; citizenship, 288, 319, 326; identity, 156
symbol, symbolism, 114, 115, 259, 263, 291. *See also* metaphors, signs
symmetry, 34, 37, 149, 181, 209, 221, 224-227, 233, 234, 240, 243

temperature, 40, 41, 42, 43, 45, 51
tensor, 101, 102, 104; Einstein, x; -scalar theory, 36
theology, xiii, 268, 282, 283
thermodynamics, xx, 14, 18, 40, 43,

44, 46, 49, 50, 52, 60, 61, 63, 66, 94, 227, 390; laws of, 65, 193; statistical, 53
thermonuclear weapons, *see* weapons
thought-experiment, 15, 17, 19, 20, 22, 68, 70, 71, 73, 123, 144, 236
time, xxiii, 3, 4, 18, 20, 27-29, 70, 89, 110, 186, 265; absolute, xviii, 4, 5, 17, 18, 29, 66, 186, 187; atomic, 89; averages, 62, 98; dilation, xviii; ephemeris, 89; mathematical, 16; physical, 16, 17, 265; plurality of, 190; psychological, 264, 265; reversal of, 41. *See also* coordinates, local time; space-time
topology, 61, 422
transference, 169
transformations, xxi, xxvi, 7, 15, 16, 101, 102, 104, 169, 422; Galilean, 16, 17; Lorentz, 17, 96, 125, 126. *See also* coordinates, transformations of; equations, transformation
transitivity, 61
truth, xv, 114, 142, 156, 124-202 passim, 206, 218, 254, 258, 259, 261, 264, 265, 266, 270, 273, 282, 283, 286, 288, 420, 422
twistor formalism, 36
two-tier thinking, 211, 213, 214-217 passim, 218, 221, 223, 224, 233, 240

unification, x, xxvi, xxvii, 5, 16, 33, 68, 87-90, 100, 164, 193, 209
unit, absolute, 87
unitariness, 213
unitary (unified) field theories, 10, 27, 32, 34-37, 225, 226, 370, 390, 405, 406, 407
United Nations, 255, 256, 380, 382-385, 418; Atomic Energy Committee, 381, 384; General Assembly, 382, 383; Security Council, 355, 356, 382, 383
United States, xvi, xviii, 109, 118, 299, 332, 335, 337, 338, 340, 341, 348, 349, 350, 355, 356, 357, 358, 359, 365, 370, 371, 372, 373, 374, 375, 377, 380, 384, 385, 390, 392, 412, 417; citizenship, 288, 387; Congress, 380; Constitution, 287; imperialism, 289; people, 381, 382
United World Federalists, 358
unity, 29, 31, 32, 38, 225, 231, 232
universal constant *k*, 44, 45, 46, 49, 51
universalism, universality, 68, 301
universe, xvii, xx, xxiii, xxiv, xxvii, 28, 30, 86-88, 108, 114, 121, 124, 131, 148, 151, 161, 162, 164, 185, 224, 228, 240, 261, 271-273, 287, 308, 330; closed, 35, 67; cylindrical model of, 86; expanding, 87, 109. *See also* cosmos
uranium, 353, 354, 373, 374, 375, 391, 392
utopia, 330, 331, 351

validity, 180, 181, 218
vector, 33, 34, 70, 94
velocity, law for the addition of, 6, 8, 17. *See also* light, speed/velocity of
Vienna Circle, 266
volume, 33, 42

waves, xxiv, 62, 63, 226; electromagnetic, 28, 67; field of, 64; function, 37, 67; gravitational, 288; length, 45, 46, 47; particle duality, 98; structure, 259. *See also* light, wave theory of; mechanics, wave
weapons: antiballistic, 360, 363; atomic, xxii, 128, 306, 375; ballistic, 363, 364; nuclear, 354, 355, 357-360, 362, 363, 366, 367, 369-393 passim. *See also* arms control
Weimar, 330, 331, 333, 337, 338
Wien's displacement law, 45, 61, 62, 63
world authority, government, 247, 257, 273, 355, 356, 357, 358, 370, 379, 380, 381, 382, 383, 384, 388, 418

Zionism, 237, 238, 249, 283-286, 289, 292, 299, 300, 302-307, 311, 334-336, 349, 418
Zurich, 40, 284, 337, 347, 370

A CATALOG OF SELECTED

DOVER BOOKS

IN ALL FIELDS OF INTEREST

BRASS INSTRUMENTS: Their History and Development, Anthony Baines. Authoritative, updated survey of the evolution of trumpets, trombones, bugles, cornets, French horns, tubas and other brass wind instruments. Over 140 illustrations and 48 music examples. Corrected and updated by author. New preface. Bibliography. 320pp. 5⅜ x 8½. 27574-4 Pa. $9.95

HOLLYWOOD GLAMOR PORTRAITS, John Kobal (ed.). 145 photos from 1926-49. Harlow, Gable, Bogart, Bacall; 94 stars in all. Full background on photographers, technical aspects. 160pp. 8⅜ x 11¼. 23352-9 Pa. $11.95

MAX AND MORITZ, Wilhelm Busch. Great humor classic in both German and English. Also 10 other works: "Cat and Mouse," "Plisch and Plumm," etc. 216pp. 5⅜ x 8½. 20181-3 Pa. $6.95

THE RAVEN AND OTHER FAVORITE POEMS, Edgar Allan Poe. Over 40 of the author's most memorable poems: "The Bells," "Ulalume," "Israfel," "To Helen," "The Conqueror Worm," "Eldorado," "Annabel Lee," many more. Alphabetic lists of titles and first lines. 64pp. 5³⁄₁₆ x 8¼. 26685-0 Pa. $1.00

PERSONAL MEMOIRS OF U. S. GRANT, Ulysses Simpson Grant. Intelligent, deeply moving firsthand account of Civil War campaigns, considered by many the finest military memoirs ever written. Includes letters, historic photographs, maps and more. 528pp. 6⅛ x 9¼. 28587-1 Pa. $11.95

AMULETS AND SUPERSTITIONS, E. A. Wallis Budge. Comprehensive discourse on origin, powers of amulets in many ancient cultures: Arab, Persian Babylonian, Assyrian, Egyptian, Gnostic, Hebrew, Phoenician, Syriac, etc. Covers cross, swastika, crucifix, seals, rings, stones, etc. 584pp. 5⅜ x 8½. 23573-4 Pa. $12.95

RUSSIAN STORIES/PYCCKNE PACCKA3bl: A Dual-Language Book, edited by Gleb Struve. Twelve tales by such masters as Chekhov, Tolstoy, Dostoevsky, Pushkin, others. Excellent word-for-word English translations on facing pages, plus teaching and study aids, Russian/English vocabulary, biographical/critical introductions, more. 416pp. 5⅜ x 8½. 26244-8 Pa. $8.95

PHILADELPHIA THEN AND NOW: 60 Sites Photographed in the Past and Present, Kenneth Finkel and Susan Oyama. Rare photographs of City Hall, Logan Square, Independence Hall, Betsy Ross House, other landmarks juxtaposed with contemporary views. Captures changing face of historic city. Introduction. Captions. 128pp. 8¼ x 11. 25790-8 Pa. $9.95

AIA ARCHITECTURAL GUIDE TO NASSAU AND SUFFOLK COUNTIES, LONG ISLAND, The American Institute of Architects, Long Island Chapter, and the Society for the Preservation of Long Island Antiquities. Comprehensive, well-researched and generously illustrated volume brings to life over three centuries of Long Island's great architectural heritage. More than 240 photographs with authoritative, extensively detailed captions. 176pp. 8¼ x 11. 26946-9 Pa. $14.95

NORTH AMERICAN INDIAN LIFE: Customs and Traditions of 23 Tribes, Elsie Clews Parsons (ed.). 27 fictionalized essays by noted anthropologists examine religion, customs, government, additional facets of life among the Winnebago, Crow, Zuni, Eskimo, other tribes. 480pp. 6⅛ x 9¼. 27377-6 Pa. $10.95

THE BEST TALES OF HOFFMANN, E. T. A. Hoffmann. 10 of Hoffmann's most important stories: "Nutcracker and the King of Mice," "The Golden Flowerpot," etc. 458pp. 5⅜ x 8½. 21793-0 Pa. $9.95

FROM FETISH TO GOD IN ANCIENT EGYPT, E. A. Wallis Budge. Rich detailed survey of Egyptian conception of "God" and gods, magic, cult of animals, Osiris, more. Also, superb English translations of hymns and legends. 240 illustrations. 545pp. 5⅜ x 8½. 25803-3 Pa. $11.95

FRENCH STORIES/CONTES FRANÇAIS: A Dual-Language Book, Wallace Fowlie. Ten stories by French masters, Voltaire to Camus: "Micromegas" by Voltaire; "The Atheist's Mass" by Balzac; "Minuet" by de Maupassant; "The Guest" by Camus, six more. Excellent English translations on facing pages. Also French-English vocabulary list, exercises, more. 352pp. 5⅜ x 8½. 26443-2 Pa. $8.95

CHICAGO AT THE TURN OF THE CENTURY IN PHOTOGRAPHS: 122 Historic Views from the Collections of the Chicago Historical Society, Larry A. Viskochil. Rare large-format prints offer detailed views of City Hall, State Street, the Loop, Hull House, Union Station, many other landmarks, circa 1904-1913. Introduction. Captions. Maps. 144pp. 9⅜ x 12¼. 24656-6 Pa. $12.95

OLD BROOKLYN IN EARLY PHOTOGRAPHS, 1865-1929, William Lee Younger. Luna Park, Gravesend race track, construction of Grand Army Plaza, moving of Hotel Brighton, etc. 157 previously unpublished photographs. 165pp. 8⅞ x 11¾. 23587-4 Pa. $13.95

THE MYTHS OF THE NORTH AMERICAN INDIANS, Lewis Spence. Rich anthology of the myths and legends of the Algonquins, Iroquois, Pawnees and Sioux, prefaced by an extensive historical and ethnological commentary. 36 illustrations. 480pp. 5⅜ x 8½. 25967-6 Pa. $8.95

AN ENCYCLOPEDIA OF BATTLES: Accounts of Over 1,560 Battles from 1479 B.C. to the Present, David Eggenberger. Essential details of every major battle in recorded history from the first battle of Megiddo in 1479 B.C. to Grenada in 1984. List of Battle Maps. New Appendix covering the years 1967-1984. Index. 99 illustrations. 544pp. 6½ x 9¼. 24913-1 Pa. $14.95

SAILING ALONE AROUND THE WORLD, Captain Joshua Slocum. First man to sail around the world, alone, in small boat. One of great feats of seamanship told in delightful manner. 67 illustrations. 294pp. 5⅜ x 8½. 20326-3 Pa. $5.95

ANARCHISM AND OTHER ESSAYS, Emma Goldman. Powerful, penetrating, prophetic essays on direct action, role of minorities, prison reform, puritan hypocrisy, violence, etc. 271pp. 5⅜ x 8½. 22484-8 Pa. $6.95

MYTHS OF THE HINDUS AND BUDDHISTS, Ananda K. Coomaraswamy and Sister Nivedita. Great stories of the epics; deeds of Krishna, Shiva, taken from puranas, Vedas, folk tales; etc. 32 illustrations. 400pp. 5⅜ x 8½. 21759-0 Pa. $10.95

BEYOND PSYCHOLOGY, Otto Rank. Fear of death, desire of immortality, nature of sexuality, social organization, creativity, according to Rankian system. 291pp. 5⅜ x 8½. 20485-5 Pa. $8.95

A THEOLOGICO-POLITICAL TREATISE, Benedict Spinoza. Also contains unfinished Political Treatise. Great classic on religious liberty, theory of government on common consent. R. Elwes translation. Total of 421pp. 5⅜ x 8½. 20249-6 Pa. $9.95